how to
know the

ferns and fern allies

P9-DJL-361

The **Pictured Key Nature Series** has been published since 1944 by the Wm. C. Brown Company. The series was initiated in 1937 by the late Dr. H. E. Jaques, Professor Emeritus of Biology at Iowa Wesleyan University. Dr. Jaques' dedication to the interest of nature lovers in every walk of life has resulted in the prominent place this series fills for all who wonder **"How to Know."**

John F. Bamrick and Edward T. Cawley
Consulting Editors

The Pictured Key Nature Series

How to Know the

- **AQUATIC INSECTS,** Lehmkuhl
- **AQUATIC PLANTS,** Prescott
- **BEETLES,** Arnett-Downie-Jaques, Second Edition
- **BUTTERFLIES,** Ehrlich
- **ECONOMIC PLANTS,** Jaques, Second Edition
- **FALL FLOWERS,** Cuthbert
- **FERNS AND FERN ALLIES,** Mickel
- **FRESHWATER ALGAE,** Prescott, Third Edition
- **FRESHWATER FISHES,** Eddy-Underhill, Third Edition
- **GILLED MUSHROOMS,** Smith-Smith-Weber
- **GRASSES,** Pohl, Third Edition
- **IMMATURE INSECTS,** Chu
- **INSECTS,** Bland-Jaques, Third Edition
- **LAND BIRDS,** Jaques
- **LICHENS,** Hale, Second Edition
- **LIVING THINGS,** Jaques, Second Edition
- **MAMMALS,** Booth, Third Edition
- **MARINE ISOPOD CRUSTACEANS,** Schultz
- **MITES AND TICKS,** McDaniel
- **MOSSES AND LIVERWORTS,** Conard-Redfearn, Third Edition
- **NON-GILLED FLESHY FUNGI,** Smith-Smith
- **PLANT FAMILIES,** Jaques
- **POLLEN AND SPORES,** Kapp
- **PROTOZOA,** Jahn, Bovee, Jahn, Third Edition
- **SEAWEEDS,** Abbott-Dawson, Second Edition
- **SEED PLANTS,** Cronquist
- **SPIDERS,** Kaston, Third Edition
- **SPRING FLOWERS,** Cuthbert, Second Edition
- **TREMATODES,** Schell
- **TREES,** Miller-Jaques, Third Edition
- **TRUE BUGS,** Slater-Baranowski
- **WATER BIRDS,** Jaques-Ollivier
- **WEEDS,** Wilkinson-Jaques, Third Edition
- **WESTERN TREES,** Baerg, Second Edition

how to
know the

ferns and fern allies

John T. Mickel
Curator of Ferns
New York Botanical Garden

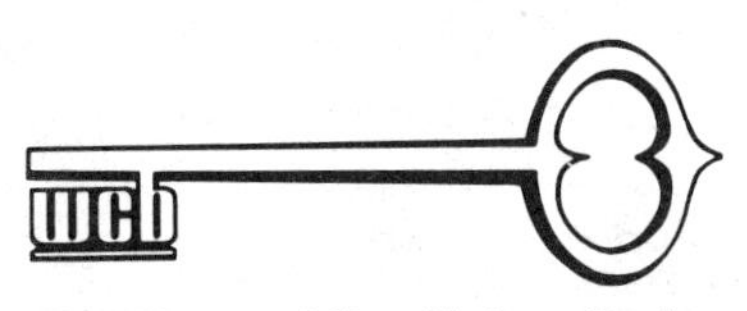

The Pictured Key Nature Series
Wm. C. Brown Company Publishers
Dubuque, Iowa

Illustrated by
Edgar M. Paulton

Copyright © 1979 by Wm. C. Brown Company Publishers

Library of Congress Catalog Card Number: 78-57876

ISBN 0—697—04771—7 (Paper)
ISBN 0—697—04770—9 (Cloth)

All rights reserved. No part of this publication may be reproduced, stored in a retrieval system, or transmitted in any form or by any means, electronic, mechanical, photocopying, recording, or otherwise, without the prior written permission of the publisher.

Printed in the United States of America
10 9 8 7 6 5 4 3

This book is dedicated to

Dr. Edgar T. Wherry and Dr. Warren H. Wagner, Jr., who through their enthusiastic teaching and research, have greatly furthered the study of the North American ferns and fern allies.

Contents

Figure 1 Fiddleheads of cinnamon fern

Bill Swan

Preface

There is an element of mystery in the ferns and fern allies. Perhaps more than any other plant group, they represent the geological past. Their origins go back four hundred million years when they emerged as the first land plants, the cornerstone of higher plant evolution and diversification.

Their reproduction is cryptic; there are no showy flowers and no seeds or fruits, no apparent reproductive similarity to the other conspicuous plants of our planet. For many years people believed the ferns must have seeds but produced them secretly, dropping them at midnight on St. John's Eve, June 24, each year; possession of such seeds would make the bearer invisible and give him other extraordinary powers. We now know that ferns reproduce by microscopic spores, but the fact that the process is inconspicuous enhances their mystique.

Ferns exhibit an amazing variety of leaf form. Their delicacy is renowned and their diversity of form, size, and texture seems endless, with leaves one-eighth of an inch to sixty feet in length, undivided straps to finely-cut lace, membranous and translucent to thick and leathery. Some ferns are so bizarre that one would hardly imagine them to be ferns at all.

The ferns and fern allies, or pteridophytes as they are called, are a manageable group for study. There are about twelve thousand species of ferns and one thousand species of fern allies in the world. Fewer than four hundred occur in North America north of Mexico, with only half of these being at all frequent, so it is not difficult to gain a passing familiarity with most of the common ones in any region.

Finding ferns can be a very challenging and rewarding pastime. Even some of the more common species are frequently overlooked, and there are some very rare species which have been found only a few times. One species of wood-fern is so rare it has not even been found yet, but its existence has been deduced from naturally-occurring hybrids.

There are excellent manuals and field guides for many regions or states. In fact, most parts of North America are covered in existing manuals, but there is no single volume that covers all of the ferns and fern allies of North America. Comparison of pteridophytes from different parts of the continent requires a shelf-full of manuals and monographs. Using these different treatments is difficult because of different approaches, different terminology, and frequently different names for the same species. Furthermore, few individuals have access to all the necessary literature.

In 1900 all the species of pteridophytes

known from North America were described in the sixth edition of Underwood's *Our Native Ferns.* This last detailed treatment of the whole area included two hundred species of ferns and seventy fern allies, as compared to the presently known three hundred ferns and eighty-five fern allies and many sterile hybrids. In 1938 Broun's very useful work, *Index to North American Ferns,* provided a catalogue of all the species, varieties, and forms known at that time, including their synonymy, habitats, and ranges, but no descriptions or other means of identification.

How to Know the Ferns and Fern Allies is intended to fill the need for a modern volume covering all the North American pteridophytes. This field book attempts to bring together, in a concise manner, keys and other pertinent information for rapid and accurate identification of all North American species. We do not want to give the impression that *all* species can be easily identified; nature is not that cooperative. Many species and genera are distinct from one another, but some (relatively few, fortunately) are so closely related that it is difficult to say categorically whether two plants are really distinct enough to be given different names. Problems of this nature will be explained so that this confusion will be understandable if not resolvable.

In preparing this book I have tried to make it clear enough for the amateur and yet complete enough to be useful for those with more technical background. There is no overall agreement among botanists regarding the number and delimitation of fern families. Their definition requires highly technical information and greater magnification than most people have available, and for species identification the family names are not essential anyway. Therefore, the keys go directly to the genera, and the genera, arranged alphabetically, have keys to their species, thus avoiding the family names altogether. Those readers interested in learning generic relationships among the ferns can refer to Fig. 2.

Certain background topics are treated in the introductory chapters, such as the general structure and life history of ferns and their propagation. References are provided for more thorough investigation by the reader as his interest warrants.

Every subject has its own vocabulary, and pteridology, the study of ferns and fern allies, is no exception. I have tried to minimize the technical terms when simpler ones will suffice, but some special terminology is necessary. Most terms are illustrated in the glossary or the introductory chapters.

A checklist of the ferns and fern allies of North America is provided at the back of the book as a record of your observations.

The illustrations are based on actual specimens and have been prepared especially for this book by Mr. Edgar M. Paulton, whose pen and wit have been a constant help. Several other individuals have been very generous with their assistance. I am indebted to Charles Anderson, Joseph M. Beitel, Dr. James G. Bruce III, Dr. Richard L. Hauke, Hugh J. Lee, Dr. James D. Montgomery, Dr. Arden W. Moyer, Clifton E. Nauman, Timothy Reeves, Dr. Alan R. Smith, Dr. David H. Wagner, and Dr. Warren H. Wagner, Jr., for reading the manuscript, testing the keys, and offering valuable suggestions. My wife Carol was especially helpful in typing, proof-reading, and general support throughout the preparation of the book.

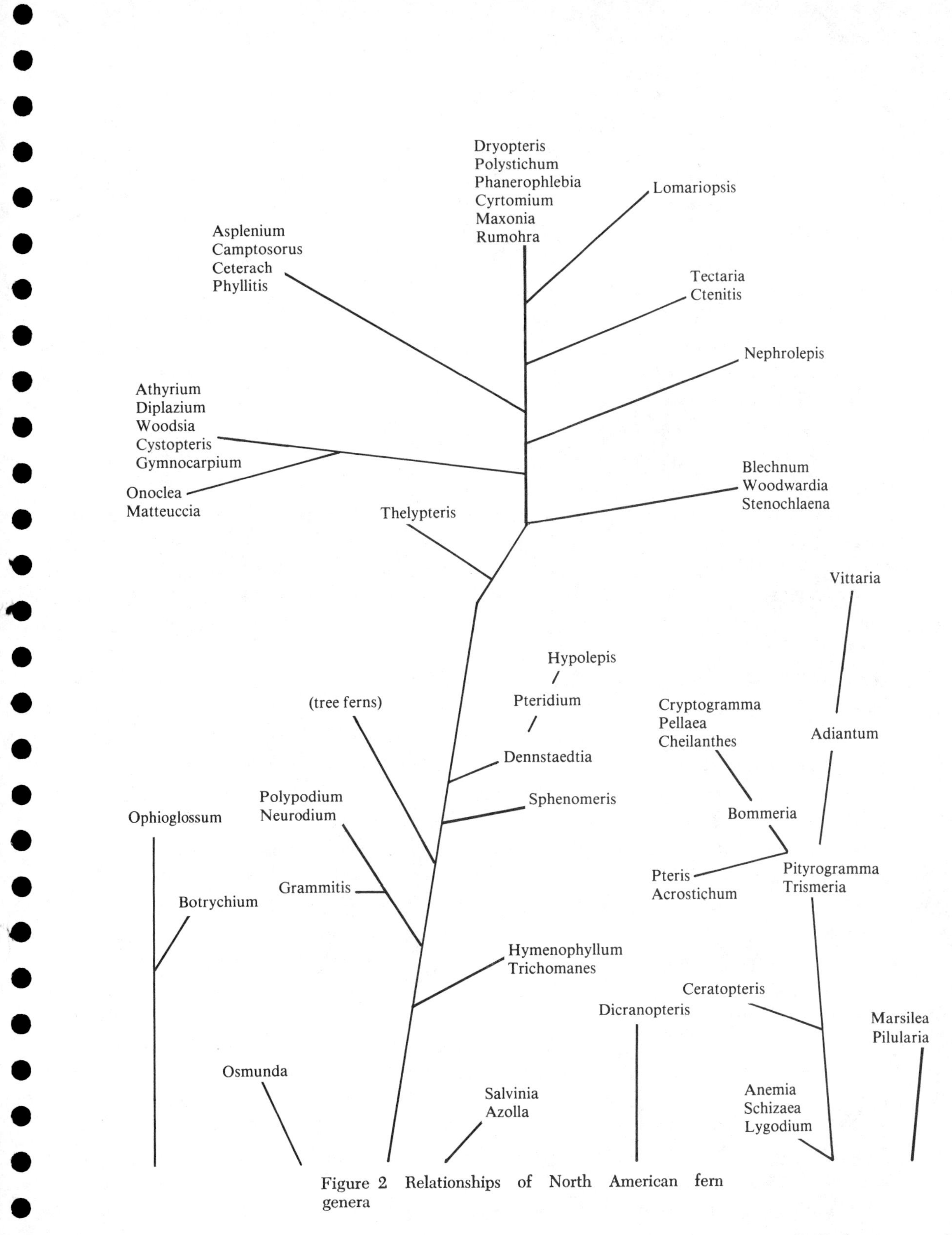

Figure 2 Relationships of North American fern genera

What Is a Fern?

Ferns are usually thought of as rather lacy plants without flowers, but in addition to the usual forms of many species, there are giant tree ferns, delicate filmy ferns, leathery strap ferns, the curious water-clover fern, and the duckweed-like floating ferns. On the other hand, many people grow plants called asparagus fern and artillery fern, which have a lacy, fern-like appearance but are not ferns at all. How then are we to tell which is and which is not a fern?

The technical definition of a fern is "a vascular plant with megaphylls that reproduces by spores."

A vascular plant is one that conducts water, minerals, and food in specialized tissues (xylem and phloem) which appear as special bundles in the roots, stems, and leaves. The mosses, liverworts, lichens, fungi, and algae are not vascular plants, being of more simple construction (Fig. 3).

Higher vascular plants—the angiosperms (flowering plants) and gymnosperms (conifers, cycads, and other naked-seeded plants)—spread by seeds rather than by spores. In the ferns and fern allies (lower vascular plants) the dispersal stage is a one-celled spore.

How do you tell a plant that produces spores? In ferns, spores are usually produced on the underside of the leaves in small fruiting

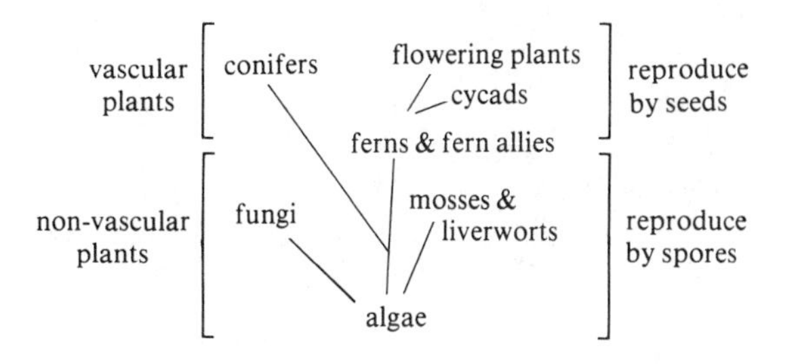

Figure 3 Relationships of major plant groups

dots, called **sori** (Fig. 4). Sometimes (e.g., cinnamon fern, sensitive fern) entire leaves are modified to produce spores, reducing the leaf tissue to almost none. In others (e.g., grape ferns, adder's-tongue) the spores are produced on special parts of the leaf that stand straight up. Often, however, there are no spores on the leaf at all, as some leaves are "sterile," or the season is wrong, or the plant too young. This presents a real problem for distinguishing some ferns at this stage.

What are **megaphylls?** Literally, this term means "large leaves," but in effect it means complex leaves with a branching vein system. The leaf is such a dominant part of the plant that its vascular tissue in leaving the stem interrupts the stem's vascular pattern at that point, leaving a "gap" (**leaf gap**) in the vascular tissue (Fig. 8). (The gap is not hollow but is filled with other kinds of cells.) Megaphylls are

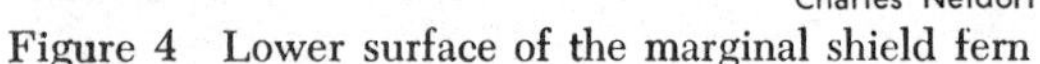

Charles Neidorf

Figure 4 Lower surface of the marginal shield fern

found in ferns, gymnosperms, and angiosperms—in other words, all vascular plants except the fern allies. On the other hand, **microphylls,** which are found in most other lower vascular plants, such as clubmosses, horsetails, quillworts, and spikemosses, have a single, unbranched vein that does not modify the vascular pattern of the stem.

The fern allies, being vascular plants that spread by spores, are of more or less the same evolutionary level as the ferns, but are not closely allied to them. They all may or may not have had a common origin, but if they are related at all, they diverged at such an early stage of evolution that we can find in the fossils very few examples that might tie them together (Fig. 5).

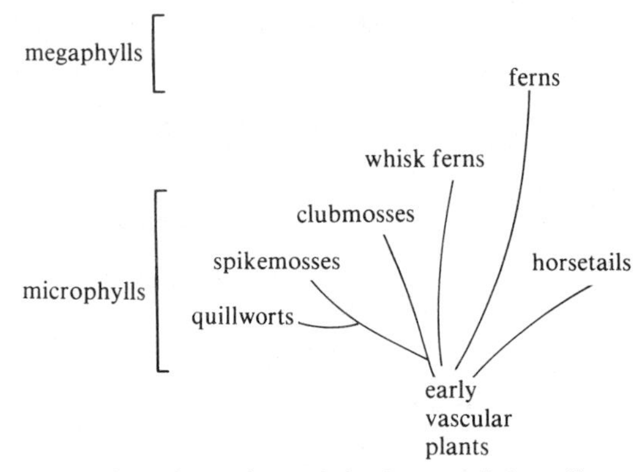

Figure 5 Relationships of the ferns and fern allies

One of the most distinctive features of the typical ferns is the manner in which they expand their leaves. Fern leaves mature from the base to the tip, gradually unrolling from a **fiddlehead,** or **crosier** (Fig. 1). Some ferns, such as the grape-ferns and adder's-tongue, do not have distinct fiddleheads, whereas a number of flowering plants, such as sundews (*Drosera*), do have them, so this character, though helpful, is not one hundred percent certain for distinguishing ferns.

Overall, then, there are several characters to look for in distinguishing a fern—vascular bundles, crosiers, elaborate leaves, spores—but when the plant is sterile, it may sometimes be difficult to tell a flowering plant from a fern, even for a professional botanist.

Fern Structure and Life History

STRUCTURE

The mature fern plant is constructed of essentially the same parts—roots, stems, and leaves—as other kinds of plants we are accustomed to seeing, but these organs are different in proportion and appearance from what we might anticipate. The part of the fern most obvious to us is the leaf, which in ferns is commonly called the **frond.** In fact, nearly everything we see of the fern above the ground is leaves (Fig. 6). The stem is a relatively inconspicuous part, its main functions being the production of leaves and roots and the storage of food. Generally the stem is horizontal, creeping along at ground level. Roots are produced by the stem and they help anchor the plant to the soil (or rocks or tree trunks) and absorb water and minerals for the plant.

Stem

The fern stem is usually called a **rhizome.** It develops horizontally on or just beneath the ground surface and may take different forms (Fig. 7). Most are short-creeping, resulting in a clump of fronds arising near each other. Many species have long-creeping rhizomes with the fronds spaced apart, forming a diffuse clump. In some, the rhizome ascends at the tip or rare-

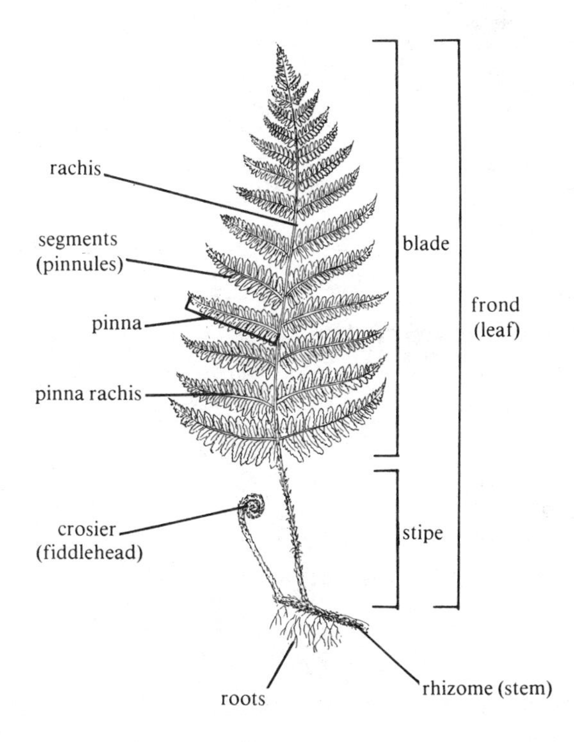

Figure 6 Parts of a fern

ly forms a short trunk. The growing tip that elongates the stem also initiates new fronds. Inside the rhizome the conducting tissue (vas-

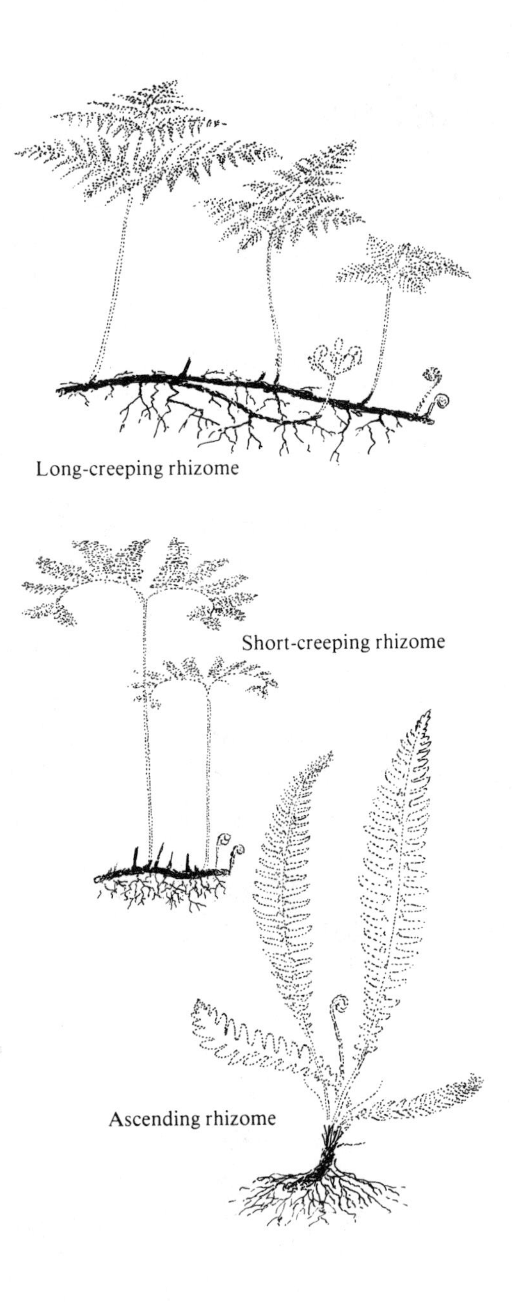

Figure 7 Rhizome habits

cular bundles) carriers water and minerals and food through the plant. The vascular bundles, which can be seen when the rhizome is cut across with a knife, are called collectively the **stele.** The stele may appear as a ring (a **siphonostele**), but more often it is discontinuous (a **dictyostele**) (Fig. 8). The **xylem** (water and mineral-conducting tissue) is surrounded

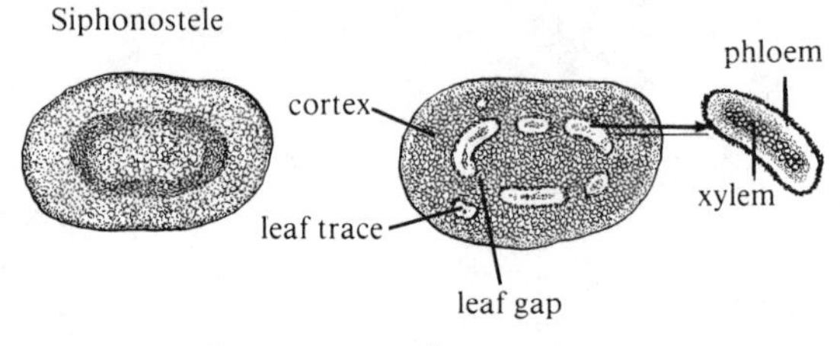

Figure 8 Rhizome internal structure

both to the inside and outside by the **phloem** (food-conducting tissue). Interruptions in the stele occur where fronds arise close to one another and portions of the vascular tissue leave the stele to serve the fronds. As the bundle passes through the outer part of the stem (cortex), it is called a **leaf trace.**

Fern roots are usually very slender and arise along the length of the stem. Those of the grape ferns and adder's-tongue, however, are somewhat fleshy, and in the leather fern the roots are quite massive.

The rhizome and often the frond are clothed with some sort of protective **indument,** either **hairs** or **scales.** Hairs are linear structures, one or more cells long and only one cell wide. Fig. 9 shows several types of hairs found in the ferns. Some long hairs have a glandular tip that secretes a volatile substance that may give the plant a distinctive smell, as in the hay-

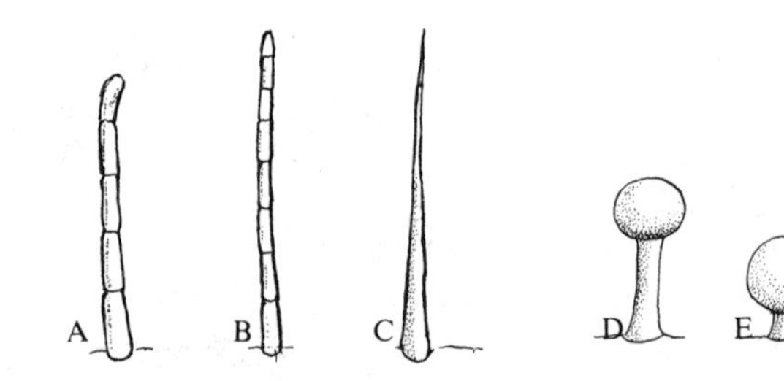

Figure 9 Fern hairs

scented fern (*Dennstaedtia*) (Fig. 9A). The hairs of *Thelypteris* are characteristically sharp-pointed and often one-celled (Fig.9C). In some ferns short, gland-tipped hairs produce a conspicuous white or yellow wax, as in the silverback and goldback ferns *(Pityrogramma* and *Cheilanthes)* (Fig. 9D, E). Scales are somewhat like hairs but broader—several cells wide, although generally only one cell thick (Fig. 10). They are often linear or lance-shaped. Usually they are uniform in color **(concolorous)** but in some species there is a dark brown or black streak down the center of each scale **(bicolorous).** Most scale variations are not named but one especially distinctive type is called the **clathrate scale** (Fig. 10C), found

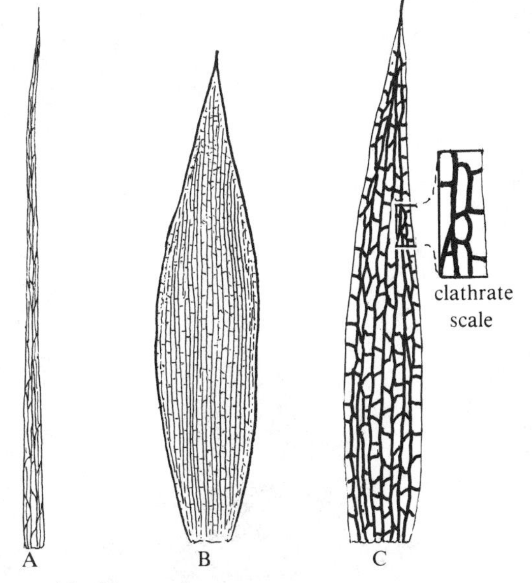

Figure 10 Fern scales

typically in the spleenworts (*Asplenium*) and their relatives and the shoestring fern (*Vittaria*). The cells of a clathrate scale are translucent and show glistening colors like stained glass church windows, and their side walls are thick like leaded glass. The marginal cells have thin outer walls, however. The structure of a clathrate scale can be seen with a 10X hand lens.

Fronds

The frond is divided into two main parts, the **stipe** (leaf stalk or petiole) and the **blade** (the leafy, expanded portion of the frond) (Fig. 6). A cut through the stipe will show the vascular bundles and provides a useful tool for distinguishing some major fern groups (Fig. 11). For example, the more primitive groups, such as *Dennstaedtia* and *Osmunda,* have a single U-shaped bundle, whereas *Thelypteris* and

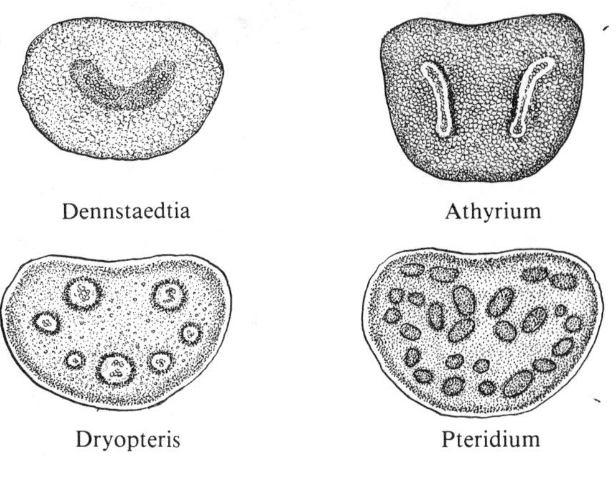

Figure 11 Stipe bundles

Athryrium have two strap-shaped bundles, and *Dryopteris* and *Polystichum* have several round bundles, still reflecting the basic U-shape. Of our native ferns the bracken (*Pteridium*) has one of the most complex patterns. At first glance it appears to have a random mess of bundles, but looking more closely and knowing the patterns of its close relatives in the tropics, we can see it has a variation of the basic U-pattern.

The blade may be undivided (**simple**) to finely cut, each degree of dissection having a specific term (Fig. 12). A **pinnatifid** blade is cut nearly to the midvein. **Pinnate** blades are divided into leaflets with each leaflet narrowly attached to the main axis. Blades more divided are designated as **bipinnate** or **tripinnate,** and for intermediate degrees of dissection the suffix "pinnatifid" is used; thus, **pinnate-pinnatifid** is not quite divided enough to be called bipin-

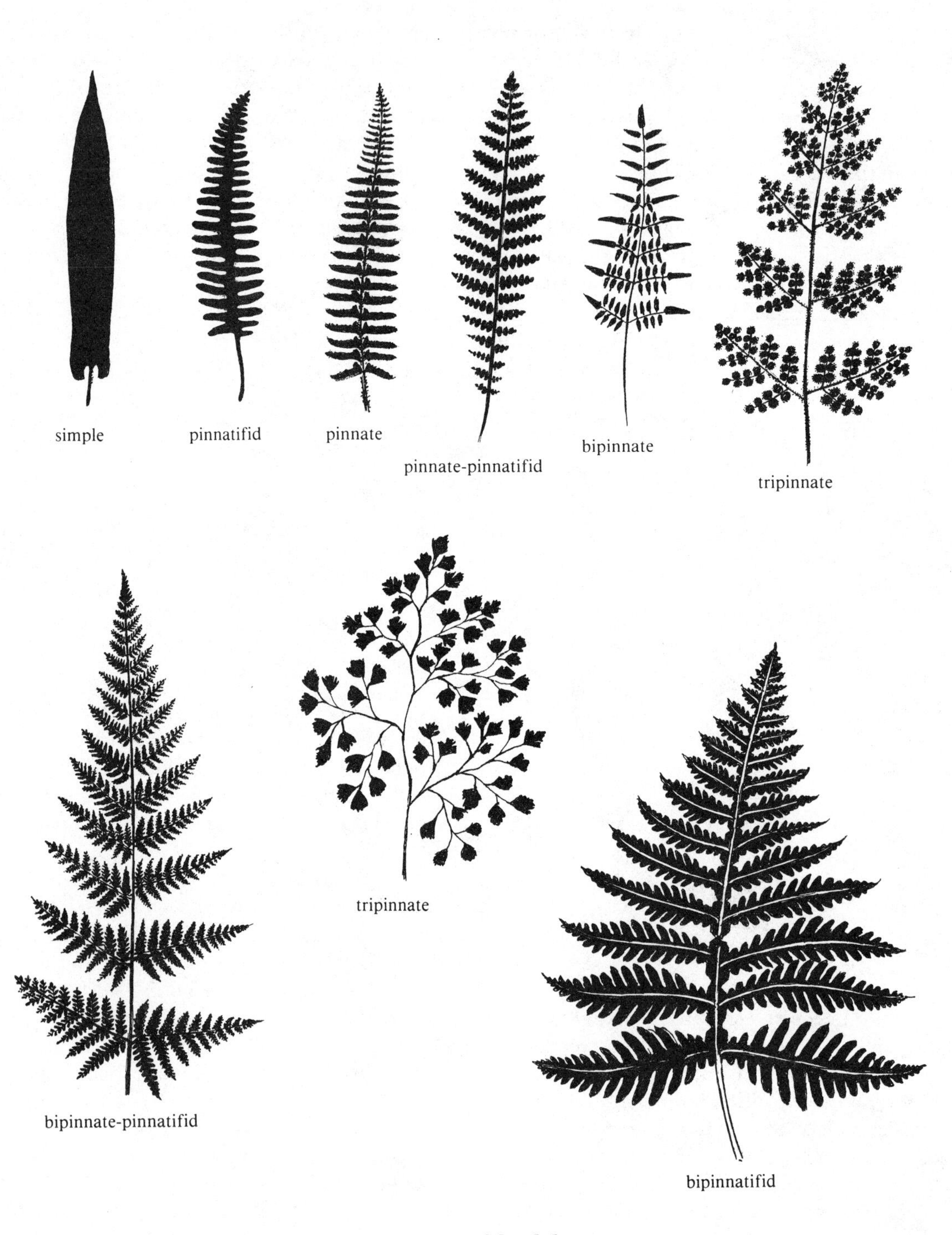

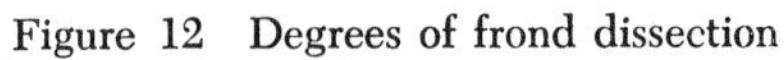
Figure 12 Degrees of frond dissection

nate, etc. The central axis of a divided leaf is called the **rachis.** The primary divisions are called pinnae, and finer units are called **pinnules** or **segments.**

In most North American ferns the veins are free, i.e., they run from the midvein to the margin without forming a network (Fig. 13). They may branch but do not unite with other veins. However, when ferns do form a vein network, it is often very distinctive and thus very helpful in identification, as in *Onoclea, Woodwardia,* and several species of *Thelypteris* and *Polypodium* (Fig. 14).

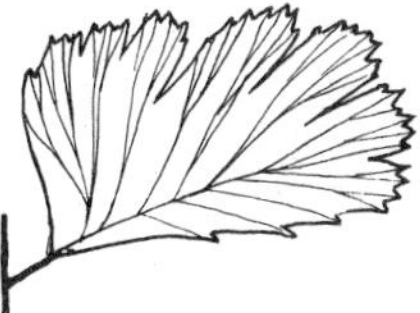

Figure 13 Examples of free veins

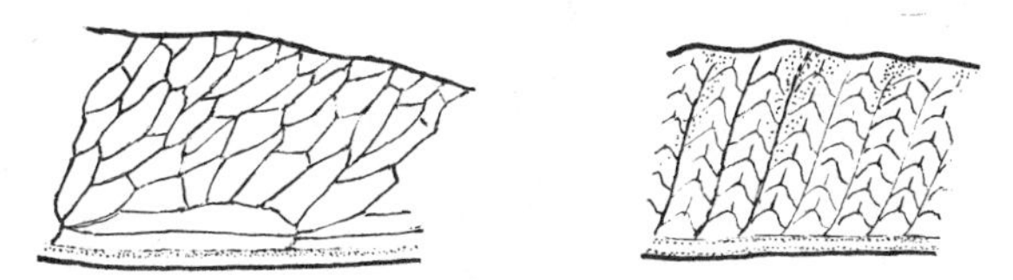

Figure 14 Examples of netted veins

The leaf surfaces often have scales and/or hairs in varying degrees of abundance. The leaves in some species are completely covered, especially the under surface, whereas others are almost completely naked.

The texture of the fern blade may be quite thin or membranous (papery when dry), firm, or leathery.

LIFE HISTORY

Ferns

When present, the reproductive structures (spore cases, or **sporangia**) are borne on the back or margin of the frond. These are usually found in small clusters, called **sori** (sorus, singular) (Fig. 15). At maturity the sori appear brown and may seem like disease pustules or small insects on the back of the leaf. This is a normal part of the development of the plant, although it may be undesirable for appearance's sake in cut fronds. The shape and arrangement of the sori are of great importance in fern identification, and ferns too immature to show these features are often quite difficult to identify. In most ferns the sori are on the un-

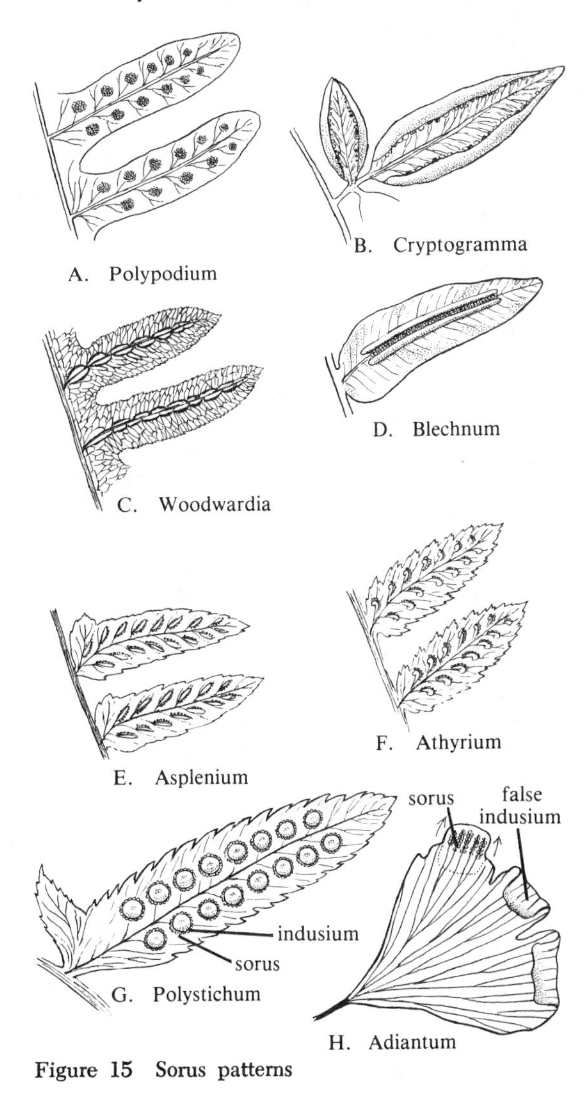

Figure 15 Sorus patterns

der surface of the leaf, about half way between the midvein and margin (medial) (Fig. 15 A, E, F, G) or near the margin (marginal) (Fig. 15 B, H) or near the midvein (Fig. 15 C, D). There is often a protective flap, the **indusium,** covering the young sorus (Fig. 15 C, D, E, F, G); marginal sori are frequently covered by the recurved leaf margin **(false indusium).** Usually the false indusium is an extension of the margin, but in *Adiantum* (Fig. 15 H) the sporangia are borne on the under side of the recurved flap. The presence and shape of the indusium (or false indusium) is often important for identification.

The sporangium of most ferns is a small, thin-walled case, usually on a stalk (Fig. 16A). Over the top there is a row of thick-walled cells, the **annulus,** that functions in opening the sporangium when the spores are mature inside. (In the more primitive fern groups the annulus is oriented diagonally or is located near the tip or on the side of the sporangium.) Within the sporangium are produced the spores, usually sixty-four in number. Most commonly, as the sporangium develops, there is a single cell in the center, the **archesporial cell,** that will divide to produce all the spores

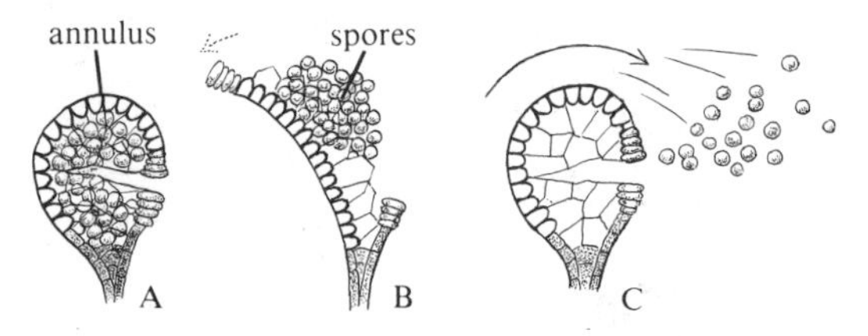

Figure 16 Release of spores from a sporangium

in that sporangium. It divides to produce two cells, which in turn divide to make four; these divide to make eight, and again to make sixteen so-called spore mother cells. At this point a special process, called **meiosis,** involving two successive nuclear divisions, takes place to make four nuclei in each of the 16 **spore mother cells.** Cell walls form between the four nuclei; these products finally separate to form a total then of sixty-four individual spores in the sporangium. (The fern plant we see is called the **sporophyte** generation, which means "plant producing spores.")

In the primitive ferns, such as the osmundas, more divisions take place before meiosis, resulting in approximately five hundred-twelve spores per sporangium. Some very primitive ferns have thousands of spores per sporangium. On the other hand, in some ferns with special reproductive quirks, as in *Ceratopteris,* one or more of the divisions is eliminated, resulting in only 32 or 16 spores.

Because of the two-step division of meiosis, the spores are produced in groups of four, each group being called a **tetrad** (Fig. 17, 18). When the spore walls form after meiosis, the resulting spores will be either bean-shaped (called **bilateral** or **monolete** because of the single scar due to its line of attachment to the other spores of its tetrad) (Fig. 17, 19B) or **tetrahedral** (or **trilete** because of its

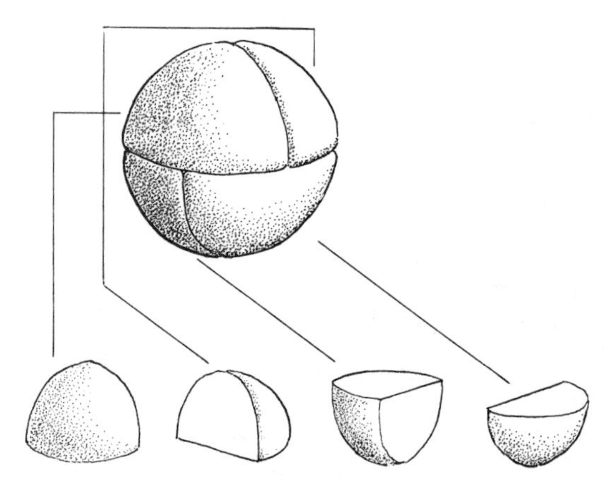

Figure 17 Monolete spores in a tetrad

three-pronged scar) (Fig. 18, 19A). Generally the spores are of the same type for all members of a genus or even family; e.g., spleenworts, wood ferns, and Christmas fern all have bilateral spores, whereas maidenhair, cliffbrakes, hay-scented fern, and the osmundas have tetrahedral spores.

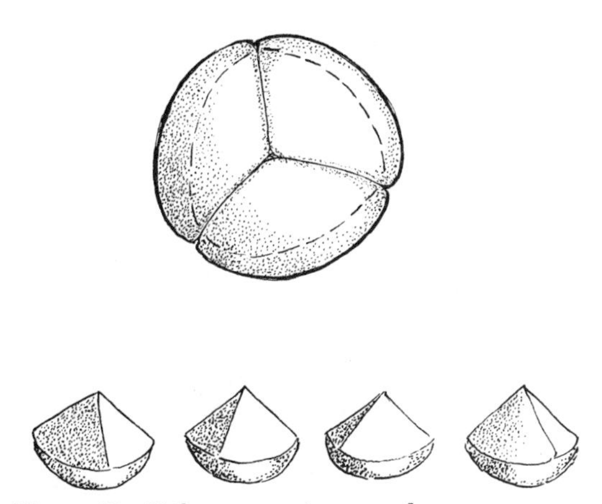

Figure 18 Trilete spores in a tetrad

After the walls form and the spores of the tetrads begin to separate, a thicker wall is formed on the outside of the original wall. It is thought that probably much of this outer wall material is laid down from the surrounding nutritive fluid (the **tapetum**), but the layers of spore walls are not well understood and developmental studies on spores are greatly needed to clarify this process.

The outer spore wall layer is very distinctive for many kinds of ferns and fern allies and is useful in some cases for identification of individual species or genera (Fig. 19). In the quillworts (*Isoetes*), for example, much of the classification is based on spore appearance.

During the spore production, the number of chromosomes (see below) is divided in half, so that each resulting spore has only half as many chromosomes as the spore mother cell or any other cell of the mature plant. This is not detrimental to the spore since the cells of the mature plant have two sets of chromosomes, one from each parent.

When the spores are mature, the sporangium slowly opens and then suddenly closes, literally throwing the spores out like a catapult (Fig. 16 B, C). Most of the spores eventually die from not finding a proper place to germinate, or they germinate and then die

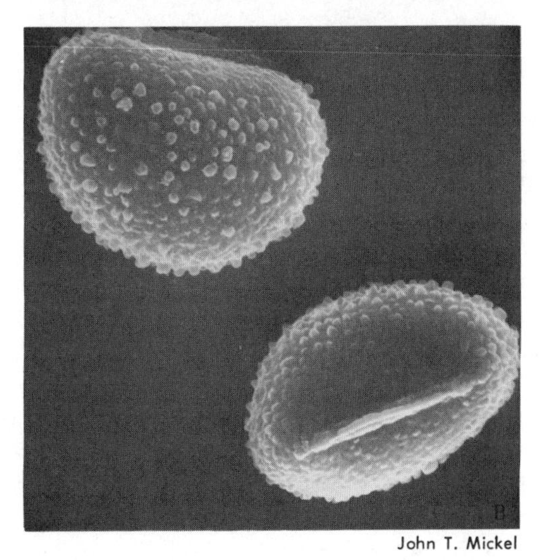

John T. Mickel

Figure 19 Spores. A. *Anemia* (trilete). B. *Polypodium* (monolete)

from dessication. The few successful ones germinate (Fig. 20) and each produces a very small prothallus. The **prothallus** is generally heart-shaped and has a slightly thickened midrib (Fig. 21). Most are less than the size of your fingernail and grow on mud or among wet moss. Toward the base (the point of the heart) there are produced many hair-like structures **(rhizoids)** that obtain water and minerals for the prothallus. The prothallus is green and

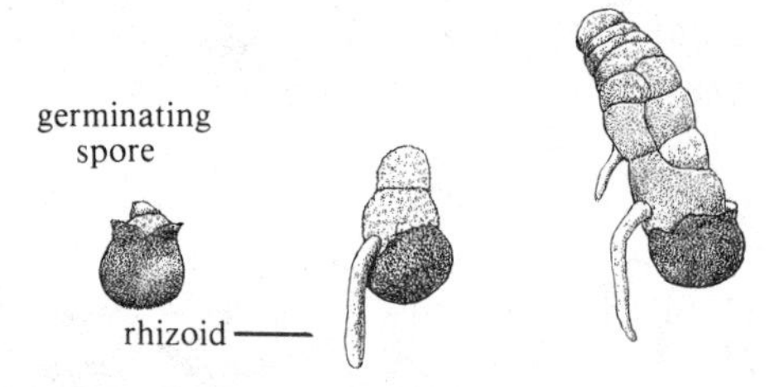

Figure 20 Stages in spore germination

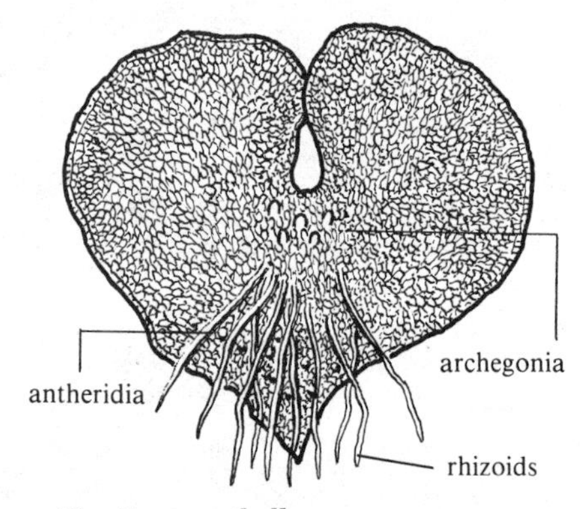

Figure 21 Fern prothallus

thus has chlorophyll and can manufacture its own food. It is on the prothallus that the sex organs are produced. (The prothallus with its sex organs is called the **gametophyte** generation, which means "plant producing gametes, or sex cells.") At the base and on the wings of the prothallus are found the male organs, the **antheridia,** tiny capsule-like sacs with sperms inside (Fig. 22). When mature, the sperms are released and swim in a spiral pathway toward the female structure, the **archegonium,** which encloses the egg. Located near the notch of the prothallus, an archegonium is a flask-shaped organ which opens at the tip like a chimney, allowing the sperms to swim down to the bottom to fertilize the egg (Fig. 22). Since the sperm cell has one set of chromosomes and the egg has one set, the resulting fertilized egg has two sets. Generally, only one fertilization takes place on each prothallus even though many archegonia may be present.

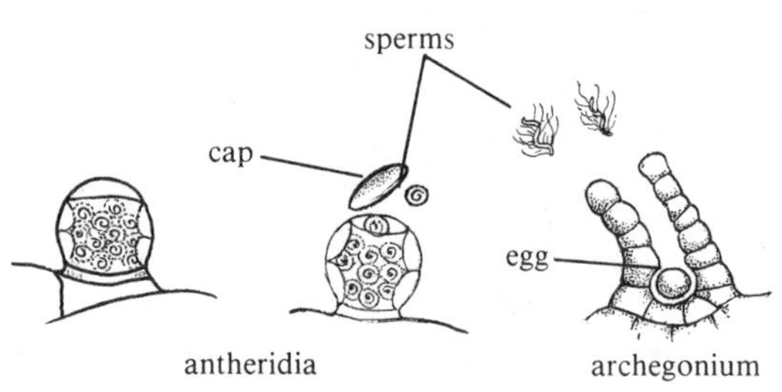

Figure 22 Sex organs of a fern

The fertilized egg begins to divide, producing first a foot that remains embedded in the prothallus for early nutrition, then a root for absorbing water and minerals, a leaf to manufacture food, and finally a stem that elongates to produce more roots and leaves (Fig. 23). As the new sporophyte becomes self-sufficient, the prothallus gradually withers.

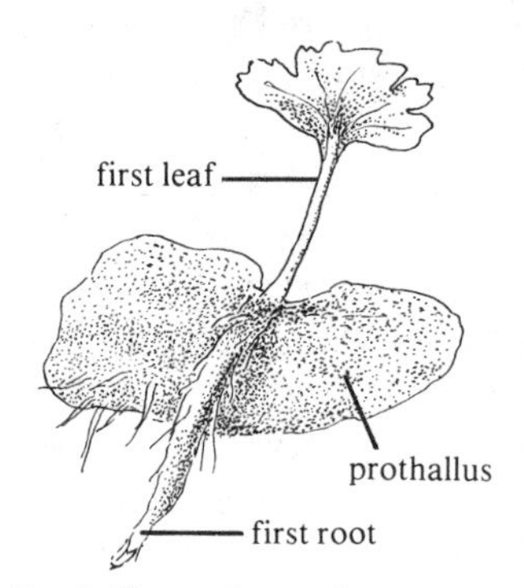

Figure 23 Prothallus and sporeling

The Fern Allies

The life history in the fern allies is essentially the same as that of the fern. A plant produces sporangia and spores, the spores germinate to form an independent prothallus with sex organs; a sperm fertilizes an egg and a new plant is produced.

In the case of the whisk fern (*Psilotum*) the plant has only rudimentary leaves and roots, but appears rather like a bundle of green, forking sticks (Fig. 24). The three-chambered sporangia are borne scattered on the stems, and each is subtended by a small forked **bract.** Its gametophyte is subterranean.

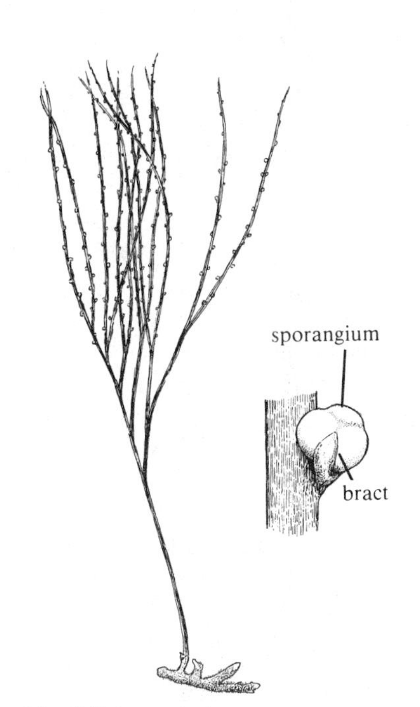

Figure 24 *Psilotum*

In the clubmosses (*Lycopodium*) the leaves are minute and undivided, appearing more like small needles or scales. The sporangia are borne on the stem at the bases of the leaves, sometimes the base of normal leaves (Fig. 25A), but more commonly the leaves with the sporangia are slightly modified and aggregated together into a **cone** (Fig. 25B). In most species the gametophyte develops under the ground and is very difficult to find.

The horsetails (*Equisetum*) differ markedly in their appearance from the clubmosses. Although they too have tiny leaves, the leaves are whorled and the stems are grooved, hollow, and jointed (Fig. 26). The narrow leaves form a sheath at the joints (**nodes**), but are usually not green. The sporangia are found in cones at the tips of the stem (Fig. 27A). Each cone is composed of many umbrella-like structures, the **sporangiophores,** under each of which there are attached about six **sporangia** (Fig. 27B).

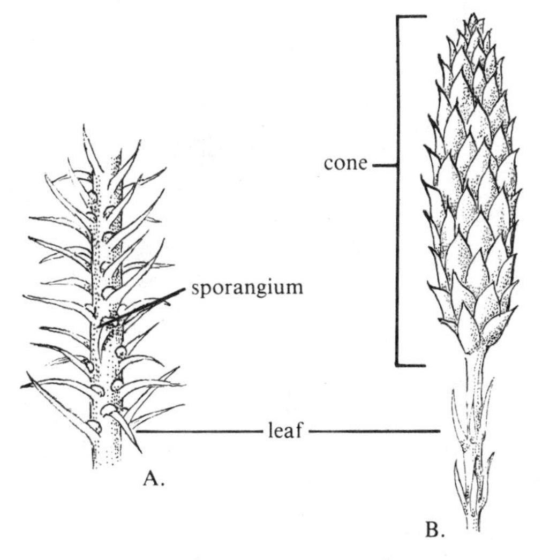

Figure 25 *Lycopodium* sporangia and cone

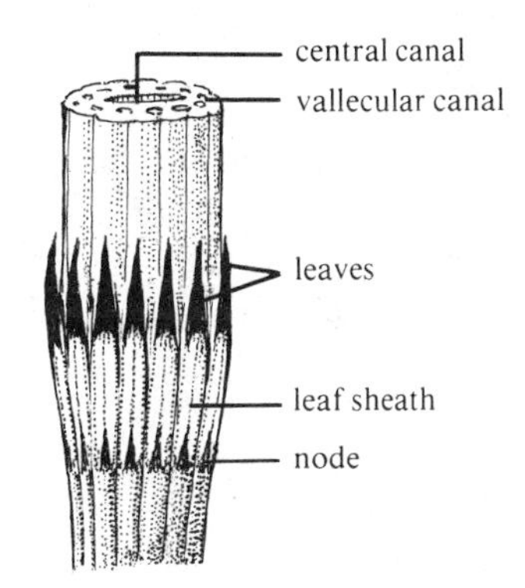

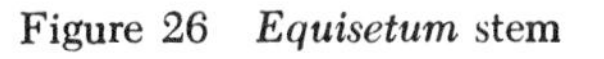
Figure 26 *Equisetum* stem

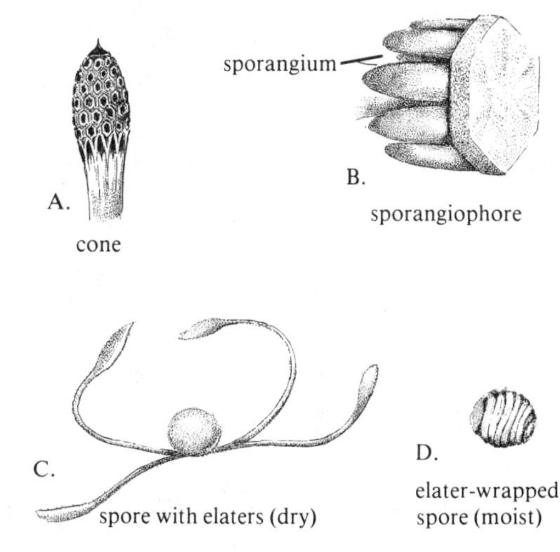

Figure 27 *Equisetum*

The spores of *Equisetum* are green and possess four long, twisting arms, or **elaters** (Fig. 27 C, D). The spores are not thrown out as in the ferns but since the elaters react to changes in humidity—coiling and straightening—it is thought that the elaters help in working the spores out of the sporangia when they open. The spores germinate on the surface of wet soil and produce prothalli that are somewhat more fleshy and irregularly shaped than those of the ferns.

Up to this point we have been discussing plants that have only one kind of spore (**homosporous**); all of the spores are alike in appearance and function. On the other hand, two distant relatives of the clubmosses, the quillworts (*Isoetes*) and spikemosses (*Selaginella*) are **heterosporous,** i.e., they produce *two* kinds of sporangia, male and female. The female, or **megasporangia,** produce relatively few large spores, the **megaspores,** that are readily seen with the naked eye (Fig. 28). The male, or

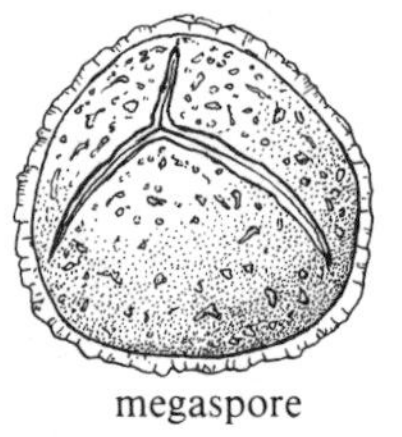

Figure 28 Relative sizes of megaspores and microspores in *Selaginella*

microsporangia, produce many tiny spores, the **microspores,** which are like powder. The megaspore cracks open but continues to grow within the original spore wall; on the surface are produced several archegonia (Fig. 29). Inside the microspore the cell divides to form a few sterile cells and several sperms. The sperms are released and swim to the archegonia. The resulting fertilized egg develops into a new plant. Heterospory, the condition of having two kinds of spores, is the first step in the evolution of seeds. (Seeds do not occur in the pteridophytes today, but in Carboniferous times there were abundant seed-producing fern-like plants that have become extinct.)

As in the clubmosses, the quillworts and spikemosses have their sporangia associated with the leaf bases. Spikemosses look like small clubmosses or mosses, but always have cones (usually four-sided) at the branch tips (Fig. 30). The quillworts look more like little tufted grasses or small onions growing submerged in shallow water or on the muddy shore (Fig. 31, p. 13). The sporangia are embedded in the bases of the leaves, and the fleshy stem appears like a small bulb or corm.

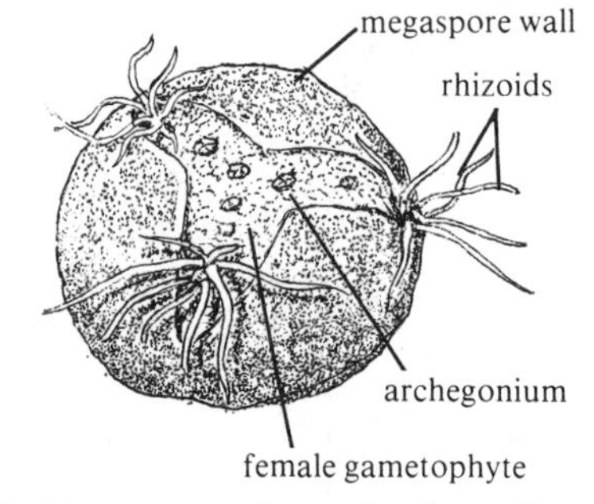

Figure 29 Megagametophyte of *Selaginella*

Figure 30 *Selaginella* cone

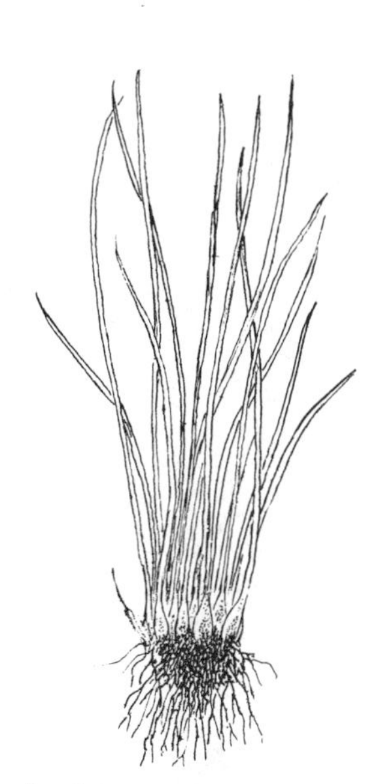

Figure 31 *Isoetes*

Water Ferns

Most of the so-called water ferns—water clover (*Marsilea*), pillwort (*Pilularia*), mosquito fern (*Azolla*), and water spangles (*Salvinia*) are true ferns but are heterosporous and are the most unfern-like in their appearance of all the ferns. (The other water fern (*Ceratopteris*) is homosporous and is a typical true fern.)

The water clover (*Marsilea*) looks like a clover with four leaflets on a long stipe (Fig. 32); the pillwort (*Pilularia*) is similar but lacks the leaflets, leaving only the stipes, and thus looks like a small grass (Fig. 33). Both have rhizomes that creep in the mud, and their sporangia are borne in special hard, nut-like structures, **sporocarps,** at the base of the leaves, usually just one or two on a stalk at the base of the stipe.

Azolla (Fig. 34) and *Salvinia* (Fig. 35, p. 14) are generally floating and have simple or lobed leaves. *Salvinia* has conspicuous hairs on the tops of the floating leaves, and its root-like structures that hang down beneath in the water are actually skeletonized leaves that resemble roots. The sporocarps in both genera are round, yellow, bead-like structures borne under the water. Unfortunately, sporocarps are rarely produced in either genus.

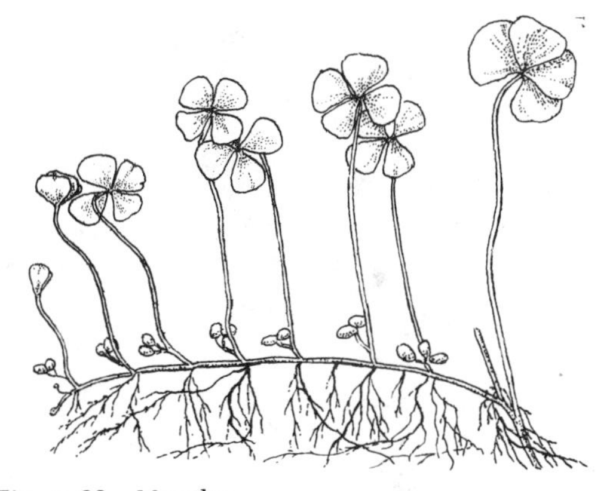

Figure 32 *Marsilea*

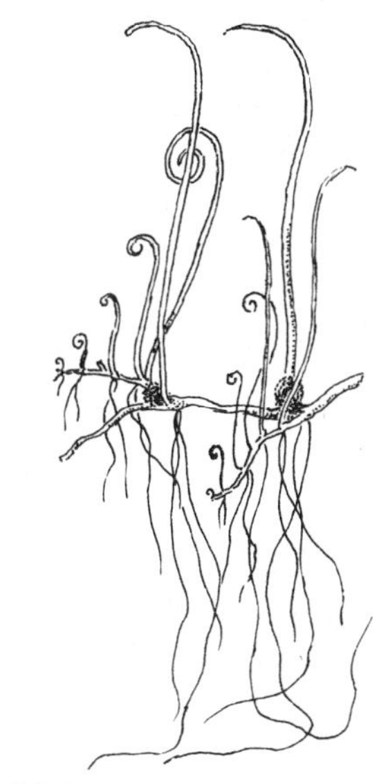

Figure 33 *Pilularia*

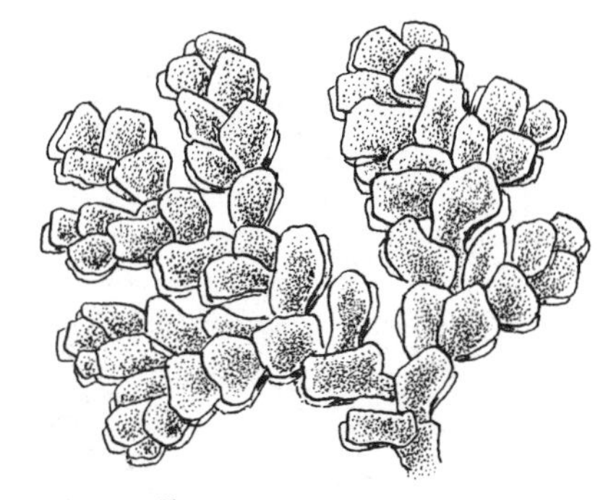

Figure 34 *Azolla*

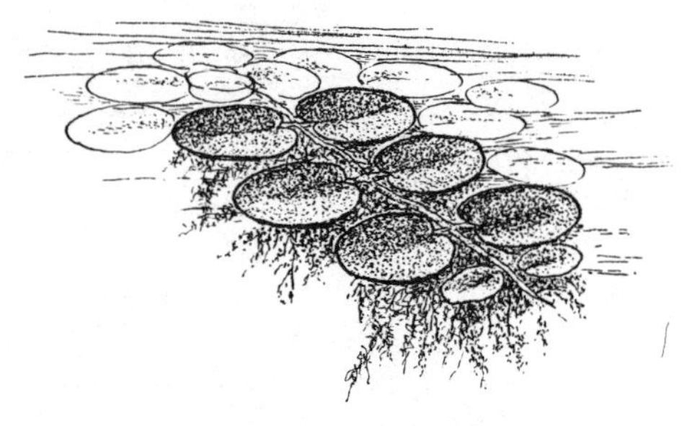

Figure 35 *Salvinia*

Apogamy

There are a few species of pteridophytes that go through the motions of a normal life history but bypass fertilization. These ferns produce spores that develop into prothalli, and although normal antheridia and even archegonia may be formed, a simple bud is formed at the edge or on the surface of the prothallus that grows directly into a new sporophyte plant, thus avoiding fertilization altogether. This process is called **apogamy** (without gametes). Apogamous ferns often produce only thirty-two spores in each sporangium.

Hybridization in the Ferns

Hybridization is the crossing of two species or varieties to form offspring that combine the characters of the two parents. This can take place only between plants that are sufficiently closely related. In the ferns this is brought about by the sperms from the prothallus of one species swimming to and fertilizing the egg in the prothallus of another species (or variety). Fern hybrids are not common, but they are frequent enough so that when two or more species of a genus are found growing together, plants with intermediate characters are often found if one searches closely.

Because hybridization plays such an important role in the study of ferns, we should consider some of the ways in which we establish that a given plant is a hybrid. For many years it was believed that ferns and their allies could not hybridize. For this reason some readers may find the following rather technical description useful. The first thing that we notice about all fern hybrids is that they are intermediate between their parents. One parent species might have simple leaves, the other one pinnate leaves. In this case the hybrid will have lobed leaves, intermediate in shape between the parents (Fig. 36). The second thing that we notice is that the hybrids are almost always in the minority. In fact, we commonly

American Fern Society

Figure 36 Scott's spleenwort, the hybrid between the walking fern and ebony spleenwort.

"Do you Asplenium platyneuron take this Camptosovus rhizophyllus to be your unlawful hybridizing ptevidophyte?"

find only a single plant, surrounded by dozens or hundreds of plants of the parents.

Hybrids are generally sterile (their spores abort; see below) and are thus reproductive deadends. If these sterile hybrids are frequent or especially interesting, they are given a species name. Often they are named as species before their hybrid origin is realized. Their hybrid nature is shown by placing a "times sign" between the genus and species name, e.g., *Dryopteris* × *triploidea*. Hybrids that have become fertile, producing viable spores, are generally treated as normal species and no "×" is put in its name. Instead of or in addition to a specific epithet, a hybrid can be expressed as a formula involving the two parents; e.g., *Dryopteris* × *triploidea* can also be written *Dryopteris intermedia* × *spinulosa*.

Frequently sterile hybrids will be inordinately common in an area. This may be due to many cases of hybridization when the parents are both especially abundant, but often one or more of the parents is rare or even lacking, and no simple explanation for the hybrid's abundance is obvious. In some hybrids, among the shrivelled abortive spores there are found large spherical spores, which may indeed be viable. Studies of such cases have not yet been made but would be helpful in solving this problem.

In at least three groups of North American ferns, the spleenworts (*Asplenium*), the wood ferns (*Dryopteris*), and the holly ferns (*Polystichum*), it has been shown that hybridization has played a major role in the origin of reproductive species. The hybrids that formed became fertile and now exist as if they were normal species. They will be discussed later under the appropriate genera, and diagrams of the species-hybrid relationships will be given.

Hybrid ferns can be artificially produced by sowing spores of different species together. The results are haphazard, though, and generally not very satisfying. However, with very careful technique and much patience, it is possible to bring together the sperm-bearing prothallia of one species with the archegonia of another species. Artificial hybridization is commonly studed by botanists but rarely used by horticulturists to form new varieties.

In order to understand which species of a fern genus are the basic species and which are of hybrid origin, pteridologists study the plants' chromosomes. The number of chromosomes per cell provides valuable information regarding the plant's ancestry and relationships with other species. Thus, fern literature often contains references to chromosome numbers and their evolutionary importance.

WHERE ARE CHROMOSOMES FOUND?

The plant is composed of millions of microscopic cells, and in each cell there is a nucleus that contains the genetic material. This material is arranged on long strings called chromosomes. Generally these structures cannot be seen, even with a microscope, but rather the nucleus appears as a homogeneous mass. The chromosomes can be seen clearly only when the nucleus is dividing, i.e., when the cell is reproducing itself to make the plant grow or to make spores.

In the gametophyte generation (prothallus), each cell contains one set of chromosomes. Each sperm cell and egg cell also contains one set of chromosomes. When the two meet in fertilization, the resulting fertilized egg then has two sets. The fertilized egg divides many thousands of times to produce the familiar mature fern plant (sporophyte), and each cell making up that plant also contains two sets of chromosomes. In the process of making spores, however, as previously mentioned, a special division, called meiosis, reduces the number of chromosomes in the resulting spores to only one set each, so the gametophytes developing from the spores again have only one set. Thus, the two major phases of the fern's life history have the same kind of chromosome material, but one phase has twice as much as the other. The condition of having one set is called **haploid** (1n); two sets is **diploid** (2n).

WHY COUNT THE CHROMOSOMES?

Each kind of plant has a fixed basic number of chromosomes in each cell, for example, twenty-nine. If the gametophyte has twenty-nine chromosomes in a cell's nucleus, the mature fern (sporophyte) would have fifty-eight. Most members of a genus or often of a family have the same basic number. Thus, we can generally tell the relationship of a plant if we know its chromosome number: the cheilanthoid ferns have a base number of twenty-nine or thirty, *Asplenium* has thirty-six, *Polypodium* has thirty-seven, *Dryopteris* and *Polystichum* have forty-one (Fig. 37). This information is handy in some difficult or especially problematic genera. The family for the tropical genus *Hyalotricha,* for instance, was not known for certain until it was shown to have a base number of thirty-seven, thus confirming its relationship to *Polypodium* rather than to *Tectaria* (forty) or *Vittaria* (thirty) which were also postulated relatives. Chromosome evidence also strengthens the argument for keeping *Thelypteris* (twenty-seven to thirty-six) separate from *Dryopteris* (forty-one).

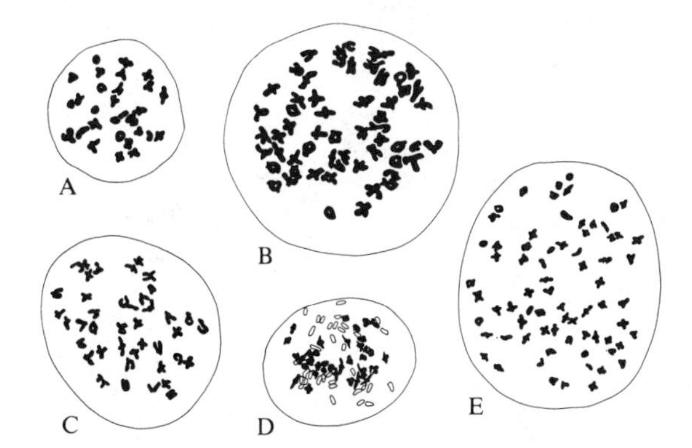

Figure 37 Chromosome squashes to show chromosomes during meiosis. A. *Adiantum* (30 pairs). B. *Adiantum* (58 pairs). C. *Diplazium* (41 pairs). D. *Thelypteris* (36 pairs + 36 singles). E. *Thelypteris* (72 pairs).

Within a genus the chromosome number can help shed light on the origin and relationships of the species. Most species generally have one set in the gametophyte, two in the sporophyte. Thus, most plants are diploids. Some species have multiples of the basic number and may be **tetraploids** (tetra = four), **hexaploids** (hexa = six), or other ploidal levels. These higher numbers may be reached in various ways, but the most common is as a result of hybridization.

In fern hybrids the resulting mature plant is nearly always sterile. The plant can exist without difficulty, but it cannot make viable spores. In the spore-making process the crucial nuclear division, meiosis, involves the pairing of similar chromosomes. In normal species there are two sets that are alike, so each chromosome of one set has one (its homolog in the other set) to pair up with. In hybrids, however, the chromosomes come from two different species and so are not exactly alike or compatible; thus, they cannot pair up, meiosis cannot be accomplished smoothly, and the spores abort.

Rarely there is a doubling of the chromosome number in the sporangium, allowing the chromosomes to pair up, and spores can then be produced. The resulting spores, though, have twice the normal complement of chromosomes (diploid) and after fertilization the plants will be tetraploids. This is essentially a new species, being different from any other one and able to reproduce itself. This, then, is how many of the **"polyploids"** are formed, through hybridization and subsequent doubling of the chromosome number.

To make the situation even more interesting, the tetraploids can hybridize with other species too, resulting in many possible ploidal levels, depending on the ploidy of the hybridizing plants; thus, a tetraploid crossing with a diploid produces a **triploid** (sterile because of lack of complete chromosome pairing), which in turn might double to make a hexaploid. If a tetraploid crosses with one of its own parent species (diploids), the chromosomes it has in common with that parent would pair and the others not—a triploid with two sets pairing and one set not. (Fig. 37D). Examination of chromosomes during meiosis reveals not only the actual number of chromosomes but also their pairing behavior, such as their degree of pairing.

HOW TO COUNT THE CHROMOSOMES

In order to see the chromosomes, it is necessary to catch them when the nucleus is dividing, either in areas of active vegetative growth or in spore production. Chromosomes are generally counted in the root tips or the young sporangia. Sporangial material is often easier to count since the chromosomes are paired in meiosis, thus making only half as many units to count. Some plants, especially when cultivated, are reluctant to become fertile, though, so it becomes necessary to use root tips in those cases.

In either case the procedure is basically the same. You must pick the part of the plant (roots or fertile segments), and place them in a fixative, a solution that will stop the nuclear division where it is and preserve it until you can examine it. After several hours a tiny amount of material may be removed from the fixative (a root tip or a sorus) and placed in a drop of chromosome stain on a microscope slide. A cover slip is added and then pressure is applied by a heavy thumb, padded pliers, or by other imaginative methods. The slide is then examined under the microscope. Chromosomes can be seen with a magnification of one hundred times but to be drawn or photographed we often increase this to one thousand times. The important points are to catch the nuclei at the proper stage and to spread the chromosomes sufficiently to count them. This is not accomplished on every slide, and often many slides are prepared before a suitable chromosome "squash" is found. Great patience is required, but the results are fascinating.

REFERENCES

Andrews, H. N. Studies in Paleobotany. New York: Wiley & Sons. 1961.

Arnold, C. A. An Introduction to Paleobotany. New York: McGraw-Hill. 1947.

Bierhorst, David W. Morphology of Vascular Plants. New York: The Macmillan Co. 1971.

Bold, H. C. Morphology of Plants (3rd ed.). New York: Harper and Row. 1973.

Bower, F. O. The Ferns (3 vols.). London: Cambridge Univ. Press. 1923-28.

Copeland, Edwin B. Genera Filicum: The Genera of Ferns. Waltham, Mass.: Chronica Botanica Co. 1947.

Delevoryas, Theodore. Morphology and Evolution of Fossil Plants. New York: Holt, Reinhart and Winston. 1962.

Eames, A. J. Morphology of Vascular Plants (Lower Groups). New York: McGraw-Hill. 1936.

Foster, Adriance S., & Ernest M. Gifford, Jr. Comparative Morphology of Vascular Plants (2nd ed.). San Francisco: W. H. Freeman. 1974.

Jermy, A. C., J. A. Crabbe, and B. A. Thomas (eds.). The Phylogeny and Classification of the Ferns. Botanical Journal of the Linnaean Society 67, Suppl. 1: 1-284. 1973.

Manton, Irene. Problems of Cytology and Evolution in the Pteridophyta. Cambridge, England: Cambridge Univ. Press. 1950.

Parihar, N. S. Introduction to the Embryophyta, Vol. II: Pteridophytes. Allahabad: Central Book Depot. 1956.

Scagel, R. F., et al. An Evolutionary Survey of the Plant Kingdom. Belmont, Calif.: Wadsworth Publ. Co. 1965.

Smith, G. M. Cryptogamic Botany, Vol. II. Bryophytes and Pteridophytes (2nd ed.). New York: McGraw-Hill. 1955.

Sporne, K. P. The Morphology of Pteridophytes. London: Hutchinson Univ. Library. 1962.

Taylor, Thomas N. (ed.) Evolution of systematic characters in the ferns. Annals of the Missouri Botanical Garden 61: 307-482. 1974.

Verdoorn, F. Manual of Pteridology. The Hague: Martinus Nijhoff. 1938.

How to Grow Ferns

SPORES

Although a mature plant can be obtained most rapidly by dividing existing plants or by other vegetative means, propagation by spores will result in larger numbers of plants, and often this method is the only way in which plants will be available.

Ferns can be grown from spores in the home, classroom, or office with a minimum of materials or mess. The prothallus that grows directly from the spore, and the subsequent juvenile fern plant make an interesting display of the life history of the fern.

Obtaining the Spores

The most direct source of fern spores, naturally, is the living plants themselves. These may be native ferns outdoors or plants in a greenhouse or even house plants. Some groups are seasonal in their spore production while some tropical ferns produce spores all year round. Remove a frond or portion of a frond bearing sori and place it sorus side down on a clean sheet of smooth white paper. Avoid coarse, rough paper because the spores will be harder to remove later. Place another sheet of paper over the frond and put a light weight on top. If the leaf is in just the right stage of development (see below), the spores will soon be released and in a few hours there will be enough spores for your use. Waiting a day or two will allow time for all the ripe sporangia to open.

Remove the leaf from the paper. Tap the paper gently until the dust (spores and empty sporangia) goes to one side of the paper. Continue tapping until the coarse debris falls off the paper; most of the spores will remain. It is important to be rid of the debris because it often harbors fungus spores. The fern spores may be stored either in envelopes or in small vials or medicine capsules.

If you do not have living plants available in the proper condition for collecting spores, there are other sources of living spores. Most fern spores will remain alive for quite some time—several months or often many years. Some spores have been germinated after fifty years; germination after seventy to one hundred years has been reported but not documented. Thus, spores can be taken from dried fern specimens with success if the plants were not dried with strong heat.

Other sources of fern spores are the spore exchanges of various fern societies. Generally this service is open only to members of the societies, but the cost of membership is modest. Many fern enthusiasts collect and donate spores to these spore banks, which include spores from around the world in addition to the ferns of North America.

How do you recognize the right stage for picking the fertile frond? At the best stage the sporangia will appear dark brown or black (or orange in the polypodies and some others); the individual sporangia will be full and round. When the spores or sporangia are too young, the sporangia (and whole sorus) will appear pale green or almost white. When the sporangia are too old, they will appear tan or brown; the sporangia will be open and the spores gone. A good hand lens (10×) may help you see this to better advantage, but with experience the naked eye is sufficient. As the ripe sporangia dry out, they begin to open, slowly cracking along a predetermined line, bend open, and suddenly, almost faster than the eye can follow, snap closed and sling the spores out.

Sowing the Spores

For sowing the spores, the type of container makes little difference. The main requirement is a tight cover to prevent drying out. Some people use pyrex dishes with glass covers. Plastic freezer cartons are quite adequate. Some growers use small flower pots, putting several in a closed plastic shoe box. The containers should be washed and sterilized by pouring boiling water over them or washing with a chlorine bleach to prevent excessive contamination from fungi, mosses, and algae.

The soil or substrate for the spores also makes little difference as long as one hundred percent humidity is maintained. A mixture of sphagnum and sand will do. If a rich soil is used, it should be sterilized by baking it for two to three hours at 300° F. Pre-sterilized soil mix, such as African violet mix, is available commercially. In using soil, some people go one step farther and pulverize pieces of broken flower pot to spread over the surface of the soil to help keep any potential fungus contamination a slight distance from the developing ferns. Actually, the fern spores can be germinated directly on distilled (or boiled) water or on sterilized pieces of pot in the water.

To sow the spores, take the paper or vial with the spores in it and gently tap it over the moist germination medium. You will probably see the dust-like spores falling to the surface. Do not sow too densely or you will have problems separating the plants later. Cover the container right away, and label with the species name, source or locality (if known), and date of sowing. Set the containers on a window sill or shelf with light. The light should be normal day length (or in winter supplemented a few extra hours) but avoid direct sunlight. Strong light will overheat the containers and cook the ferns. If the containers are properly covered, further watering will be necessary rarely, if at all.

The next requirement is patience. The spores will not germinate for at least a week or two. Even then growth is relatively slow and you may not see any tint of green for three to four weeks or longer. The prothalli may mature in one to three months. If the prothalli seem to reach mature size and produce no young ferns by themselves, it may be necessary to drip some distilled water on them to provide some free water in which the sperms can swim to the egg.

Transplanting

Even when the young sporophytes are visible, be patient a while longer. Let them grow a little, and don't transplant them until they have a few small leaves. It may be wise to place them in rows in another container for a while, but it is still imperative to provide the plants with high humidity. Plant them in a terrarium or a plastic-lined box with a plastic bag over it. A clear plastic shoe box makes a good small terrarium. The soil must be loose with good drainage. This can be the African violet soil mix, or a mixture of peat, vermiculite, soil, and sand in equal quantities.

When the plants have finally reached a

size of two to three inches, transplant them to individual two-inch pots. After they outgrow those pots, they can be planted as you wish, keeping in mind that most ferns do best with high humidity. They can be gradually placed outdoors during the warm months, and if they are hardy, they can be mulched well and left outside for the winter.

The main points to remember throughout are that the ferns require very high humidity in the air, good drainage in the soil, relatively cool temperatures, and much patience.

VEGETATIVE PROPAGATION

Vegetative propagation of ferns is much more rapid and more certain than reproduction by spores, and in nature this serves as a very important means of reproduction. Similarly, in cultivation the surest way of getting mature plants is by dividing existing plants. Ferns that produce crowns will usually send out side branches, so that one or more small plants will appear next to the parent plant. When the offshoot has reached one-fifth to one-third the size of the parent, it is safe to separate it. Carefully dig down beside the plant and cut the young plant off with a sharp knife, being sure not to damage the parent plant and to include the root system of the new plant. In some crown-formers, such as the ostrich fern, the side branches are rather long (**stolons**), resulting in offspring several inches away from the parent.

Ferns with creeping rhizomes can be divided when the plant has several branches and forms a clump. Break or cut the clump into several pieces and replant separately. Be sure that each clump has at least one growing tip, the tender rhizome tip, from which the new leaves develop. If that is damaged or missing, there will usually be no new growth.

In addition to division of the branching system, some ferns have special ways of producing new plants. In the walking fern, the prolonged leaf tip has a bud that develops into a new plant when it touches the ground. Similarly, Anderson's holly-fern and the European chain-fern have a large bud near the tip of each leaf. As their leaves get old and bend over to the ground, the buds take root and grow into new plants. The bulblet bladder-fern, on the other hand, produces several buds (called bulblets) on the under side of the rachis which readily fall off the plant and roll to suitable habitats, serving as the principal means of reproduction in this species.

An uncommon fern of the Southwest, *Asplenium exiguum,* has a bud at the tip of each pinna. Each bud can, with very high humidity, develop into a new plant.

The water fern, *Ceratopteris,* produces buds along the margin of the leaf. When the leaf is in water or lying on mud, these buds develop into new plants. This is sold as an aquarium plant, and it can quickly proliferate new plantlets to fill the tank.

ESTABLISHING A FERN GARDEN

If you have a shaded part of your yard, you might consider establishing a fern garden. Many of our native ferns can readily be transplanted and grow well under cultivation. The plants of your area are adapted to your local climate; plants from other areas should be selected with consideration of their hardiness and the cold of the winter or heat of the summer. Basically, you should try to simulate the natural habitat of the ferns you wish to transplant, but many ferns have broad latitude and can withstand a wide range of conditions.

For a general fern garden, start with a shaded area. This might be the north side of a house, open to the sky, or an area shaded by trees. It is important that the shade not be dense; few ferns survive in very dense shade. Rather, they should be exposed to two to three hours of sunshine each day. Dappled shade is ideal. Avoid strong, direct, midday sun.

The second general requirement is consistent moisture. The garden need not be swampy but it should have moisture throughout the year. Dry soil in summer will require watering.

Ferns do best in a rich, humusy soil, but the garden can begin with almost any kind of soil that is well drained. The regular addition of compost and mulch, and perhaps wood chips, will soon convert the area into a fern haven.

When transplanting a fern to your garden, first select the part of your garden that most closely approaches the plant's natural habitat. With a shovel, dig a hole larger than the root ball of the plant. Add compost or peat and a little bonemeal to the bottom of the hole. Plant the fern with the crown at the surface of the soil, never under the ground level. Fill around the plant with more compost, firm it in place, and water well. Add more compost, dry leaves, or other mulch around the plant and make sure the plant stays moist until it is well established. Each spring and fall add more mulch or compost and occasionally some bonemeal. If you want to remove unwanted leaves or branches, do it by hand, never with a rake for fear of damaging the crosiers.

It is best to space ferns apart to best show off their form. Be sure to take into consideration the plant's mature size. Some plants are crown-formers, growing slowly and forming a vertical crown of fronds. These may stay short but some may reach three to four feet in height. Many ferns, on the other hand, have wide-creeping rhizomes that may spread quickly. Leave space for them to expand and be ready to thin them periodically. A few species are so aggressive that some gardeners prefer not to get them started in the first place. Among the worst offenders are the bracken, sensitive fern, New York fern, and hay-scented fern, but all of these are known to make good garden plants on occasion.

Logs, stones, and other points of interest make effective additions to the fern garden and help to show off the ferns to better advantage. In addition, closely-set rocks will provide a habitat for many of the rock-loving ferns. It is important to know whether the ferns prefer acidic rocks, such as sandstone or granite, or calcareous rocks, such as limestone, and plant accordingly. Most ferns prefer rocks that are mossy and moist. The walking fern, bulblet bladder-fern, and cliff-brakes require limestone, whereas the ebony spleenwort, fragile fern and polypody can grow either on limestone or acidic rocks. The majority of Appalachian spleenworts grow only on acidic rocks. In the drier parts of the country there are many lip ferns, cloak ferns, and spikemosses that can be used. In the southern states the maidenhairs, halberd ferns, and ladder brake thrive on limestone rocks.

If you have a swampy area by a pond or stream, or a seepage area, you can use the ferns that naturally do well with their "feet wet." These include the royal fern, cinnamon fern, the chain ferns, ostrich fern, and sensitive fern. Some of the shield ferns do well there too, such as the crested and spinulose shield-ferns.

A few ferns can survive in full sun most of the day. The best of these is the hay-scented fern, which, although it creeps widely, is very effective among rocks to soften the harsh lines of the stones.

Some enthusiasts may use ferns alone in a shady border, mixing species of different sizes, shapes of frond, and shades of green. However, they also make a good foil for shade-loving flowering plants. Bleeding heart and astilbe mix well with the larger ferns, such as the shield ferns and osmundas. Wild flowers, such as wild geranium and violets, add just a touch of color to the shades of green of the ferns. Columbines come in a variety of sizes and colors and can be interspersed very effectively. Some small flowering shrubs, such as azaleas, add a larger mass of color in the springtime.

Some of the finest foliage plants to associate with fern fronds are hostas, with their variegated leaves striped with green and white, or the multicolored coleuses. The pale green-and-white leaves of aegopodium also contrast well with the darker green of the ferns.

Ferns can be used to edge walkways made of woodchips. Spreading rhizomatous species are best for this as they fill in the space quickly. Maidenhairs are especially good. Even the much-maligned bracken has been used at times in more open areas as a walk border.

In addition to your local ferns, nurseries often carry species from other parts of the world that will suit your climate. Of course, the parts of the country that lack frost can take advantage of the many tropical species, but even the more temperate parts of the continent have a wider selection of possible species than is often realized. There are many species in the Northeast and Northwest but some hardy species from northern Europe and Japan are also commercially available. The autumn fern (*Dryopteris erythrosora*) and the Japanese painted fern (*Athyrium niponicum*) from Japan are especially good; the hart's-tongue varieties and several crested forms of lady fern and the male fern from Europe are sometimes available.

Some of the fern allies can also be introduced to the garden, but most of them are either not well adapted for cultivation or they are undesirable. Best are the shining clubmoss (*Lycopodium lucidulum*) and the tree clubmoss (*L. obscurum*). Most of the other clubmosses do not transplant well, so the success rate is rather low. Some of the horsetails might be introduced, especially if the area is very wet and fairly sunny, but they may spread too rapidly. Horsetails provide interesting texture and lines not found in other pteridophytes, and more could be done with them. The woodland horsetail (*Equisetum sylvaticum*), variegated scouring-rush (*E. variegatum*), and dwarf scouring-rush (*E. scirpoides*), are especially interesting garden plants; they are less likely to take over than the field horsetail (*E. arvense*) and common scouring-rush (*E. hyemale*).

REFERENCES

Davenport, Elaine. Ferns for Modern Living. Kalamazoo, MI: Merchants Publ. Co. 1977.

Handbook on Ferns. Brooklyn Botanic Garden Record Vol. 25 (1)(Spring) 1969.

Foster, F. Gordon. Ferns to Know and Grow. New York: Hawthorn Books. 1971.

Hoshizaki, Barbara Joe. Fern Growers Manual. New York: Knopf. 1975.

Kaye, Reginald. Hardy Ferns. London: Faber & Faber. 1968.

Mickel, John T. The Home Gardener's Book of Ferns. Ridge Press/Holt, Rinehart & Winston. 1979.

How to Collect Ferns

Fern plants are often collected to be used for comparison, study, teaching, identification, and as records or vouchers of local species. Although a living collection has definite advantages, dried specimens are better in many ways for permanent study: they require less care, less space, and will last indefinitely. In order to prepare dried specimens, you need some means of pressing the plants and extracting their moisture. Usually this is done with a **plant press.** A plant press consists of a series of corrugated cardboards or blotters or both to dry the plants, sheets of newspaper to hold the plants, and a sturdy top and bottom frame (Fig. 38) or two pieces of plywood twelve by eighteen inches. Ropes or belts around the press are used to pull it tight.

To begin, the plant must be removed from the soil. Dig the entire fern in order to have the rhizome and roots for identification. It may be necessary to dig under the rhizome with a trowel, rock hammer, or shovel, but some fern rhizomes can be removed by hand. Shake or wash all the soil from the roots.

Open the press and lay the first corrugate on the bottom frame. Open a single sheet of newspaper and lay the fern on half the paper (Fig. 39). Be sure all of the plant parts are spread flat so there is little overlap. Turn over one frond or frond part so the finished specimen will show both top and bottom surfaces. It is best if the frond is fertile, i.e., with sporangia, since that is the basis of most identification. If the frond is longer than the newspaper,

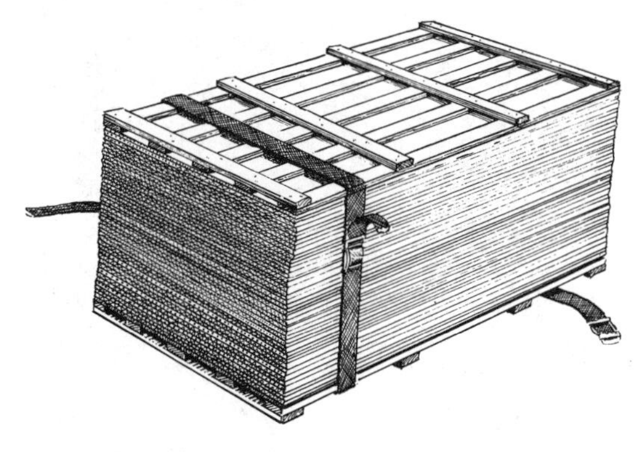

Figure 38 Plant press

Figure 39 Specimen in newspaper sheet

fold it in a V or N (Fig. 42), with as little overlap as possible, or cut the frond into parts and press them separately. With large fronds you may wish to remove the pinnae from part of one side to minimize the overlap. Close the newspaper, put a corrugate on top and you are ready to press the next specimen. Make notes on the margin of the newspaper and in your notebook on details not shown by the specimen alone, such as locality, collector, date, habitat, abundance, size variation, and color.

When you have finished placing your plants in the press, put the rest of the corrugates on top, add the top frame, and tighten the belts or ropes around it to apply maximum pressure to the press. With a full press it is often helpful to stand on the press while pulling the ropes tight.

John T. Mickel

Figure 40 The author preparing specimens in the field

Blotters may be used instead of corrugated cardboards to absorb moisture. (Blotters can be made from builders deadening felt which can be obtained in rolls from lumber yards.) Some people use both, in the sequence of: corrugate, blotter, newspaper with specimen, blotter, corrugate, blotter, newspaper with specimen, etc. Corrugates without blotters are often preferred for simplicity.

The completed press can then be set aside for a few days for the plants to dry. If corrugates are used, heat may be used to speed the drying process. Though the press can be set out in the sun, it is best to have a heat source under the press to allow the heat to rise through the air spaces of the corrugates, carrying off the moisture. Lanterns, camp stove, electric lights, catalytic heaters, or any other source of heat can be used. To direct the maximum heat through the press, it should be placed on a rack or frame with the heat source beneath and the rack surrounded by a piece of canvas or asbestos (Fig. 41). The distance between the heat and the press is dictated by the strength of the heat; i.e., catalytic heaters may require a foot or more of space below the press to avoid burning the corrugates, whereas light bulbs may be only a few inches below the press. (In Mexico I once placed a press over an open fire to dry, but it was too close and the press caught fire. We nearly burned down the thatched hut we were working in.) Generally with strong heat, a press of ferns will be completely dried in eight to twelve hours. If a practical heat source is not available, drying can be speeded by replacing the corrugates or blotters with dry ones each day, the moist ones being dried in turn by placing them in the sun or by a fire. If you are travelling, another method of drying is to place the press on top of the car with the corrugate openings facing into the wind. As the plants dry, the ropes or straps often loosen, so it is wise to check and retighten them periodically.

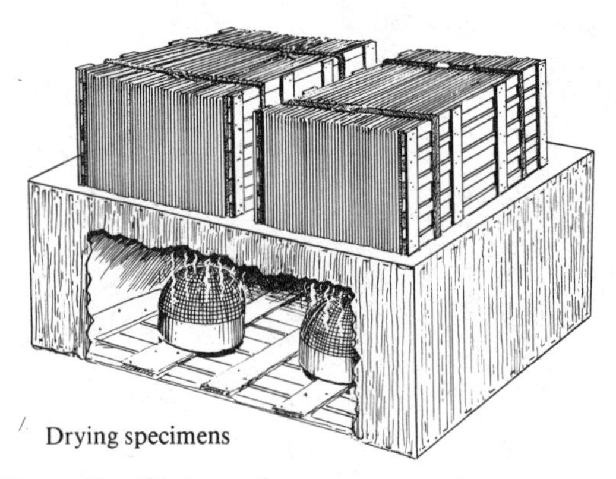

Figure 41 Drying rack

For smaller specimens, it is not always necessary to use a press for drying. In fact, it may actually be preferable to use old telephone directories or magazines with absorbant (not slick) pages for drying leaves. You can also put the plant in a newspaper with other newspapers above and below as blotters and a book or other weight on top. Many variations can be improvised, the main elements being some sort of blotting action accompanied by pressure to keep the specimen flat.

Once the specimens have dried, the next step depends on the purpose you have for them. For formal herbarium specimens, the plant is mounted on a standard-sized sheet of heavy white paper eleven-and-one-half by sixteen-and-one-half inches, obtainable from biological supply houses. The specimen may be attached to the sheet with Elmer's glue or special mounting plastic. Generally, the fern specimens are mounted with the bottom side of the frond up to show the sori (although part of the frond should be turned over to show the top side as well). Dribble glue or mounting plastic on the back side of the frond, not to completely cover the back surface, but to have streaks of glue. Make heavier application on the thicker axes and rhizome. Then place the specimen in an esthetically pleasing position on the sheet, pressing it lightly. Add small weights to the specimen to hold it down while the glue dries. Add an extra dribble of glue as a strap over the larger axes and the rhizome. In place of glue or plastic, some people prefer to use gummed linen straps to hold the plant on the sheet. Never use cellophane tape since this discolors, dries out, and comes off after a few years.

In the lower right corner you should affix a label to give the name of the plant, where and when and by whom it was collected, and any other pertinent information, such as habitat conditions, from your notes (Fig. 42).

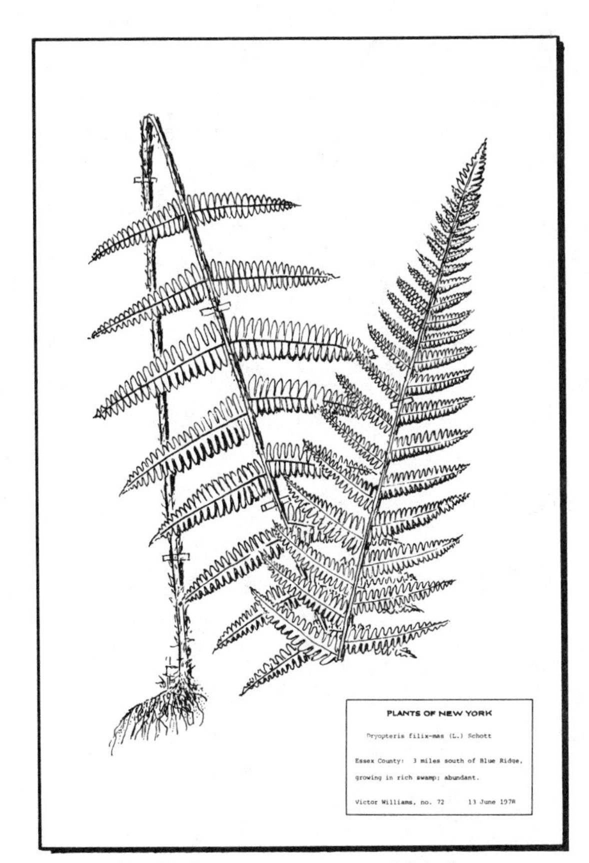

Figure 42 Herbarium specimen and label

You may want to use the dried specimens for a small teaching collection. Some people mount the specimens on sheets in a notebook, protecting the specimen with clear contact paper. Others use the specimens to make dried floral pictures, bookmarks, and other craft items. You may want to show the developmental sequence of fronds of one kind of plant, from juvenile to adult. Some people like to make a collection of specimens of local ferns and to exchange specimens with botanists from other areas.

In collecting specimens it is important that you take only species that are in no way rare or endangered. When in doubt, leave the plant alone or carefully remove only an individual frond (if the plant is robust), but even this must be done with utmost care and a sharp knife since pulling off a leaf usually results in disturbing the entire plant. Grape ferns are best collected in this way since they rarely have identification characters on the underground stems, and removal of a leaf does not seriously harm the plant. In many states ferns are protected by law, so it is best to acquaint yourself with the laws of your state.

How Ferns Are Named

The botanical name of a plant is in Latin or Greek and consists of two parts—the genus name and the specific epithet, usually a noun and an adjective respectively. The genus name is a group name, like oak or spleenwort and is capitalized—*Quercus, Asplenium.* The specific epithet, or species name, applies to a particular kind in that group, like white oak, ebony spleenwort. The epithet may be descriptive, or tell where it was found, or it may honor its collector or other appropriate person (e.g., *Asplenium platyneuron, Polypodium virginianum, Cheilanthes fendleri*). The epithet is generally in lower case, although personal names may be capitalized. To be complete, the botanical name is followed by its author's name, the botanist who named the species and described it. Often one author's name is in parentheses and there is a second author's name. The one in parentheses is the one who originally described that species but under a different genus name or at a different classification level, such as variety of a species. The second author placed it under its present genus or brought it up from varietal rank; e.g., *Dryopteris marginalis* (L.) A. Gray was first described by Linnaeus under *Polypodium* and later placed in *Dryopteris* by Asa Gray. Similarly *Botrychium oneidense* (Gilbert) House was first named by Gilbert as a variety of *B. ternatum* and raised to species rank by House.

The variety name that repeats the species name (the so-called type variety) is not followed by an author's name, since it is automatically the same as that of the species.

One problem that disturbs many people is that the names do not seem fixed or permanent. A frequent accusation is that botanists are continually changing the plant names. Actually scientific names are intended to be stable, and there is an International Code of Botanical Nomenclature to direct the naming of plants. The prime rule is that the earliest name for a species must be used.

Name changes come about either to correct mistakes or to reflect changes in botanical opinion, usually as a result of new knowledge. In some cases botanists in the past unknowingly did not use the earliest name for a species. Sometimes new information makes it clear that some species do not belong in the genus they previously were placed in; for example, many species of *Dryopteris* were transferred to *Thelypteris* when several lines of evidence showed that these groups were quite distinct and not at all related to one another.

On the other hand, as in most bodies of knowledge, there are differences of opinion among the experts. Some botanists do not believe that certain genera are distinct enough to warrant separate status and therefore place the species of those genera in other genera. Other botanists may have a more restricted view and place higher value on certain distinguishing charactristics, recognizing a higher number of genera with fewer species in each. For example, some believe that *Phyllitis* and *Camptosorus* should be lumped with *Asplenium* since their species often hybridize, suggesting a very close relationship. Others recognize the close relationship but prefer to keep the genera separate because they appear so distinct. Thus, different authors may apply different correct names to the same plant. This is understandably confusing to the amateur botanist.

Much to the relief of many people, most of our native ferns have been given colloquial English names, or so-called "common names." These have the advantage of being more easily learned and pronounced, but they are not entirely satisfactory. With few exceptions, there is no general agreement on which common names to use. An outstanding example is *Lycopodium clavatum,* which has twenty-four common names. Similarly, the same common name is often applied to different species; "ground pine" applies to five species of *Lycopodium,* "shield fern" is used for species of both *Dryopteris* and *Polystichum.* Furthermore, common names are of no use in communicating with botanists from other countries, where they have their own sets of common names. This is not to say we should avoid the common names; a few have been more stable than the scientific name (e.g., ostrich fern). Common names are often convenient and useful, but when accuracy is necessary, as in botanical research, the scientific names must be used.

How to Use This Book

In order that you might fully use the resources of this book, I would strongly urge that before using the keys, you familiarize yourself with the general structure of the ferns as described in the introductory chapters. Otherwise, some of the characters mentioned in the keys and species diagnoses will be unfamiliar to you and you will tend to overlook some very useful information. These features are illustrated in the glossary and/or the introductory chapters.

For ease of identification it is important to look at the entire plant, including the underground parts and fertile fronds. Many parts of the keys refer to the rhizome, the rhizome scales, and the disposition of the sori. If these are lacking, it makes identification difficult, if not impossible.

For better observation some sort of magnification is helpful. A 10-power hand lens should be adequate for most observations; in fact, the keys are specifically designed for use with a hand lens and the unaided eye, ignoring, for the most part, characters that require a microscope. There are a few groups that require stronger magnification, however. The classification of *Isoetes,* for example, is based primarily on the sculpturing of the megaspores, which requires a dissecting microscope with magnification of thirty times, and *Azolla* glochidia require even greater magnification.

If you do not know the general group your fern belongs to, go first to the generic key (p. 34), which will lead you to the genus. Because of the great range of variation within several of the fern genera, there are problems in developing a completely fool-proof generic key without using very technical characters or becoming so wordy and complex that the key loses its usefulness. Hopefully it will be only the exceptional plant that will not key out easily to the genus. Reference to the species illustrations should be of great help in finding the correct genus. Each genus in turn will have a key to its species. If you know the genus, go directly to it and key out the species.

In using any key, you may have difficulty at a particular choice and cannot decide which way to go. In that case, try both leads of the key; the wrong way will probably become evident from the wrong geographical range, the use of characters that do not exist on your plant, or the species illustrations will not be of your plant.

Obviously this book has been prepared around the keys, and the intention is that they be as useful as possible. However, I realize that most people do not like to read any more than necessary (myself included), and there are several suggestions for saving time. First, I have included a list of especially distinctive species or genera that can be identified readily without wading through a long key. Second, careful browsing through the illustrations of the book will familiarize you with the diversity of form and assist you in rapid identification. Third, the geographic ranges will, in many cases, help eliminate a great many possibilities. Fourth, the notes on relative frequency will also be a rough guide to your chances of finding the species in the wild. The abundance of a species, however, is not uniform throughout its range; it may well be abundant in some areas, but be a rarity on the fringes of its range. For example, the sensitive fern is abundant in the Northeast but is a rarity in Florida.

Comparison of your plant with correctly identified specimens is also a great help. These reference collections might be your own or those of a friend or even a botanical museum.

Best of all, go out with people who know the local ferns. Learn the common species first. Compare them with the illustrations and text of this book and run them through the key. This will make the key easier to use for other species. Try to learn to genera of your area so you can go directly to the species keys.

Note the checklist at the end of the book as an aid in keeping a record of your finds. In warmer parts of the United States, especially along the Gulf coast, there is some likelihood of finding species of ferns that are not in this book because cultivated species escape and can become naturalized. There are also some species blown in as spores from the West Indies which become established, some lasting for a single generation whereas others may reproduce themselves and spread.

FERN SOCIETIES

There are several fern societies in North America, ranging from local groups to those international in scope. The society with the broadest interest in native North American pteridophytes is the American Fern Society. Organized in 1893, it now has nearly two thousand members, mostly amateurs, who are interested in all aspects of fern study. The Society publishes the American Fern Journal, which contains largely technical articles and notes, and a bulletin, Fiddlehead Forum, which contains articles, notices, and reports of a less technical nature. An international spore exchange, including spores of over seven hundred species, is maintained for Society members. Meetings, field trips, and several regional chapters facilitate interchange of ideas. Further information can be obtained from the author or from:

American Fern Society
Department of Botany
Smithsonian Institution
Washington, D.C. 20560

Quick Recognition Groups

Certain genera have such distinctive features that they can be recognized easily without using the more technical characters in the key. Some of these genera and their more obvious features are listed here. Space is left below for you to add your own short-cut notes for groups particularly obvious to you.

Silverback or goldback: some *Cheilanthes, Pityrogramma, Trismeria*
Floating ferns: *Azolla, Ceratopteris, Salvinia*
Four-leaved clover: *Marsilea*
Shoestring fern: *Polypodium angustifolium, Vittaria*
Climbing, vine-like leaf: *Lygodium*
Climbing rhizome: *Lomariopsis,* some *Polypodium, Stenochlaena*
Grass-like leaves: *Asplenium septentrionale, Isoetes, Pilularia, Schizaea*
Filmy ferns (leaves one cell thick): *Hymenophyllum, Trichomanes*
Erect spore-bearing portion of leaf: *Anemia, Botrychium, Ophioglossum*
Upright forking green stems with no leaves at all: *Psilotum*
Moss-like: *Azolla, Lycopodium, Selaginella*
Spiny stipe: *Hypolepis*
Rooting leaf tip: *Camptosorus, Thelypteris reptans*
Bulblets (buds) on rachis: *Ceratopteris, Cystopteris bulbifera, Polystichum andersonii, Woodwardia radicans*
Black, wiry stipe and rachis: *Adiantum,* some *Asplenium,* some *Cheilanthes,* some *Pellaea*

Keys to the North American Ferns and Fern Allies

1a Leaves small (needle- or scale-like) or grass-like, with only one unbranching vein; leaves rarely absent altogether; sporangia borne on stem or in axils of leaves, often in cones. .... 2 (FERN ALLIES)

1b Leaves broader, with branching veins; sporangia on leaves or in nut-like sporocarps, never borne in cones or in axils of leaves. .... 6 (FERNS)

2a Stems branching into equal divisions, erect, green; leaves and roots lacking; sporangia three-chambered. .... (p. 180) *Psilotum*

2b Stems branching variously; leaves and roots present; sporangia one-chambered. 3

3a Leaves whorled, scale-like, non-green; branches (if present) whorled; stems jointed, grooved, mostly hollow; sporangia borne in terminal cones composed of polygonal, umbrella-like structures with sporangia beneath. .. (p. 114) *Equisetum*

3b Leaves spirally arranged, green; stems not jointed, grooved, or hollow; sporangia borne in leaf axils or in leaf bases, or if in a cone, in axils of minute modified leaves. 4

4a Plants in or near water; leaves long, grass-like; sporangia embedded in leaf bases. .... (p. 124) *Isoetes*

4b Plants terrestrial; leaves short, scale- or needle-like; sporangia in axils of leaves or modified leaves of cone. 5

5a Plants homosporous; cones cylindrical or sporangia borne in axils of vegetative leaves. .... (p. 130) *Lycopodium*

5b Plants heterosporous; cones four-sided in cross-section (only rarely cylindrical). .... (p. 186) *Selaginella*

6a (1b) Plants heterosporous, spores borne in sporocarps; unfernlike in appearance; aquatic, floating or rooting in mud. 7

6b Plants homosporous; typical ferns, occupying various habitats. 10

7a Plants floating; fronds very small (one-half inch across or less), lacking stipes, opposite or whorled. 8

7b Plants rooting in mud (rarely floating free); fronds with stipes two inches or more long; blade one inch or more across or lacking; fronds spirally arranged. 9

8a Fronds about one-half inch across, round or oval, the upper surface covered with erect hairs. .... (p. 185) *Salvinia*

8b Fronds about one-eighth inch across, two-lobed; hairs lacking. .... (p. 60) *Azolla*

9a Pinnae four, wedge-shaped, crowded at the tip of the stipe, looking like a four-leaved clover. .... (p. 142) *Marsilea*

9b Pinnae lacking, fronds consisting of only green stipes. .... (p. 165) *Pilularia*

10a (6b) Frond divided near the base into a vegetative portion and one or two erect fertile spikes or dissected parts. 11

10b Frond lacking a distinct erect fertile portion arising from the stipe. 13

11a Erect fertile parts two; stem a creeping rhizome at or near the ground surface. .... (p. 44) *Anemia*

11b Erect fertile portion single (rarely several if an epiphyte); plant soft; sporangia massive; stem erect, tuberous, subterranean, an inch or more beneath ground surface (or buried in leaf bases of palms). 12

12a Sterile blade undivided or rarely a hand-shaped epiphyte; veins netted; fertile portion an undivided spike or spikes. .... (p. 150) *Ophioglossum*

12b Sterile blade divided; veins free; fertile portion much branched. .... (p. 63) *Botrychium*

13a (10b) Fronds extremely delicate and filmy, only one cell thick, translucent, less than six inches tall; sporangia in cups on leaf margin; rhizome fine and thread-like. 14

13b Fronds more than one cell thick, not translucent, more than six inches tall, or if smaller, then with sori not in marginal cups; rhizome one-eighth inch or more thick. 15

14a Marginal soral cups each consisting of two flaps. ...... (p. 122) *Hymenophyllum*

14b Marginal soral cups tubular. .... (p. 208) *Trichomanes*

15a Fronds regularly dividing into equal branches; pinnae deeply pinnatifid. .... (p. 104) *Dicranopteris*

15b Fronds not regularly dividing this way. 16

16a Individual fronds several feet long, climbing, twining, slender. .... (p. 141) *Lygodium*

16b Individual fronds not climbing (although whole plant may climb). 17

17a Vegetative fronds like grass, twisted spirally or rarely straight; fertile fronds with long, slender stipe and tiny pinnae at the end to bear the sporangia. .... (p. 185) *Schizaea*

17b Fronds not grass-like; fertile frond not just long naked stipe with tiny pinnae at the tip. 18

18a Fertile fronds or pinnae strikingly different from the sterile, not just slightly narrower. 19

18b Fertile fronds or pinnae about the same as the sterile, sometimes somewhat narrower. 28

19a Fertile fronds or pinnae totally lacking leafy tissue, but consisting of masses of sporangia which are dark green when young, brown after spores are shed. .... (p. 154) *Osmunda*

19b Fertile fronds or pinnae possessing leafy tissue, though occasionally small in amount or turning brown with age. 20

20a Fertile fronds tough and woody, with the pinnules tightly curled and enclosing the sori; remaining erect through the winter, releasing spores in late winter or spring. 21

20b Fertile fronds of same texture as the sterile, with the pinnules not tightly curled; releasing spores during summer, shortly after being formed. 22

21a Fronds tapering toward the base; veins free; blade pinnate-pinnatifid. .... (p. 145) *Matteuccia*

21b Fronds broadest near base of blade; veins netted; blade pinnatifid, the pinnae with wavy margin. .... (p. 149) *Onoclea*

22a Fertile fronds once pinnate. 23

22b Fertile fronds more than once divided. 25

23a Veins of sterile blade netted. .... (p. 214) *Woodwardia*

23b Veins of sterile blade free. 24

24a Rhizome erect, terrestrial; sterile leaves lying on ground, fertile erect; northwestern. .... (p. 61) *Blechnum*

24b Rhizome long-creeping, climbing trees; leaves spreading; Florida. .... (p. 129) *Lomariopsis*

25a (22b) Sterile fronds once pinnate; rhizome long-creeping, climbing trees. .... (p. 196) *Stenochlaena*

25b Sterile fronds more than once pinnate; rhizomes short-creeping or ascending. 26

26a Fronds less than six inches tall; northern and western. ...... (p. 98) *Cryptogramma*

26b Fronds 5-36 inches tall; southern. 27

27a Rhizome stout, woody, climbing trees; sori round, medial. .... (p. 146) *Maxonia*

27b Rhizome slender, soft, largely aquatic or of muddy banks; sori elongate along the margin. .... (p. 72) *Ceratopteris*

28a (18b) Sporangia on the under surface of the frond, in discrete sori or densely spread over the surface; never with a white or yellow waxy lower surface. .. 29

28b Sporangia along the margin, or if on the under surface, then running along the veins without an indusium; often with a white or yellow waxy lower surface. .. 52

29a Sporangia spread over the entire under surface; blade once pinnate. .... (p. 40) *Acrostichum*

29b Sporangia not spread across the under surface, but located in discrete sori. 30

30a Sori with an indusium. 31

30b Sori lacking an indusium. 48

31a Sori elongate along both sides of the midvein. 32

31b Sori not elongate along the midvein. .. 33

32a Sori continuous, only one sorus on each side of pinna midvein. .... (p. 61) *Blechnum*

32b Sori interrupted, chain-like, with several sori on each side of the pinna midvein. .... (p. 214) *Woodwardia*

33a Sori round. 34

33b Sori elongate along veins. 43

34a Veins netted with intricate network; pinnae three pairs or fewer (rarely more). .. .. (p. 197) *Tectaria*

34b Veins free (rarely netted); pinnae four or more pairs. 35

35a Indusium attached in the middle, umbrella-like, rigid. 36

35b Indusium attached laterally (hood-like) or below the sporangia (napkin-like or nearly cup-like) or kidney-shaped. 38

36a Sori usually in a single row on each side of the midvein; blade pinnate to bipinnate, pinnate species with pinnae about one-half inch wide. .... (p. 175) *Polystichum*

36b Sori in two or more rows on each side of the midvein; blade pinnate; pinnae three-fourths to 1 inch wide. 37

37a Veins free; plants of the Southwest. .... (p. 163) *Phanerophlebia*

37b Veins netted; escape from cultivation in the Southeast and California. .... (p. 100) *Cyrtomium*

38a (35b) Indusium attached beneath the sporangia or at one side of the sorus and covering it, hood-like. 39

38b Indusium kidney-shaped, attached at its sinus. 40

39a Indusium hood-like. .... (p. 101) *Cystopteris*

39b Indusium attached under the sorus and coming up around it on all sides. .... (p. 210) *Woodsia*

40a Stipe with two strap-shaped vascular bundles near the base; blade with numerous tiny, needle-shaped hairs. .... (p. 198) *Thelypteris*

40b Stipe with several round vascular bundles near the base; blade naked or with hairs but hairs not sharp-pointed. 41

41a Frond once pinnate with the pinnae articulate to the rachis. .... (p. 146) *Nephrolepis*

41b Frond more than once divided; pinnae not articulate to the rachis. 42

42a Frond with many very small hairs (scales usually present too); southern Florida. .... (p. 99) *Ctenitis*

42b Frond lacking hairs, although scales and minute glands may be present; widespread. .... (p. 106) *Dryopteris*

43a (33b) Frond undivided. 44

43b Frond divided. 45

44a Frond slender, four to eight inches long, rooting at tip; veins netted; sori scattered in all directions. .... (p. 71) *Camptosorus*

44b Frond strap-shaped, about one inch wide, eight to twelve inches long; veins free; sori long, parallel, paired and facing each other. .... (p. 164) *Phyllitis*

45a Fronds two to eighteen inches tall; rhizome scales clathrate; stipe with one x-shaped bundle. 46

45b Fronds twelve to sixty inches tall; rhizome scales not clathrate; stipe with two discrete ribbon-shaped bundles. 47

46a Blade undivided to four times divided; if pinnatifid, plants of eastern North America. .... (p. 46) *Asplenium* and *Asplenosorus*

46b Blade pinnatifid; Arizona. .... (p. 73) *Ceterach*

47a Sori at least partly in pairs back to back on the same vein. .... (p. 104) *Diplazium*

47b Sori all single, often hooking over the vein like a buttonhook. .... (p. 57) *Athyrium*

48a (30b) Blade undivided to once pinnate. 49

48b Blade pinnate-pinnatifid to tripinnate. 50

49a Fronds three to twenty-four inches long, naked or scaly, not hairy; widespread. .... (p. 167) *Polypodium*

49b Fronds one to two inches long, hairy, soft; North Carolina. .... (p. 121) *Grammitis*

50a Fronds sixteen to forty-eight inches tall; blade oval to lance-shaped. .... (p. 57) *Athyrium*

50b Fronds six to twelve inches (rarely to eighteen inches) tall; blade broadly triangular. 51

51a Blade pinnate-pinnatifid or bipinnatifid, with tiny, sharp-pointed hairs. .... (p. 198) *Thelypteris*

51b Blade bipinnate to tripinnate, naked or with minute glands. .... (p. 121) *Gymnocarpium*

52a (28b) Blade undivided and unlobed; epiphytes. 53

52b Blade pinnatifid or more divided; terrestrial. 54

53a Blade one-eighth inch broad, like a shoestring. .... (p. 210) *Vittaria*

53b Blade one-half to one inch broad, narrowly lance-shaped. .... (p. 149) *Neurodium*

54a Sori in marginal cups. 55

54b Sori along the veins or along the margin, not in cups. 56

55a Fronds eighteen to forty-eight inches tall; sori on margins of segments. .... (p. 103) *Dennstaedtia*

55b Fronds less than twelve inches tall; sori at tips of very narrow, wedge-shaped segments. .... (p. 196) *Sphenomeris*

56a Sori elongate along the veins. 57

56b Sori close to the margin. 61

57a Pinnae naked. .... (p. 156) *Pallaea (bridgesii)*

57b Pinnae scaly, hairy, or waxy. 58

58a Rhizome long-creeping; blade pentagonal, densely stiff-hairy. .... (p. 63) *Bommeria*

58b Rhizome short-creeping or ascending (or if long-creeping, blade not pentagonal), variously hairy or scaly or waxy, but not densely stiff-hairy. 59

59a Blade with tiny, needle-like hairs, not waxy; sori distinct, oblong. .... (p. 198) *Thelypteris (pilosa)*

59b Blade waxy on under surface, not with sharp, needle-like hairs; sori diffuse, sporangia along veins and not in distinct sori. 60

60a Pinnae pinnately lobed or dissected. .... (p. 165) *Pityrogramma*

60b Pinnae (at least the lower ones) divided at their bases into three nearly equal narrowly oblong parts. .. (p. 209) *Trismeria*

61a (56b) Sporangia borne on under surface of recurved portion of segments, i.e., the under side of the false indusium, not just protected by them; segments fan-shaped or rhomboid, without strong midvein; blade not scaly or long-hairy. .... (p. 41) *Adiantum*

61b Sporangia borne at or near the margin of the segments, the margin slightly to strongly recurved but not bearing the sporangia on their under surfaces; segments with distinct midvein; blade naked or more often hairy or scaly. 62

62a Fronds three to eighteen inches (rarely to twenty-four inches) tall; segments less than one inch long. 63

62b Fronds mostly twenty-four to forty-eight inches (or more) tall; if smaller, then with segments over two inches long. .. 64

63a Fronds naked or with hairy stipe and rachis; margin recurved; rhizome scales usually bicolorous; blade pinnate to bipinnate, or if tripinnate, then with light-colored stipe. .... (p. 156) *Pellaea*

63b Fronds with hairs, scales, or wax, or if naked, blade tripinnate at base and rhizome scales concolorous; margin recurved or not; rhizome scales bicolorous or concolorous; blade pinnate to five times pinnate; stipe light or dark. .... (p. 74) *Cheilanthes*

64a Sori very short, protected by small indusia; stipe with small prickles. .... (p. 123) *Hypolepis*

64b Sori continuous along the margin; stipe lacking prickles. 65

65a Rhizome long-creeping, deep in the ground; rhizome and stipe hairy; blade broadly triangular. .. (p. 181) *Pteridium*

65b Rhizome ascending or very short-creeping at surface of ground; rhizome and stipe base scaly; blade somewhat lance-shaped, not triangular. .... (p. 182) *Pteris*

ACROSTICHUM

Leather Fern

Rhizome massive, creeping; fronds large and coarse, pinnate; pinnae smooth-margined, leathery; veins netted; sporangia not in discrete sori but spread across the under surface of the pinnae. Pantropical genus of three species inhabiting swamps and brackish marshes.

1a Pinna areoles divergent from pinna midvein at an angle of forty-five to sixty degrees; fronds fertile only in the upper one-fourth to one-third; blade smooth

beneath; fronds to ten feet tall. (Fig. 43A). .... LEATHER FERN, *Acrostichum aureum* L.

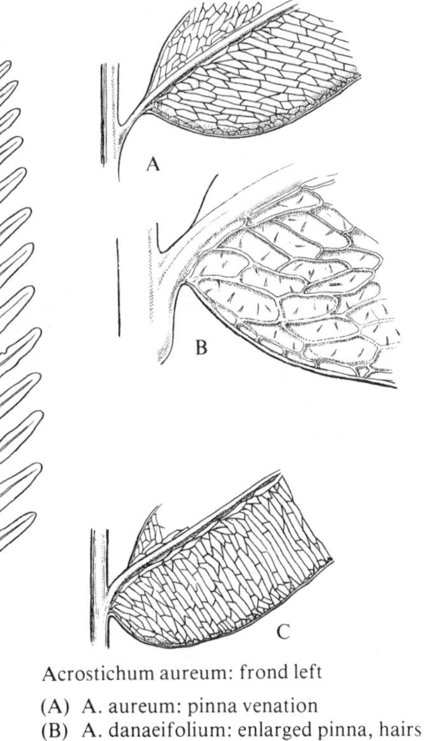

Acrostichum aureum: frond left

(A) A. aureum: pinna venation
(B) A. danaeifolium: enlarged pinna, hairs
(C) A. danaeifolium: pinna venation

Figure 43

Figure 43 *Acrostichum*

Brackish water, salt marshes and freshwater swamps. Rare. Peninsular Florida; pantropical.

1b Pinna areoles divergent from pinna mid-vein at an angle of sixty to eighty-five degrees; fronds fertile in the upper one-half to two-thirds; blade often with small, stiff hairs on the under surface; fronds to sixteen feet tall. (Fig. 43B, C). .... GIANT LEATHER-FERN, *Acrostichum danaeifolium* Langsd. & Fisch.

Fresh-water swamps. Common. Peninsular Florida; West Indies, Mexico to South America.

ADIANTUM

Maidenhair

Rhizome short-creeping, scaly; fronds small to medium-sized with black wiry stipes and rachises, and with thin blade texture; sori borne on the underside of the reflexed margin; often lime-loving. About three hundred species, largely of tropical regions.

1a Stipe and rachis smooth; segments about the same size throughout the pinna length. 2

1b Stipe and rachis hairy; segments progressively smaller toward the pinna tips. (Fig. 44). .... ROUGH MAIDENHAIR, ROSY MAIDENHAIR, *Adiantum hispidulum* Sw.

2a Stipe and rachis straight or slightly wavy, branching pinnately. 3

2b Stipe forking in half to form two subequal curving rachises, each bearing pinnae on one side. (Fig. 45). .... NORTHERN MAIDENHAIR, *Adiantum pedatum* L.

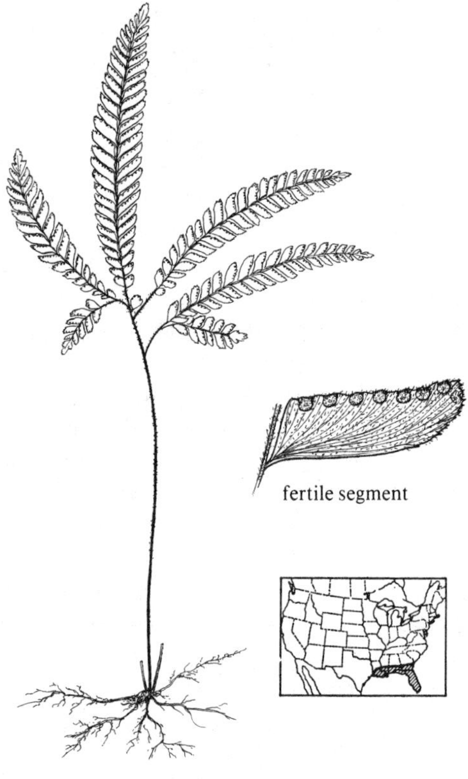

Figure 44

Figure 44 *Adiantum hispidulum*

Fronds about twelve inches tall, bipinnate to tripinnate; largest segments about three-eighth inch long, hairy below; pinnae taper toward the tip; sori round. On banks and old walls. Rare. Louisiana to southern Georgia and northern Florida, established in the wild from cultivation; native of Asia and Australia.

Another species with hairy stipe and rachis, the FRAGANT MAIDENHAIR (*A. melanoleucum* Willd.), has been found rarely on limestone in Everglades hammocks of Dade Co., Florida. It is larger (eighteen to twenty-four inches tall), has larger segments (about three-fourths inch long), is smooth below, and has oblong sori; native of the Caribbean region.

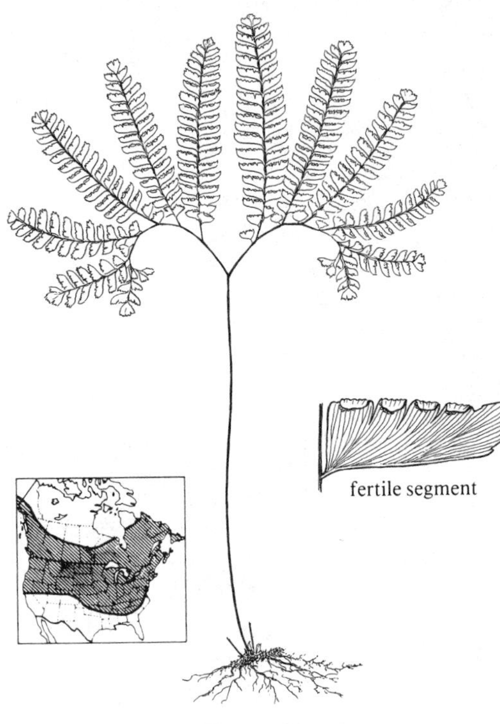

Figure 45

Figure 45 *Adiantum pedatum*

Fronds up to twenty-eight inches tall, the stipe apparently forking in half with the two branches each curving to form a semicircle; the pinnae spreading to form a fan-shaped blade; tripinnate; segments oblong, naked, about one-half inch long; sori oblong. Moist, well-drained woods. Common. Much of North America.

Var. *aleuticum* Rupr. is found in the Northeast (Vermont northward) and the Northwest and differs from var. *pedatum* in being more compact with fronds more crowded and pinnae fewer and more upright.

3a Segments often smooth or wavy-margined; sterile segments not sharp-toothed, or if so, fertile segments not deeply incised. 4

3b Both fertile and sterile segments usually incised; sterile segments sharp-toothed. (Fig. 46). .... SOUTHERN MAIDENHAIR, *Adiantum capillus-veneris* L.

Figure 46 *Adiantum capillus-veneris*

Fronds to twenty-two inches tall, bipinnate to tripinnate; segments fan-shaped to rhomboid, one-fourth to three-fourths inch long; sori oblong. Limestone rocks and walls. Frequent. Southern and western North America; widespread in tropical and warm-temperate regions of the world.

4a Segments articulate, the black color of the segment stalk stopping abruptly in a swelling at the base of the segment; Florida. (Fig. 47). .... BRITTLE MAIDENHAIR, *Adiantum tenerum* Sw.

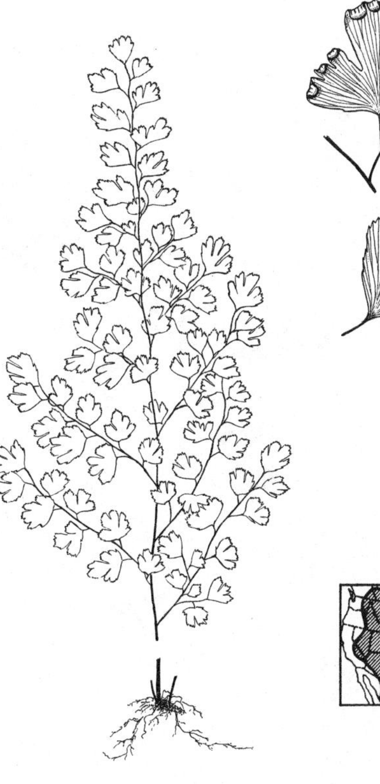

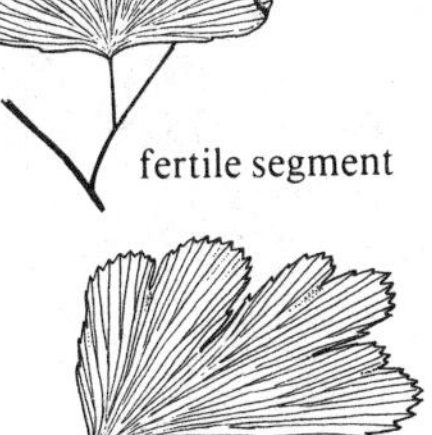

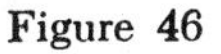

Figure 46

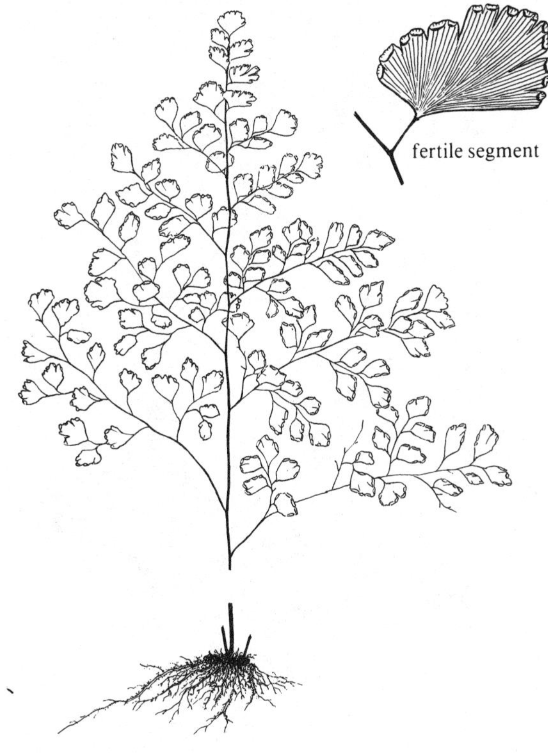

Figure 47

Figure 47 *Adiantum tenerum*

Fronds to twenty-six inches tall, tripinnate; segments rhomboid, one-half to three-fourths inch long, naked, jointed at base; sori short, several per segment. Shaded limestone rocks. Frequent. Peninsular Florida; West Indies, Mexico to South America.

4b Segments not articulate, the dark pigment of the segment stalk running slightly into the segment base; far western. (Fig. 48). ...... CALIFORNIA MAIDENHAIR, *Adiantum jordanii* C. Müll.

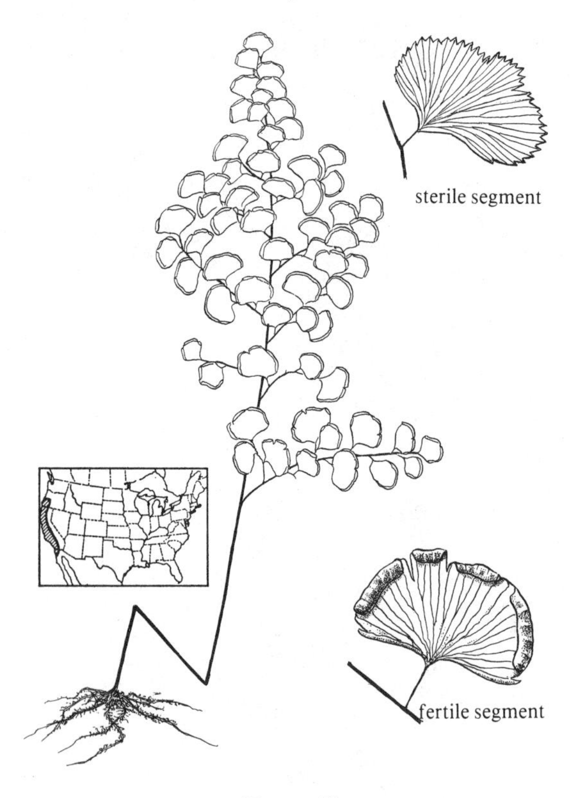

Figure 48

Figure 48 *Adiantum jordanii*

Fronds to eighteen inches tall, bipinnate to tripinnate; segments quite variable in size, three-eighths to one-and-one-fourth inches wide, fan-shaped; sterile segments toothed; sori long, one to four per segment. Shaded canyon ledges. Frequent. Far western United States; Baja California.

Adiantum jordanii hybridizes with *A. pedatum* in California to form *A.* × *tracyi* C. C. Hall ex W. H. Wagner, a sterile hybrid.

The HAIRY MAIDENHAIR, *A. tricholepis* Fée, is a rare fern in Texas but more common in Mexico. It has stiff hairs on both the top and bottom surfaces of the segments, which are nearly round, and the sterile margins are untoothed.

ANEMIA

Rhizome creeping, clothed with dark hairs; fronds pinnate to tripinnate, leathery; veins free; sporangia borne on two erect dissected pinnae arising from the stipe, just below the sterile part of the blade. One hundred species, mostly Latin American.

1a Blade once pinnate; Texas. (Fig. 49). .... *Anemia mexicana* Kl.

Figure 49

Figure 49 *Anemia mexicana*

Frond to twenty-four inches tall; blade broadly oblong, once pinnate; pinnae four to seven pairs, finely toothed, leathery, lance-shaped. Limestone ledges. Frequent. Edwards Plateau of Texas; south to southern Mexico.

1b Blade bipinnate-pinnatifid to tripinnate; Florida. (Fig. 50). .... PINE FERN, *Anemia adiantifolia* (L.) Sw.

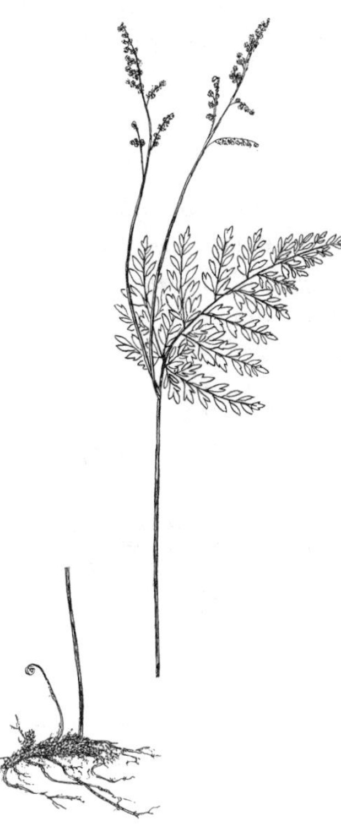

Figure 50

Figure 50 *Anemia adiantifolia*

Frond to twenty inches tall; blade triangular, bipinnate-pinnatifid to tripinnate; segments wedge-shaped, coarsely toothed, leathery. Shaded limestone. Frequent. Peninsular Florida; West Indies, Mexico, northern South America.

Two small Bahamian and West Indian species have been found recently on limestone in Dade Co., Florida. Both have strongly dimorphic fronds with sterile fronds two to five inches tall and the fertile fronds usually twice as tall. *Anemia wrightii* Baker is bipinnate and virtually naked, whereas *A. cicutaria* Kunze ex Spreng. is bipinnate to tripinnate and somewhat hairy.

ASPLENIUM AND ASPLENOSORUS

Spleenwort

Rhizome short-creeping, clothed with clathrate scales; fronds small to medium-sized, undivided to tripinnate, generally naked; sori elongate on one side of the vein, covered with an indusium. *Asplenium* is a genus of about seven hundred species, largely of the tropics.

Hybridization frequently takes place between species of *Asplenium.* Hybrids between the walking fern, *Camptosorus rhizophyllus,* and species of *Asplenium* are placed in the hybrid genus *Asplenosorus*[1]. Fertile hybrids crossing with their parents and other species make an interesting complex, as shown in Fig. 51.

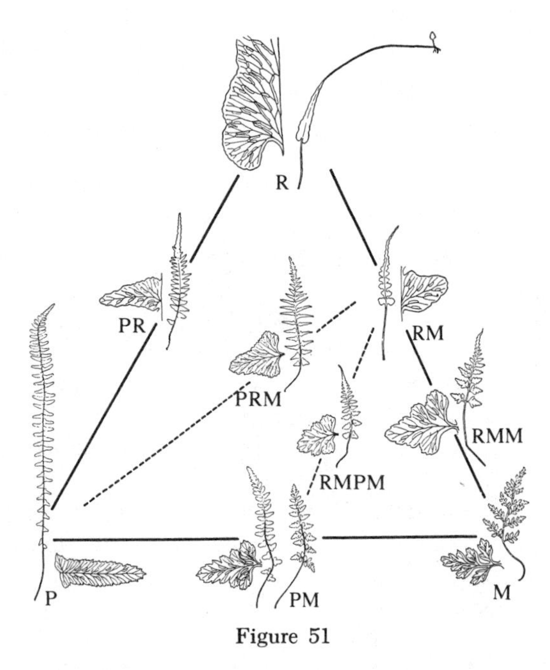

Figure 51

Figure 51 Relationships of some Appalachian spleenworts. R = *Camptosorus rhizophyllus;* P = *Asplenium platyneuron;* M = *Asplenium montanum;* PM = *Asplenium bradleyi;* PR = *Asplenosorus* × *ebenoides;* RM = *Asplenosorus pinnatifidus;* RMM = *Asplenosorus* × *trudellii;* RMPM = *Asplenosorus* × *gravesii;* PRM =*Asplenosorus* × *kentuckiensis.* (From W. H. Wagner, Jr., Reticulate evolution in the Appalachian Aspleniums, Evolution 8: 103-118 (1954), with permission of the author.)

1a **Blade undivided although the stipe may fork.** **2**

1b **Blade divided (pinnatifid to tripinnate).** **3**

2a **Leaf twelve to thirty inches long, two to four inches broad; stipe short and undivided; Florida. (Fig. 52).** .... **AMERICAN BIRD'S-NEST FERN,** *Asplenium serratum* **L.**

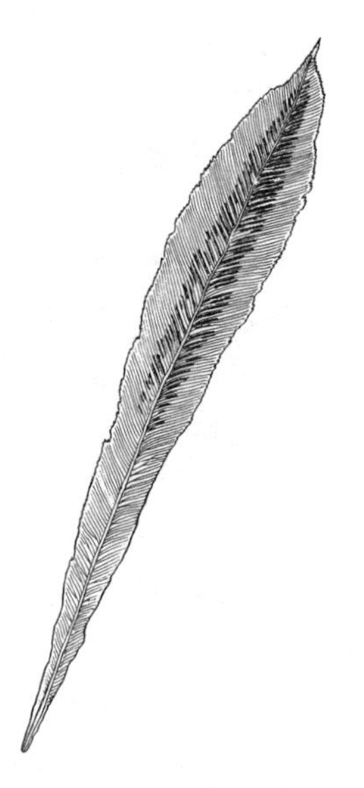

Figure 52

1. Technically this hybrid genus name is writen with a times sign before it (x *Asplenosorus*), but since some of these hybrids are fertile and cross with other species to form sterile hybrids, it becomes confusing to have x's before and/or after the genus name. Thus, for simplicity I have dropped the "x" before *Asplenosorus* and left it after only when the plant is a sterile hybrid.

Figure 52 *Asplenium serratum*

Rhizome stout; fronds undivided, strap-shaped; stipe very short (about one inch long); blade margin finely toothed toward the tip. On ground, logs, and tree trunks. Rare. Southern tip of Florida; West Indies, Middle and South America.

2b Leaf less than six inches long, one-eighth inch wide, grass-like; stipe longer than the blade, often divided in half; northern and western. (Fig. 53). .... FORKED SPLEENWORT, *Asplenium septentrionale* (L.) Hoffm.

Figure 53

Figure 53 *Asplenium septentrionale*

Rhizome slender; blades undivided but stipes often forked, slender, grass-like; stipe much longer than the blade. Crevices of non-calcareous cliffs. Rare. Western United States and West Virginia; Europe, Asia.

Very rarely *A. septentrionale* hybridizes with *A. trichomanes* to form *A.* × *clermontiae* Syme and with *Camptosorus rhizophyllus* to form *Asplenosorus* × *inexpectatus* E. L. Braun ex Friesn.

3a Rachis dark brown to black and shiny throughout its length, even to the tip; fronds linear, once pinnate; pinnae many (fifteen to forty pairs); stipe very short (one-fifth or less of the frond). 4

3b Rachis green, or if dark, not dark and shiny to the tip; fronds linear to triangular, once to tripinnate; pinnae few to many (two to twenty pairs); stipe in most species one-third or more of the frond length. 9

4a Rachis and stipe black. 5

4b Rachis brown, of if partly black, pinnae incised. 6

5a Pinna margins smooth or nearly so with slight basal auricle above and below most pinnae; lower pinnae somewhat triangular and pointing downward; leathery. (Fig. 54). .... BLACK-STEMMED SPLEENWORT, *Asplenium resiliens* Kunze

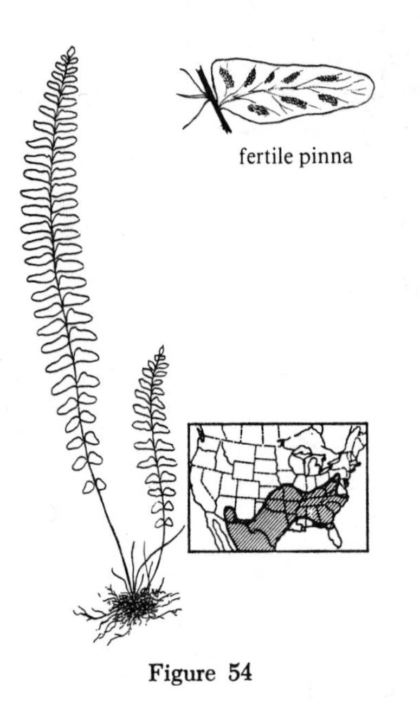

Figure 54

Figure 54 *Asplenium resiliens*

Fronds to twelve inches long; stipe one-sixth of the frond; blade pinnate, less than one inch wide. Calcareous rocks. Frequent. Southern United States; West Indies, Middle and South America.

5b Pinnae finely toothed, not auricled or only on the upper side; lower pinnae oblong and perpendicular to the rachis; thin-textured. (Fig. 55). .... ***Asplenium heterochroum*** **Kunze**

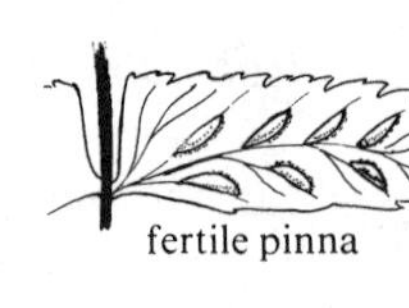

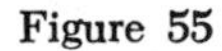

Figure 55

Figure 55 *Asplenium heterochroum*

Fronds to eleven inches long; stipe one-sixth of the frond; blade pinnate, one inch wide. Calcareous rocks. Frequent. Florida; West Indies.

The presumed hybrid between the above two species, *A. heteroresiliens* W. H. Wagner, is intermediate in form and its spores fertile. It occurs from North Carolina to northern Florida.

Asplenium palmeri Maxon, of Arizona, New Mexico, and Mexico, is much like the above two species but has a rooting rachis tip.

6a Pinnae not lobed, the margin smooth or finely cut; mostly northern or eastern. 7

6b Pinnae lobed, cut nearly one-half way to the midvein; California and Arizona. (Fig. 56). .... WESTERN SPLEENWORT, *Asplenium vespertinum* Maxon

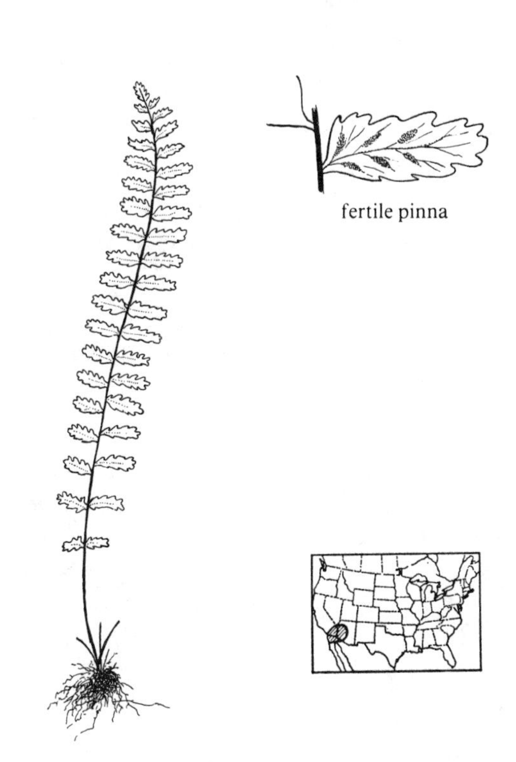

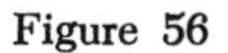
Figure 56

Figure 56 *Asplenium vespertinum*

Fronds to twelve inches long; stipe one-third of the frond; blade pinnate, one-half to one inch wide. Rocky crevices. Frequent. Arizona and southern California.

7a Segments elongate, three-eighths to one inch long, three to five times long as wide; fronds mostly eight to eighteen inches long, erect; on ground or among rocks. 8

7b Segments roundish, about one-fourth inch long, less than two times long as wide; fronds three to seven inches long; spreading on rocks. (Fig. 57). .... MAIDENHAIR SPLEENWORT, *Asplenium trichomanes* L.

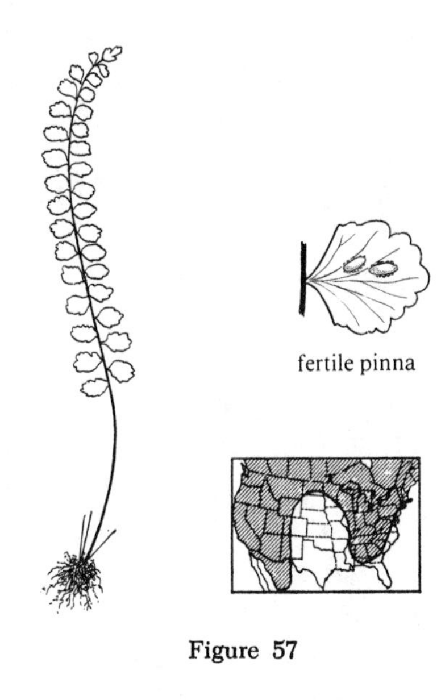

Figure 57

Figure 57 *Asplenium trichomanes*

Fronds to seven inches long; stipe one-fifth of the frond; blade pinnate, about one-half inch wide. Shaded rock crevices. Common. Widespread in North America; Europe, Asia.

8a Sori mostly toward the lower side of each pinna; usually only one sorus per pinna; pinnae not auricled. (Fig. 58). .... *Asplenium monanthes* L.

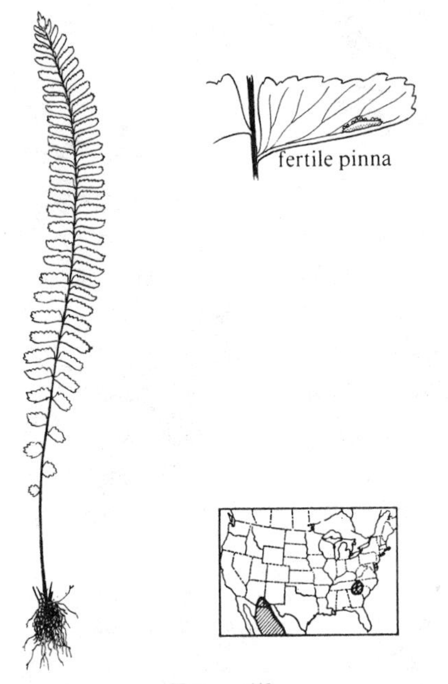

Figure 58

Figure 58 *Asplenium monanthes*

Fronds to twelve inches long; stipe one-third to one-half the frond length; blade pinnate, one-half to one inch wide. Shaded rocks. Rare. Southwestern and southeastern United States; tropical America, Africa, Hawaii.

8b Sori on each pinna about the same number in each of two rows, one row on each side of the midvein; pinnae auricled. (Fig. 59). ...... EBONY SPLEENWORT, *Asplenium platyneuron* (L.) Oakes ex D. C. Eaton

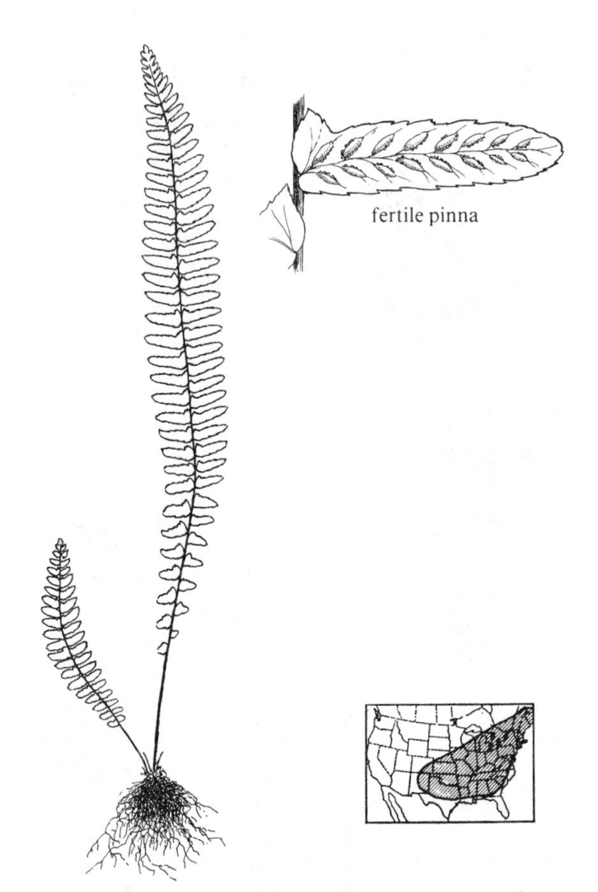

Figure 59

Figure 59 *Asplenium platyneuron*

Fronds to eighteen inches long; stipe one-sixth of the frond; blade pinnate, one to two inches wide; pinnae oblong, auricled; fronds somewhat dimorphic, the sterile ones short and spreading, the fertile ones erect; pinnae sometimes deeply incised with several forms or varieties. Dryish soil and rocks in open woods. Abundant. Eastern North America.

The ebony spleenwort hybridizes with the mountain spleenwort to form *Asplenium bradleyi* D. C. Eaton (Fig. 51), which is fertile and occurs from Pennsylvania to Georgia and Oklahoma. *Asplenium platyneuron* also crosses with *A. trichomanes* to form *A.* × *virginicum* Maxon.

9a **Blade pinnatifid to pinnate-pinnatifid. 10**

9b **Blade bipinnate or more divided. 16**

10a **Blade linear; pinnae many if pinnate (twelve to thirty pairs). 11**

10b **Blade triangular, pinnae few (two to eight pairs). 15**

11a **Blade fully pinnate. 12**

11b **Blade only pinnatifid, the tip long and slender. (Fig. 60). .... LOBED SPLEENWORT, *Asplenosorus pinnatifidus* (Nutt.) Mickel**

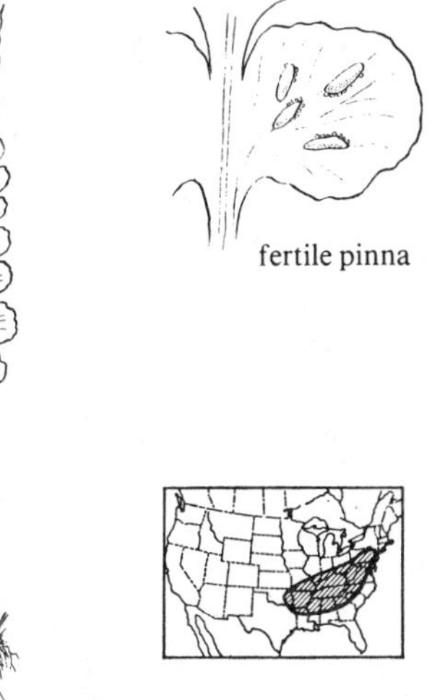

Figure 60

Figure 60 *Asplenosorus pinnatifidus*

Fronds to seven inches long; stipe one-fourth the frond length; blade pinnately lobed, occasionally rooting at the tip; evergreen. Shaded non-calcareous rocks. Frequent. Eastern United States.

The lobed spleenwort is the fertile hybrid between *Asplenium montanum* and *Camptosorus rhizophyllus*. It in turn backcrosses with *A. montanum* to form *Asplenosorus* × *trudellii* (Wherry) Mickel, and with *Asplenium bradleyi* to form *Asplenosorus* × *gravesii* (Maxon) Mickel, and with *Asplenium platyneuron* to form the very rare *Asplenosorus* × *kentuckiensis* (McCoy) Mickel (Fig. 51). The lobed spleenwort also crosses with *Asplenium trichomanes* to form *Asplenosorus* × *herb-wagneri* (Taylor and Mohlenb.) Mickel.

12a **Pinnae oval to oblong, two to four times as long as wide; blade about one-half inch wide; stipe one-tenth to one-fourth of the frond. 13**

12b **Pinnae long, four to seven times as long as wide; blade generally two to four inches wide; stipe one-fourth to one-half the frond length. 14**

13a **Stipe and rachis with scattered hair-like black scales; stipe base dark; pinnae about twice as long as wide; northern. (Fig. 61). ...... GREEN SPLEENWORT, *Asplenium viride Huds.***

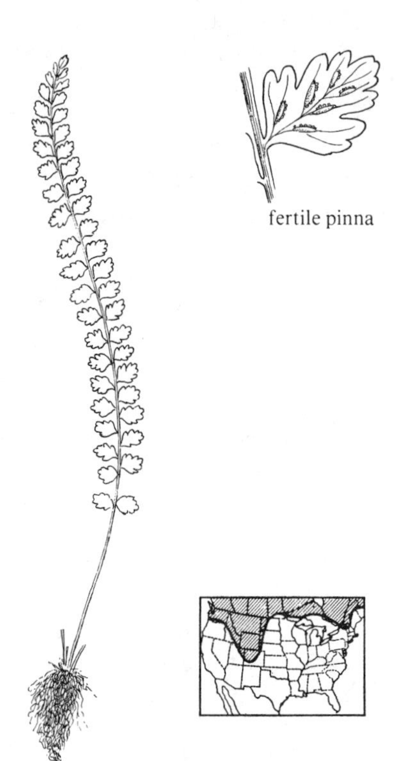

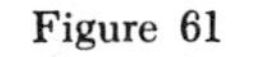

Figure 61

Figure 61 *Asplenium viride*

Fronds to seven inches long; stipe one-third the frond length; blade pinnate, linear; texture thin; evergreen. Shaded calcareous rocks. Rare. Northern North America; Europe, Asia.

13b Stipe and rachis naked; green to base of stipe; pinnae three to four times as long as wide; Florida. (Fig. 62). .... TOOTHED SPLEENWORT, *Asplenium dentatum* L.

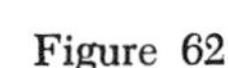

Figure 62

Figure 62 *Asplenium dentatum*

Fronds to six inches long; stipe one-sixth the frond length; blade pinnate, linear; texture thin. Calcareous rocks. Rare. Southernmost Florida; West Indies, Middle America.

14a Stipe green; pinnae progressively smaller from base to tip, distinct from one another, auricled; Florida. (Fig. 63). .... EARED SPLEENWORT, *Asplenium auritum* Sw.

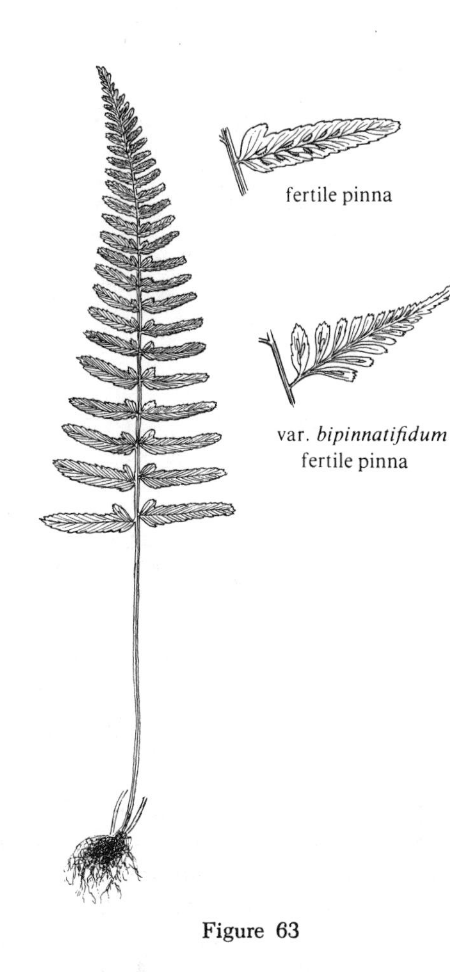

Figure 63

Figure 63 *Asplenium auritum*

Fronds to twelve inches long; stipe about one-third the frond length; blade pinnate to pinnate-pinnatifid, narrowly triangular, leathery. On trees and rocks. Rare. Southwestern peninsular Florida; widespread in tropical America, Africa, and southern Asia.

Var. *bipinnatifidum* Kunze is also known in Florida.

14b Stipe brown, shiny; pinnae of very irregular lengths, not auricled, distinct from one another only toward the base of the frond; eastern United States. (Fig. 64). .... SCOTT'S SPLEENWORT, *Asplenosorus* × *ebe-*

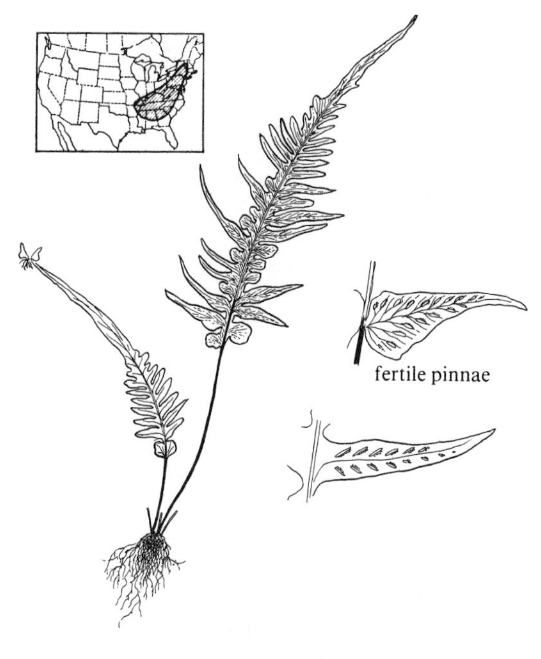

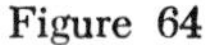
Figure 64

Figure 64 *Asplenosorus* × *ebenoides*

Fronds to twenty-one inches long; stipe one-third the frond length; blade pinnate at base, pinnatifid toward the tip, irregular in outline, of thin texture. Shaded, moist calcareous rocks. Rare. Eastern United States.

Scott's spleenwort is the hybrid between *Asplenium platyneuron* and *Camptosorus rhizophyllus*. Although it is generally sterile, rare fertile populations are known to occur.

15a Pinnae not lobed or auricled. (Fig. 65). .... *Asplenium abscissum* Willd.

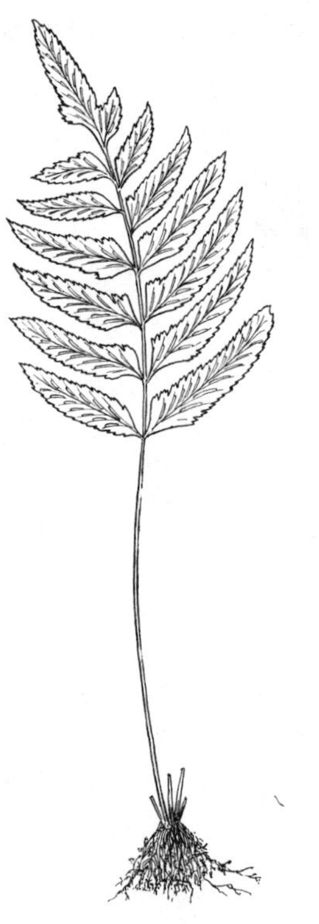

Figure 65

Figure 65 *Asplenium abscissum*

Frond to sixteen inches long; stipe one-half the frond length; blade pinnate, triangular, of thin texture. Terrestrial or on calcareous rocks. Frequent. Peninsular Florida; West Indies, Middle and South America.

15b At least the basal pinnae divided, one to three pairs of pinnae, bearing small hairs on surfaces and margin. (Fig. 66). .... DWARF SPLEENWORT, *Asplenium pumilum* Sw.

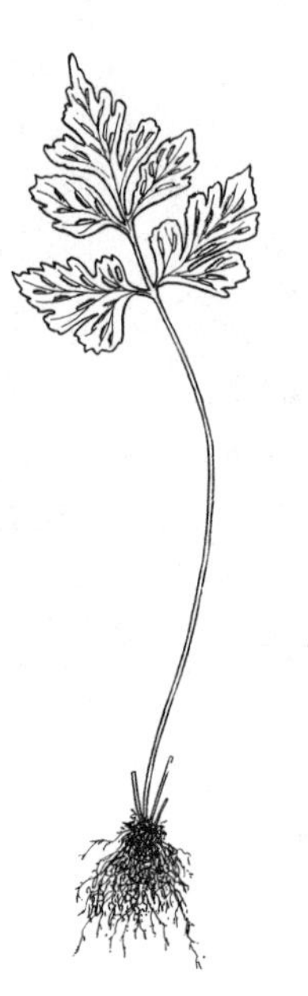

Figure 66

Figure 66 *Asplenium pumilum*

Frond to six inches long; stipe one-half the frond length; blade pinnate to pinnate-pinnatifid, broadly triangular, of thin texture, lightly hairy on both surfaces and margin. Calcareous rocks. Frequent. West peninsular Florida; West Indies, Middle and South America.

16a (9b) Fronds small (two to six inches long); medium texture; northern or southwestern. 17

16b Fronds larger (six to eighteen inches long); very thin texture; Florida. 18

17a Segments oblong, broadly attached. (Fig. 67). .... MOUNTAIN SPLEENWORT, *Asplenium montanum* Willd.

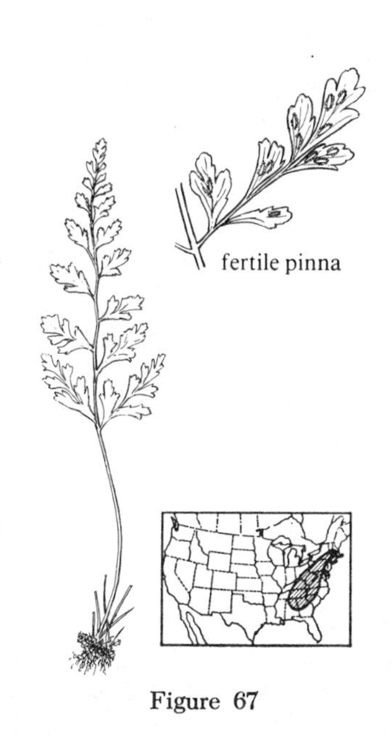

Figure 67

Figure 67 *Asplenium montanum*

Fronds to five inches long; stipe about one-half the frond length; blade bipinnate at the base, pinnate-pinnatifid in upper parts; oblong-triangular, shiny, smooth. Noncalcareous rocks. Frequent. Eastern United States.

In addition to the hybrids shown in Fig. 51, *Asplenium montanum* crosses with *A. bradleyi* to form *A.* × *wherryi* D. M. Smith.

Asplenium exiguum Bedd. is somewhat similar, though the fronds are more oblong in shape and slightly less divided, and occurs in Arizona, Mexico, and the Himalayan Mountains. It is especially distinctive in having a bud at the tip of each pinna, and the rachis may also root at the tip.

Asplenium fontanum (L.) Bernh. of Europe and Asia was reported from Ohio and Pennsylvania about 1870 but there is serious question as to whether the plants were really found in America; quite possibly the labels were erroneous.

17b Segments wedge-shaped, narrowed at the base. (Fig. 68). .... WALL RUE, *Asplenium ruta-muraria* L.

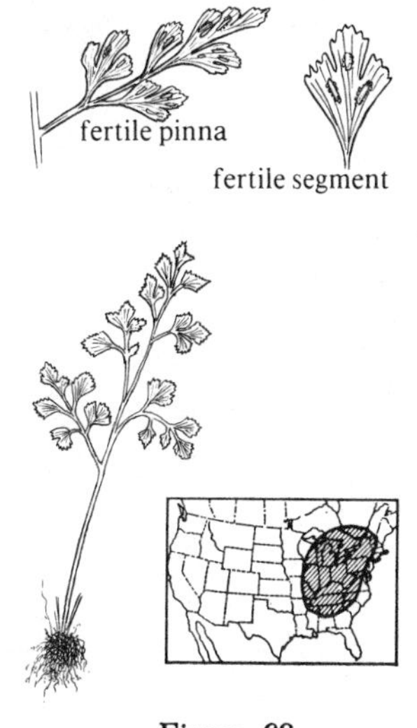

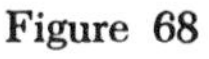
Figure 68

Figure 68 *Asplenium ruta-muraria*

Fronds to four inches long; stipe one-half the frond length; blade bipinnate; segments wedge-shaped, narrowing at the base; blade oval-triangular, dull. Calcareous cliffs. Frequent. Northern North America; Europe, Asia.

Asplenium adiantum-nigrum L. is a rare spleenwort from Arizona, Utah, and Colorado, though widespread in Europe, Asia, and Africa, and can be distinguished by its triangular blade and dark, shiny stipe.

18a Stipe much shorter than the blade, about one-sixth of the frond length; blade linear-lance-shaped, tapering to the base; stipe nearly lacking; pinnae short, not curving upward; rachis mostly brown; blade bipinnate to tripinnate, the segments oblong and smooth-margined, about one-eighth inch long; basal pin-

nules of each pinna not generally overlapping the rachis. (Fig. 69). ....
.... ***Asplenium myriophyllum*** **(Sw.) Presl**
[*A. verecundum* Chapm.]

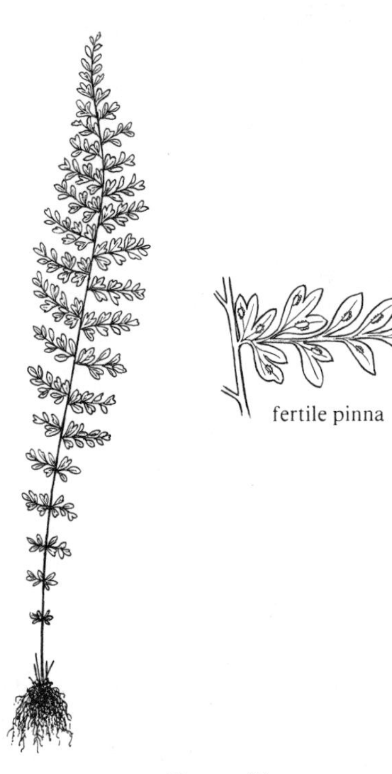

Figure 69

Figure 69 *Asplenium myriophyllum*

Fronds to sixteen inches long, though usually much shorter; stipe one-sixth the frond length; blade oval-lance-shaped. Calcareous rocks. Frequent. Peninsular Florida; West Indies, Middle and South America.

The Florida material commonly goes under the name of *A. verecundum,* but I can find no characters to separate the two.

Asplenium × *curtissii* Underw. is the hybrid between *A. myriophyllum* and *A. abscissum. Asplenium myriophyllum* also crosses with *A. dentatum* to form *A.* × *biscayneanum* (D. C. Eaton) A. A. Eaton.

18b Stipe one-third to one-half of the frond length; blade narrowly triangular, not tapering at the base; pinnae often curving slightly upward; rachis mostly green; blade bipinnate to bipinnate-pinnatifid, the segments about one-fourth inch long, deeply incised with several teeth; the basal pinnules of each pinna overlapping the rachis. (Fig. 70). ....
.... HEMLOCK SPLEENWORT,
***Asplenium cristatum* Lam.**

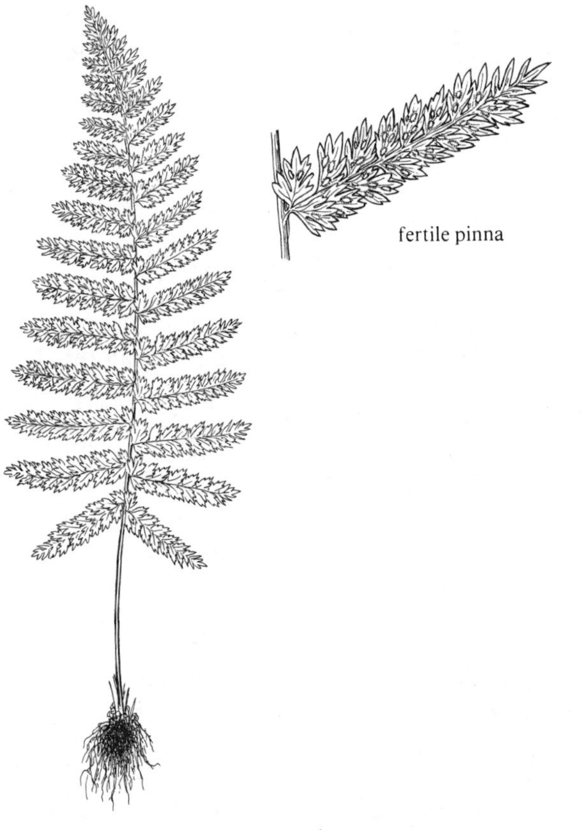

Figure 70

Figure 70 *Asplenium cristatum*

Fronds to fourteen inches long; stipe one-third to one-half the frond length; blade bipinnate to bipinnate-pinnatifid, oblong-triangular, basal pinnules of pinnae overlapping the rachis. Calcareous rocks. Rare. Peninsular Florida; West Indies, Mexico to South America.

The very rare *Asplenium* × *plenum* E. P. St. John ex Small is a sterile hybrid, apparently of complex parentage. Two possible parents are *A. cristatum* and *A. abscissum.*

ATHYRIUM

Lady Fern, Glade Fern

Rhizome short-creeping or ascending, clothed with scales; fronds thin-textured, medium to large size, pinnate to tripinnate; veins free; sori elongate along the veins, with some hooked over the veins; indusium usually present. One hundred-fifty species, largely of temperate regions.

1a **Blade pinnate to pinnate-pinnatifid; sori linear, elongate, straight, not hooked over the veins. 2**

1b **Blade bipinnate to tripinnate; sori round to elongate with one end of the sorus often hooked over the vein. 3**

2a **Pinnae undivided, long-tapering; stipe and rachis naked or sparsely hairy. (Fig. 71). .... NARROW-LEAVED SPLEENWORT, GLADE FERN, *Athyrium pycnocarpon* (Spreng.) Tidestrom**

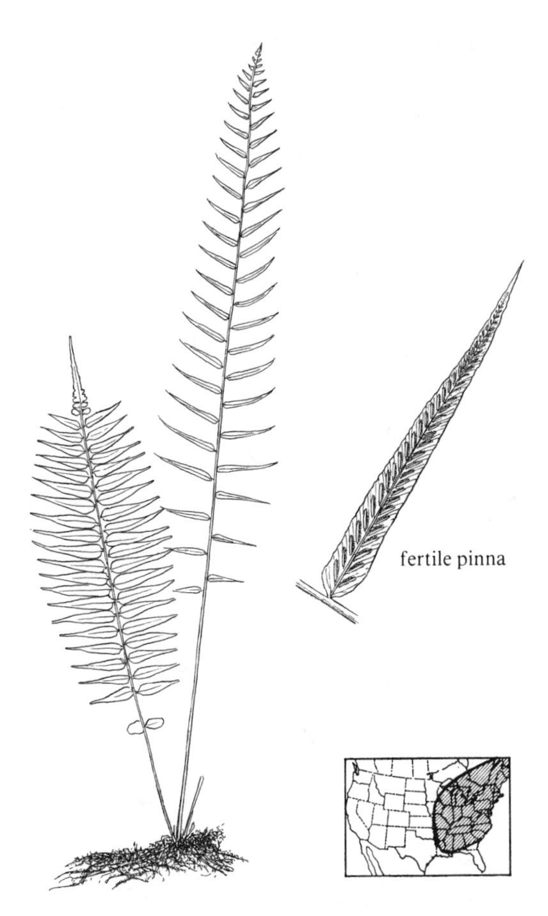

Figure 71

Figure 71 *Athyrium pycnocarpon*

Rhizome short-creeping; fronds twelve to twenty-four inches tall, six to twelve inches wide; stipe one-third of the frond length; blade pinnate, tapering slightly at the bottom of the blade; pinnae one-half inches wide, smooth-margined; stipe and rachis green, naked or only sparsely hairy; sori long, straight; fertile pinnae narrower than the sterile. Moist woodlands. Frequent. Eastern North America.

2b **Pinnae deeply pinnatifid; stipe and rachis clothed with many narrow scales and hairs. (Fig. 72). .... SILVERY SPLEENWORT,**

SILVERY GLADE-FERN, *Athyrium thelypterioides* (Michx.) Desv.

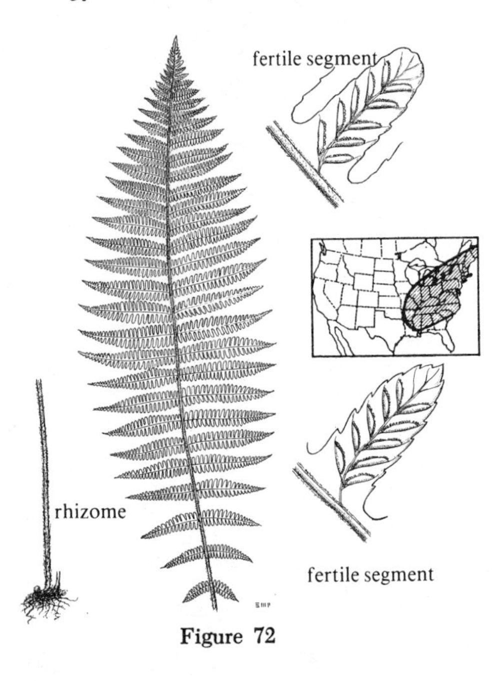

Figure 72

Figure 72 *Athryrium thelypterioides*

Rhizome short-creeping; fronds twenty-four to thirty-six inches tall, six to eight inches wide; stipe one-fourth of the frond length; blade tapering markedly at base; stipe and rachis green, appearing hairy due to abundant narrow scales and some fine hairs; young indusia silvery; sori long, straight. Damp woods, often near water. Common. Eastern North America; Asia.

There are two forms, one with the segments nearly smooth-margined and round-tipped, the other has segments toothed and slightly pointed at the tip.

3a Sori elongate or kidney-shaped, midway between midvein and margin; indusium present. (Fig. 73). .... LADY FERN, *Athyrium filix-femina* (L.) Roth

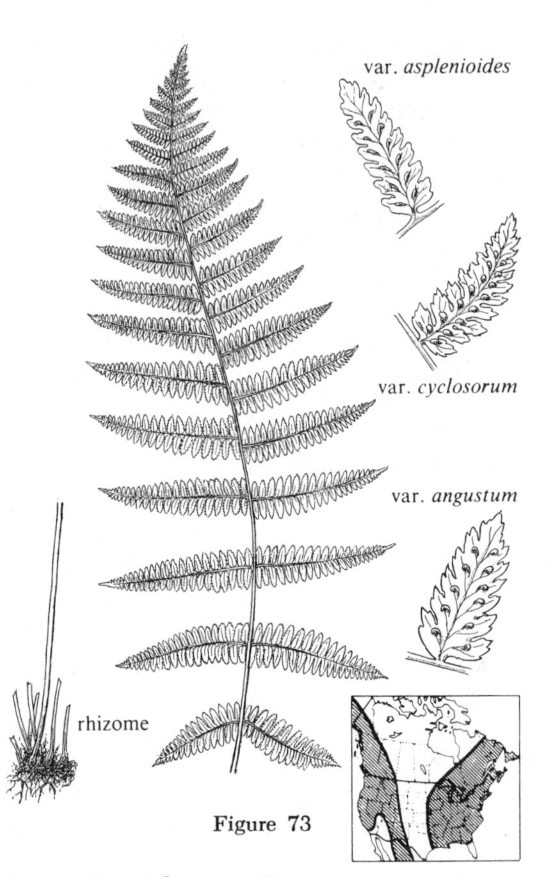

Figure 73

Figure 73 *Athyrium filix-femina*

Rhizome short-creeping, ascending or erect; fronds in a vase-like crown or slightly more diffuse clump, twelve to sixty inches tall; stipe about one-fourth of the frond length; blade bipinnate to tripinnate, spreading or tapering at base, naked; sori elongate, some hooked over the vein. Moist woods. Abundant. North America; Europe, Asia, tropical America, Africa.

There are four varieties in North America. Var. *angustum* (Willd.) Moore is the northeastern variety, two to three feet tall, somewhat tapering toward the base, the indusium short-ciliate and glandless, the spores yellow and the rhizome short-creeping. Its stipe and rachis may be wine-red in forma *rubellum*. Var. *asplenioides* (Michx.) Farwell in the Southeast has as ascending rhizome, blade broadest near the base, indusium glandular and ciliate, the spores black. Var. *filix-femina* in the West and Northwest is the largest, with fronds three to five feet tall, tapering toward the base, has an erect rhizome, indusium long-ciliate and glandless, and yellow spores. Of our varieties it most closely resembles the European lady-fern. Var. *cyclosorum* Rupr. [var. *sitchense* Rupr.] of the Northwest (and rarely in the Northeast) is of higher altitude or latitude, is of medium size, and has kidney-shaped sori with the indusium neither ciliate nor glandular.

3b Sori round, submarginal; indusium lacking or minute. (Fig. 74). .... ALPINE LADY-FERN, *Athyrium distentifolium* Tausch ex Opiz [*A. alpestre* (Hoppe) Rylands]

Figure 74

Figure 74 *Athyrium distentifolium*

Rhizome stout, ascending; fronds clumped into a vase-like crown, sixteen to thirty inches tall, four to six inches wide; stipe about one-fourth of the frond length; blade bipinnate-pinnatifid, not tapering much at the base, naked; sori round, very small, indusium minute or lacking. Moist wooded or alpine slopes. Frequent. Northwestern North America, eastern Canada; Iceland, Europe, Asia.

The Eurasian specimens are shorter and more compact, so the North American plants are considered a separate variety, var. *americanum* (Butters) Cronq.

AZOLLA

Mosquito Fern

Plants small, aquatic, free-floating; stems hair-like, branched; roots thread-like; leaves minute (about one-thirty-second inch long), borne in two rows; each leaf composed of two round lobes, the upper one green, the lower one colorless; heterosporous, male and female sporangia produced in separate round structures (sporocarps) located in the leaf axils; the male spores in masses with protruding anchor-shaped hairs, called glochidia. Six species, found in quiet streams and ponds in warm temperate and tropical regions.

Species identification is based on details of the glochidia, but since these plants are rarely fertile, identification is usually very difficult or impossible.

1a **Glochidia one-celled, or rarely with one or two cross-walls near the tip. 2**

1b **Glochidia several-celled. (Fig. 75A). .... *Azolla mexicana* Presl**

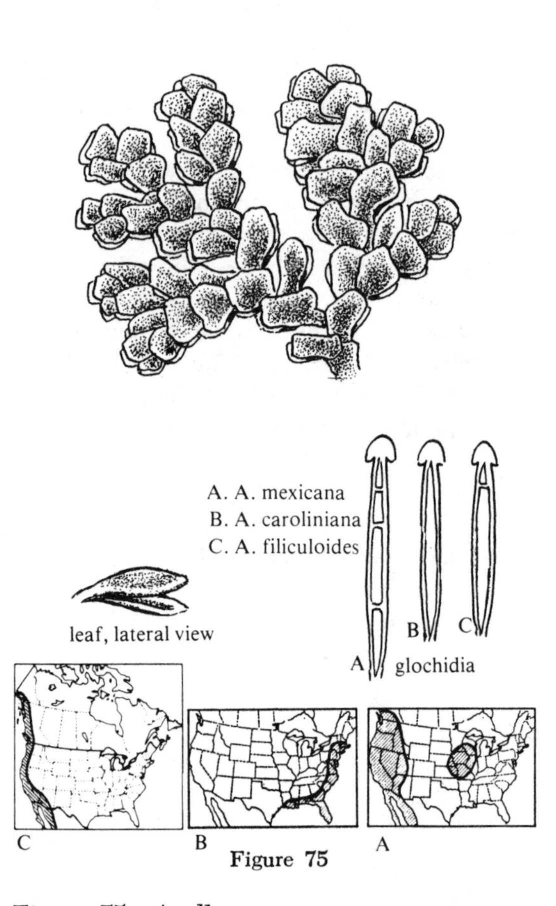

Figure 75 *Azolla*

Plants one-half to one inch long; branching a mixture of equal forkings and lateral branches. Frequent. Western and central North America; Mexico to South America.

2a **Leaves smooth; plants small, mostly less than one-half inch long; branching by equal forkings of the stem. (Fig. 75B). .... *Azolla caroliniana* Willd.**

Glochidia always one-celled. Frequent. Eastern United States; West Indies.

2b **Leaves with minute, short hairs; plants often one to two inches long; branching generally by lateral branching of the stem. (Fig. 75C). ...... *Azolla filiculoides* Lam.**

Glochidia mostly one-celled, but rarely with one or two cross-walls near the tip. Frequent. Western North America and New York; Mexico to South America.

BLECHNUM

Rhizome short- or long-creeping or ascending, scaly; fronds close together, up to seventy-two inches tall, pinnatifid to once pinnate, often pink when young; fertile and sterile fronds similar or different; sori elongate along both sides of the length of the midrib; indusium present. Two hundred species, largely of tropical and south-temperate regions.

1a **Fertile and sterile leaves the same, pinnate; Florida. 2**

1b **Fertile and sterile leaves different; pinnatifid; Northwest. (Fig. 76). .... DEER FERN, HARD FERN, *Blechnum spicant* (L.) J. E. Smith**

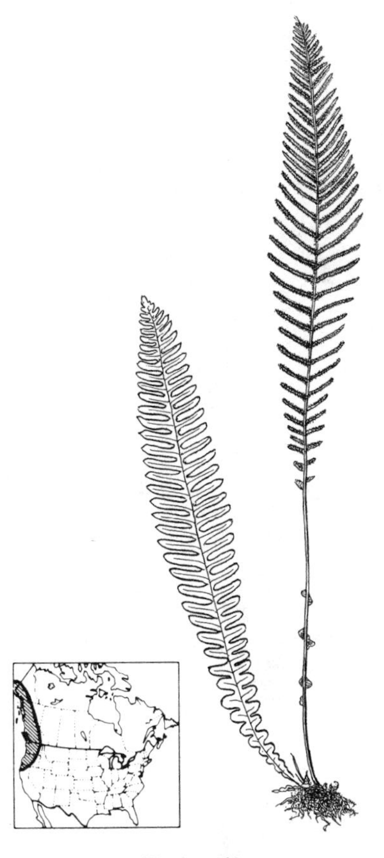

Figure 76

Figure 76 *Blechnum spicant*

Rhizome short-creeping; fronds forming a rosette on the ground; sterile fronds evergreen, hard and stiff, numerous, linear, pinnatifid, eight to ten inches long, tapering to the base; fertile fronds erect at center crown, taller than the sterile fronds. Moist coniferous forests. Common. Northwestern North America; Europe, Asia.

2a **Pinnae all narrowed at their base, toothed margined, later deciduous; blade tapering only slightly toward the tip with a pinna-like terminal segment; twenty-four to thirty-six inches tall. (Fig.**

77). .... SWAMP FERN, *Blechnum serrulatum* L. C. Richard

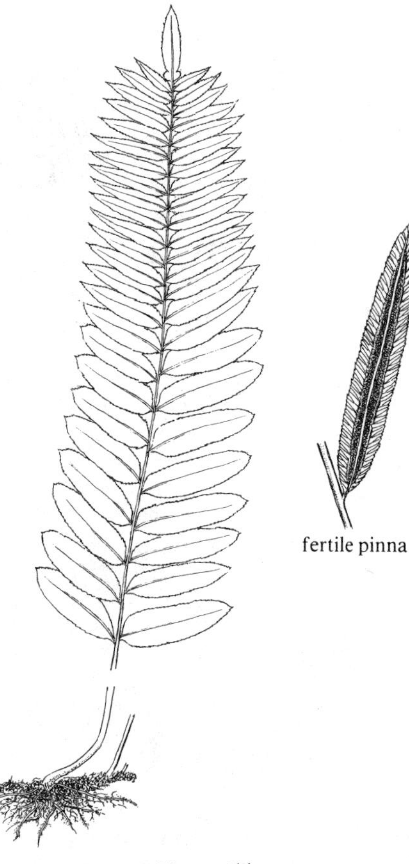

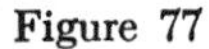
Figure 77

Figure 77 *Blechnum serrulatum*

Rhizome long-creeping; fronds scattered along rhizome, twenty-four to seventy-two inches tall, four to six inches wide; stipe one-third to one-half of the frond length; pinnate throughout; pinnae with fine teeth along margin; one pinna-like terminal segment; blade not tapering at base of frond. Swamps, ditches and other wet areas. Common. Peninsular Florida; tropical America.

2b Pinnae mostly broadly attached at base, smooth-margined, or only very finely toothed, persistent; blade diminishing gradually to the apex, with no distinct terminal segment; eight to sixteen inches tall. (Fig. 78). .... HAMMOCK FERN, *Blechnum occidentale* L.

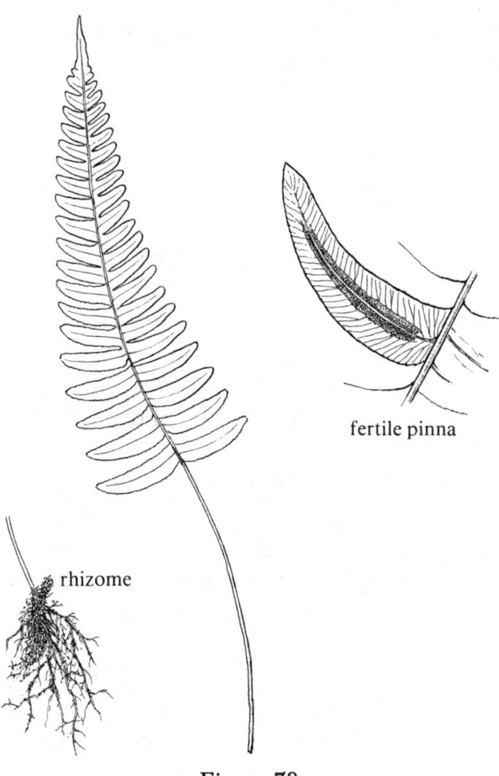

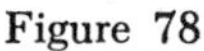
Figure 78

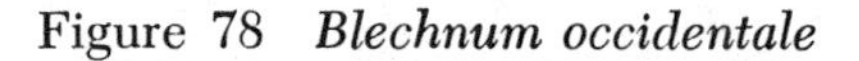
Figure 78 *Blechnum occidentale*

Rhizome ascending; fronds clustered near the tip of the rhizome, eight to sixteen inches tall, two to three inches wide; stipe one-third to one-half of the frond length; blade narrowly triangular, tapering gradually to the tip, pinnate below, pinnatifid toward the tip. Damp woods. Rare. West peninsular Florida and southern Texas; one of the most common ferns in the American tropics.

BOMMERIA

Copper Fern

Rhizome subterranean, slender, much branched, wide-creeping, scaly; blade small, pentagonal, hairy; sori running along veins near their ends without an indusium. Four species, largely of Mexico. (Fig. 79) ....
COPPER FERN, *Bommeria hispida* (Mett.) Underw.

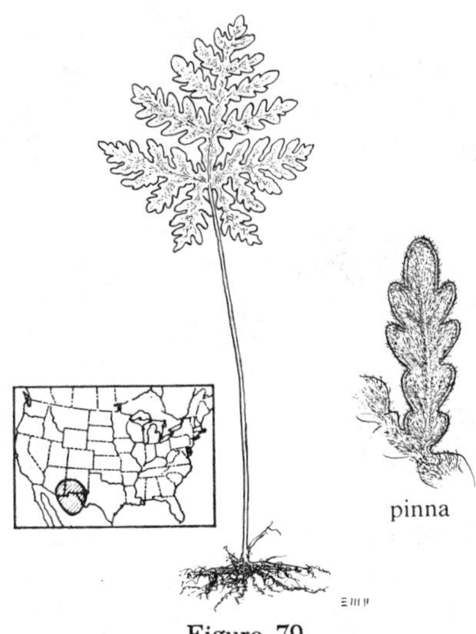

Figure 79

Figure 79 *Bommeria hispida*

Fronds four to ten inches tall, pinnate-pinnatifid, the basal pinnae with the downward pointing pinnules greatly elongated; stipe three-fourths of the leaf length. Shaded, dry, rocky ledges. Frequent. Southwestern United States; northern Mexico.

BOTRYCHIUM

Grape Fern

Coarse, fleshy stem deep in the ground; fronds only one or two per plant at one time, papery to leathery, pinnate to quadripinnate; veins free; sporangia large, round, borne on a branched, erect stalk that arises from the stipe; leaf bud for the following year ensheathed by the base of the stipe. Ferns of meadows and open second-growth woodlands. Twenty-five species, largely of temperate regions.

Some species are quite variable, and differences between species are often subtle. Dwarf and giant plants are occasionally found. Frequently several species grow together. Proper identification of this group requires great care and a keen eye. The plants are not killed if only the leaf is taken at or shortly below ground level, since the rhizome and bud are deep in the ground. The key is based upon mature leaf forms. Immature fronds are common at the same time.

1a **Sterile blades various, usually small (one to twelve inches), mostly leathery or fleshy, broadly triangular to lance-shaped; fertile stalk arising either from near the base of blade or farther down the stipe near ground level; never with the combination of characters of 1b. 2**

1b **Sterile blade ten to twenty-four inches tall, broadly triangular, tripinnate, the segments papery; fertile stalk arising from base of blade. (Fig. 80). ....
.... RATTLESNAKE FERN, *Botrychium virginianum* (L.) Sw.**

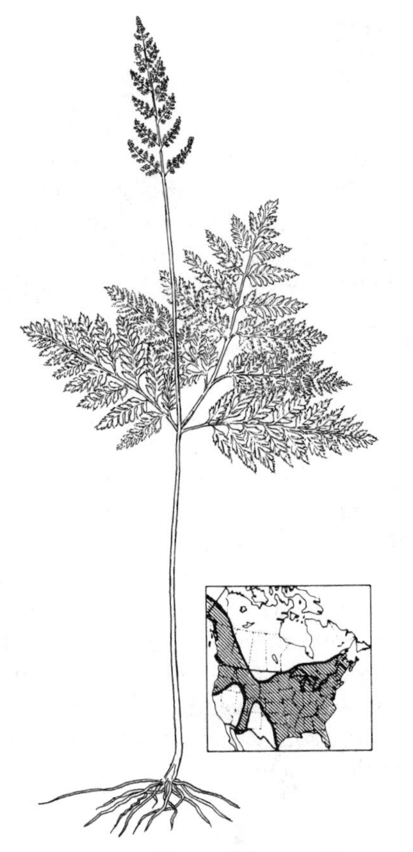

Figure 80

Figure 80 *Botrychium virginianum*

Fronds up to twenty-four inches tall, broadly triangular, bipinnate to tripinnate; segments pointed, toothed, usually papery when dry; deciduous. Moist woods. Abundant. Much of North America; Europe, Asia.

The most widespread variety is var. *virginianum,* but another so-called variety, var. *europaeum* Ångstr., is found in coniferous woods and meadows in the more northern parts of the range. It is more compact, leathery, and somewhat less divided (bipinnate), and is probably a mere environmental form of cold and exposed places.

2a **Blade bipinnate to quadripinnate; basal pinnae with downward pointing pinnules longer than those pointing upward; fertile stalk arising near base of the stipe, often below ground level. 3**

2b **Blade pinnatifid to pinnate-pinnatifid; basal pinnae, if divided, not longer on downward pointing side than the upward pointing; fertile stalk arising at or near the base of the blade. 8**

3a **Blade segments about the same size and shape; terminal segments not elongate. 4**

3b **Blade segments not all the same; terminal segments elongate and less divided. 6**

4a **Ultimate segments fan-shaped; southeastern United States. 5**

4b **Ultimate segments not fan-shaped; northern United States and Canada. (Fig. 81). .... LEATHERY GRAPE-FERN, *Botrychium multifidum* (Gmel.) Rupr.**

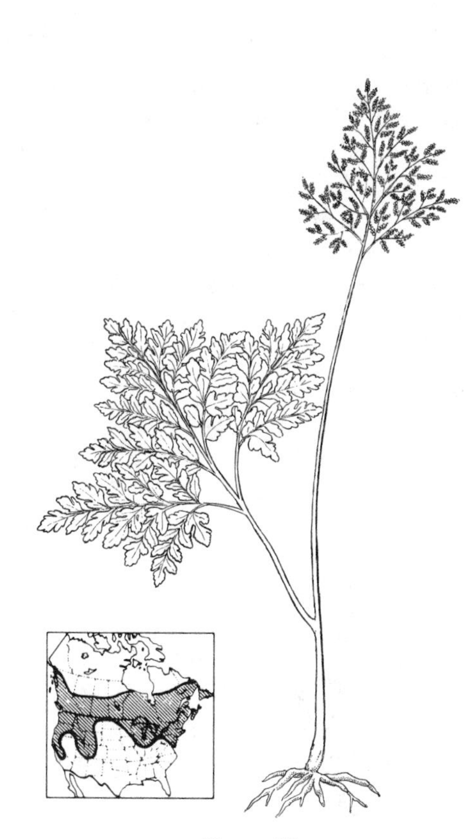

Figure 81

Figure 81 *Botrychium multifidum*

Fronds usually stout, fleshy, leathery, broadly triangular, bipinnate to quadripinnate, six to twelve inches tall; segments blunt or only slightly pointed, sparsely hairy, margins smooth to slightly wavy; fertile stalk much divided and spreading like a fountain. New fronds appear in early summer; evergreen. Frequent. Meadows and open woods. Northern North America; Europe, Asia.

A much less common species in northeastern United States is the TERNATE GRAPE-FERN, *Botrychium ternatum* (Thunb.) Sw., which differs generally from the leathery grape-fern in having smaller leaves (generally six to eight inches tall), thinner texture, and segments that are finely toothed and somewhat pointed. Rare.

5a Stipe short (one to two inches long); blade nearly flat on the ground, mostly only one to two inches long and broad. (Fig. 82). ...... WINTER GRAPE-FERN, *Botrychium lunarioides* (Michx.) Sw.

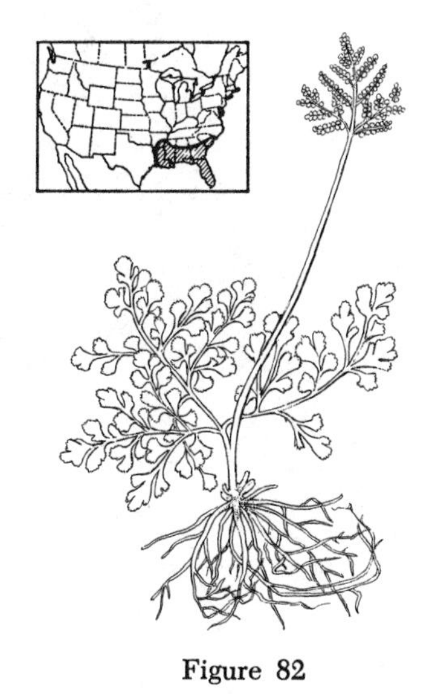

Figure 82

Figure 82 *Botrychium lunarioides*

Fronds two to four inches long, bipinnate to tripinnate; segments rounded, small, finely toothed. New fronds appear in late November; evergreen. Dryish open woods, old fields. Rare and local. Southeastern United States.

5b Stipe longer (two to four inches long); blade two to five inches long and broad. (Fig. 83). .. ALABAMA GRAPE-FERN, *Botrychium alabamense* Maxon

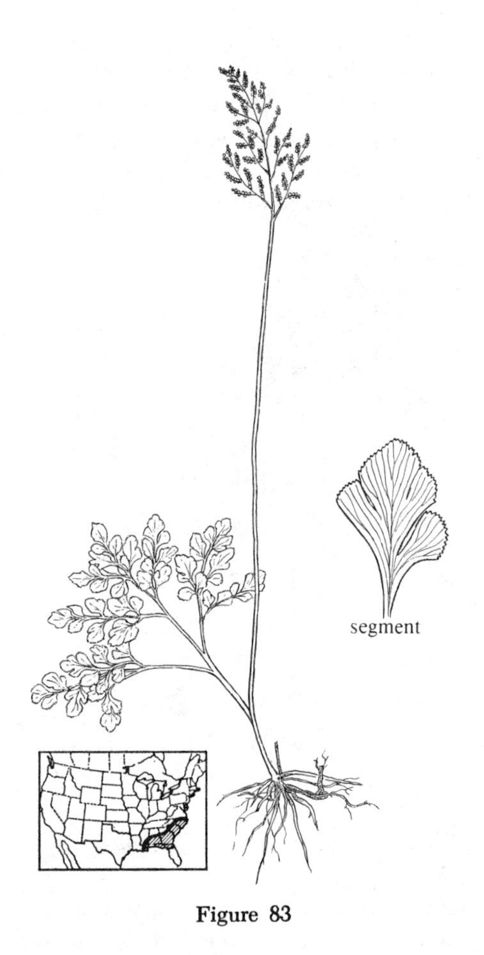

Figure 83

Figure 83 *Botrychium alabamense*

Fronds five to ten inches long, tripinnate; segments distant, small, rounded, finely toothed, fruiting in late summer. New fronds appear in summer; evergreen. Moist woods and old fields. Rare. Southeastern United States.

This species is thought to have originated as a hybrid between *B. biternatum* and *B. lunarioides.*

6a Terminal segments somewhat pointed at their tips; often bronze-colored in winter; roots dark brown to gray. 7

6b Terminal segments rounded at the tip; generally green in winter; roots pale gray to tan. (Fig. 84). .... BLUNT-LOBED GRAPE-FERN, *Botrychium oneidense* (Gilbert) House

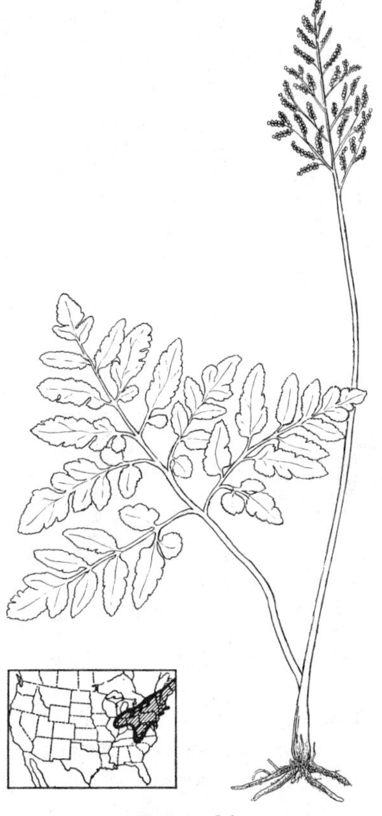

Figure 84

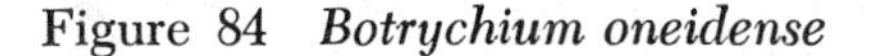

Figure 84 *Botrychium oneidense*

Fronds six to twelve inches long, triangular; segments blunt-tipped, rounded. New fronds appear in summer; evergreen. Low woods and swamps, never in open fields. Rare. Northeastern North America.

7a Blade leathery, bipinnate to bipinnate-pinnatifid; margin smooth to slightly

wavy. (Figs. 85, 86). ....
.... DISSECTED GRAPE-FERN,
***Botrychium dissectum* Spreng. [*Botrychium obliquum* Muhl.]**

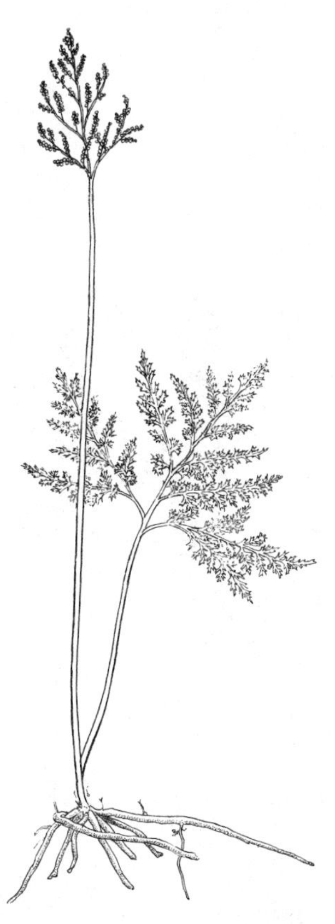

Figure 85

Figure 85 *Botrychium dissectum* f. *dissectum*

Fronds four to eight inches long, triangular, bipinnate, tripinnate at base; segments long, somewhat pointed, several times long as wide. New fronds appear in summer; evergreen, often turning bronze-colored in winter. Moist woods and meadows. Common to abundant. Eastern North America.

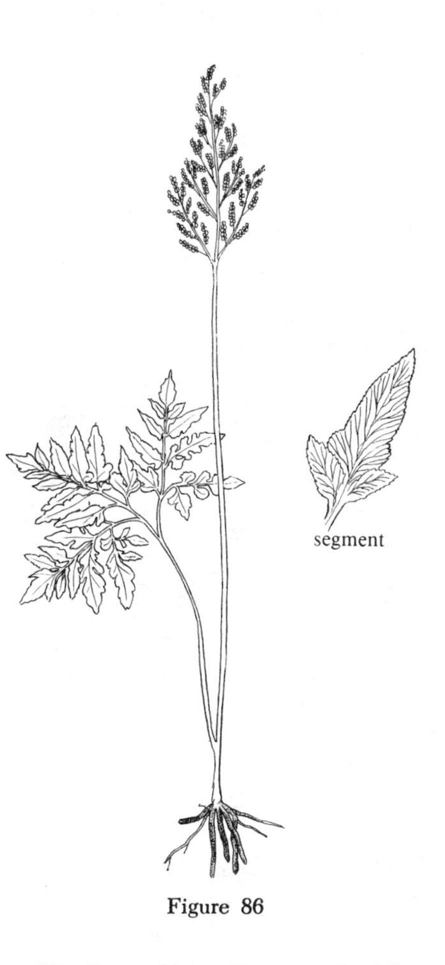

Figure 86

Figure 86 *Botrychium dissectum* f. *obliquum*

There are two quite distinct extreme forms of this species. The more unusual, f. *dissectum,* is extremely finely cut with a fine, lacy appearance (Fig. 85). The more common form is f. *obliquum* (Fig. 86). The two forms may be found growing together, and there is every intergradation between them.

7b Blade papery, pinnate to bipinnate; margin toothed. (Fig. 87). ....
.... SPARSE-LOBED GRAPE-FERN,
***Botrychium biternatum* (Sav.) Underw.**

Figure 87

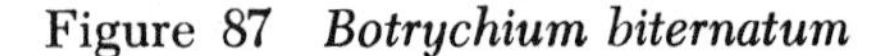

Figure 87 *Botrychium biternatum*

Fronds five to ten inches long, triangular, only two to three pairs of pinnae, the basal pinnae each with only one to two pairs of pinnules. New fronds appear in summer; evergreen, somewhat bronze-colored in winter. Wet woods and swamps. Frequent. Southeastern United States.

8a (2b) Sterile blade oblong or oval; fertile stalk arising slightly down the stipe from the blade, rarely at base of the blade. .. 9

8b Sterile blade broadly triangular; fertile stalk arising at base of blade. (Fig. 88). .... LANCE-LEAVED GRAPE-FERN, *Botrychium lanceolatum* (Gmel.) Ångs.

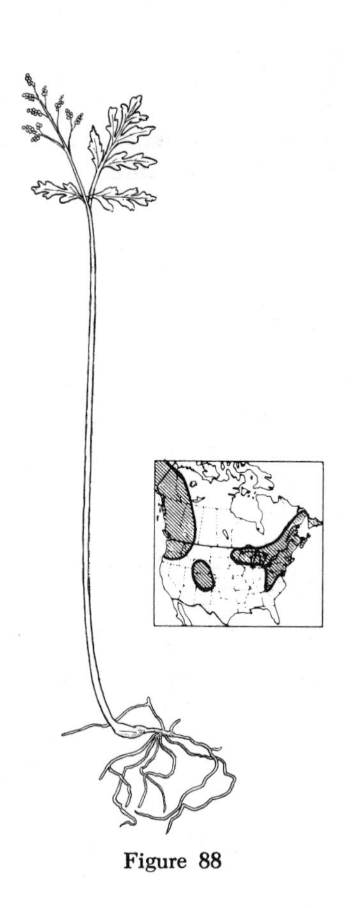

Figure 88

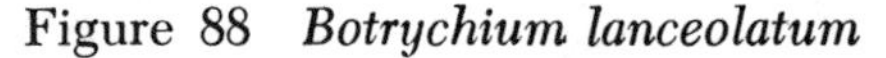

Figure 88 *Botrychium lanceolatum*

Fronds three to eight inches long, naked, triangular, pinnate-pinnatifid; pinnae few, four to five pairs; segments sharp-pointed or blunt; stipe much longer than the blade. New fronds appear in spring, lasting through the summer; deciduous. Meadows, swamps, and barrens. Frequent to rare. Northeastern and northwestern North America; Europe, Asia.

Most of the material in the Northeast is var. *angustisegmentum* Pease & Moore, which is thinner in texture and has lobes narrower at the base and more pointed at the tip. The more widespread variety is var. *lanceolatum*.

9a **Blade undivided to once pinnate, rarely bipinnate at base. 10**

9b **Blade pinnate-pinnatifid to bipinnate. 11**

10a **Pinnae broadly fan-shaped, not divided. (Fig. 89). .... MOONWORT, *Botrychium lunaria* (L.) Sw.**

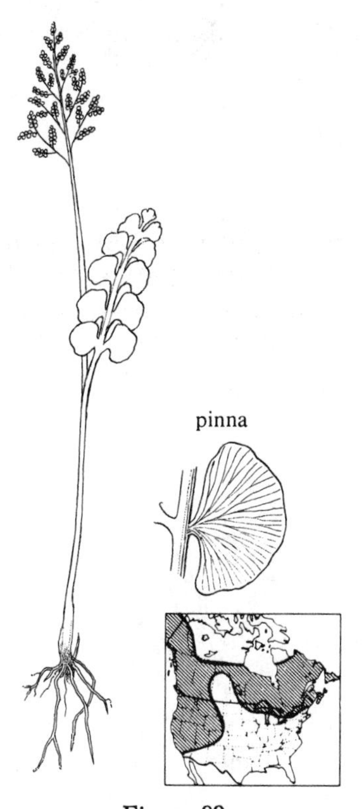

Figure 89

Figure 89 *Botrychium lunaria*

Fronds four to eight inches tall, naked; blade oblong, leathery; pinnae three to five pairs, broadly fan-shaped, the pinna base making an angle of ninety to one hundred eighty degrees. New fronds appear in spring; deciduous. Meadows, mossy rocks in open woods. Rare. Northern and western North America; northern Europe, Asia.

Botrychium minganense Victorin (Fig. 90) is closely related and often difficult to distinguish from the moonwort. It is about the same size but has pinnae narrowly fan-shaped, the pinna base making an angle of about or less than ninety degrees. Rare. Northwestern and northeastern North America.

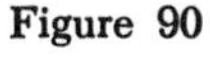

Figure 90

Figure 91 *Botrychium minganense*

In California and Arizona to British Columbia there is another variant, *B. dusenii* (Christ) Alston, which occurs also in Chile. It is very small with pinnae only four to five pairs.

10b **Segments not usually broadly fan-shaped; usually very small, or larger forms have divided lower pinnae. (Fig. 91). .... LITTLE GRAPE-FERN, *Botrychium simplex* E. Hitchc.**

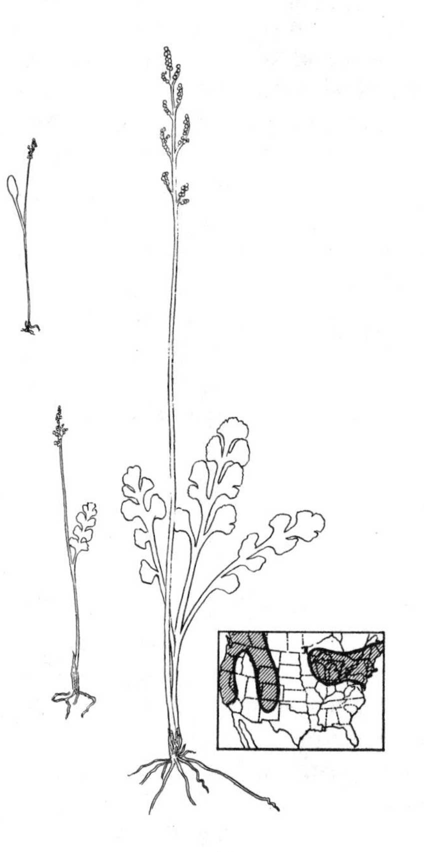

Figure 91

Figure 91 *Botrychium simplex*

Fronds one to seven inches tall; blade oblong, fleshy. New fronds appear in early spring; deciduous. Meadows, moist woods, gravelly slopes. Plants seemingly rare, but easily overlooked. Northeastern, western, and northwestern North America; Europe, eastern Asia.

Var. *simplex* is the most widespread, somewhat leathery, the fertile stalk arising at the base of the blade. Var. *compositum* (Lasch) Milde has the blade divided into three parts, each part deeply lobed, the lobes rounded and the fertile stalk arising at the base of the blade. It occurs in the western states. Var. *tenebrosum* (A. A. Eat.) Clausen of northeastern North America is smaller, one to four inches tall, and the blade is merely pinnately lobed, and the fertile stalk arising below the blade. There is a very small undescribed variety or closely related species in the upper Great Lakes region and Minnesota, called the little goblin. It is paler, more fleshy, and holds its spores until fall, rather than releasing them in the summer.

A very rare species, the PUMICE GRAPE-FERN, *Botrychium pumicola* Coville, looks like a very compact *B. simplex* var. *compositum* and is known only from the volcanic gravel at Crater Lake in southern Oregon. Its fronds are up to four inches tall, and divided into three nearly equal parts, the segments fan-shaped and overlapping.

11a Fertile stalk arising from base of blade; northwestern. (Fig. 92). .... NORTHERN GRAPE-FERN, *Botrychium boreale* (Fries) Milde

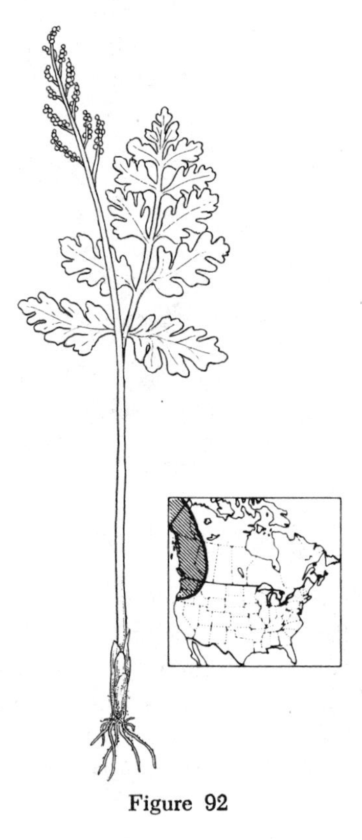

Figure 92

Figure 92 *Botrychium boreale*

Fronds to ten inches tall, pinnate-pinnatifid to bipinnate; blade broadly oblong; stipe longer than the blade. New fronds appear in spring; deciduous. Meadows. Rare and local. Northwestern North America; Europe, Asia.

11b Fertile stalk arising somewhat below the blade; northeastern. (Fig. 93). .... DAISY-LEAVED GRAPE-FERN, ***Botrychium matricariifolium*** **(Döll) A. Braun**

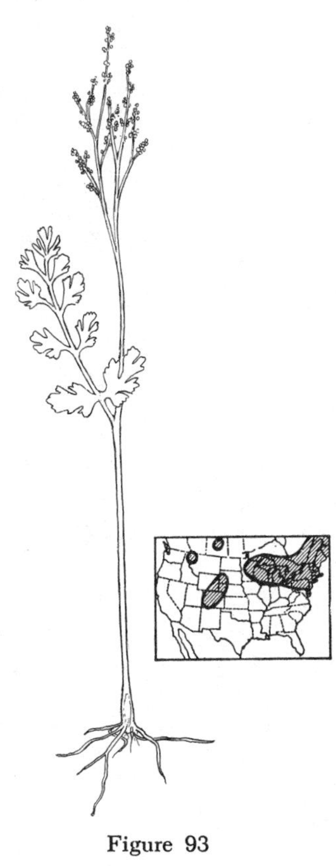

Figure 93

Figure 93 *Botrychium matricariifolium*

Fronds two to six inches tall, narrowly triangular or oblong, pinnate-pinnatifid. New fronds appear in spring and die down in midsummer; deciduous. Moist woods. Frequent. Central and northeastern North America. The most common of the "little grapeferns."

CAMPTOSORUS

Walking Fern

Rhizome ascending, clothed with small clathrate scales; frond undivided, slender, small, long-triangular in shape, rooting at the tip to form new plants; sori elongate along netted veins in an irregular pattern; indusium present. Two species of north temperate regions, one in North America, the other in eastern Asia. (Fig. 94) .... WALKING FERN, *Camptosorus rhizophyllus* (L.) Link

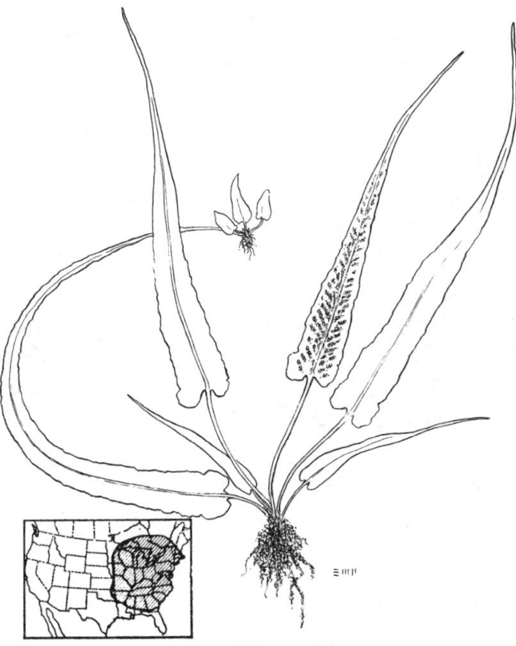

Figure 94

Figure 94 *Camptosorus rhizophyllus*

Fronds five to ten inches long, one-half to one inch wide, the base usually heart-shaped. Moist, mossy, shaded limestone or sandstone rocks. Frequent. Eastern North America.

This species hybridizes with several species of *Asplenium*, demonstrating a close relationship with that genus. Because of this, some botanists treat the walking fern as a species of *Asplenium*. Hybrids with *Asplenium* are placed in a hybrid genus, *Asplenosorus*.

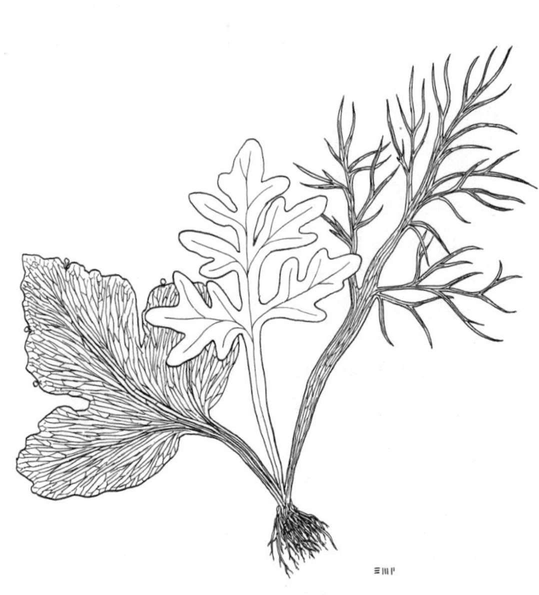

Figure 95

Figure 95 *Ceratopteris pteridoides*

Sterile fronds five to ten inches long, six inches wide; annulus cells few. On mud or quiet water. Rare. Southern Florida, Louisiana; tropical America.

CERATOPTERIS

Water Fern

Plants rooting in mud or free-floating; rhizome short-creeping, sparsely scaly; fronds dimorphic; sterile fronds pinnatifid to tripinnate, naked; veins netted; buds occur in pinna sinuses and serve as a major method of reproduction; sori marginal on more slender fronds, with the margin recurved to protect the sori. Four species of tropical regions.

1a **Stipe inflated; sterile blade pinnatifid with three to seven broad lobes. (Fig. 95) .... WATER FERN, *Ceratopteris pteridoides* (Hooker) Hieron.**

1b **Stipe not inflated; sterile blade bipinnate. (Fig. 96). .... WATER FERN, *Ceratopteris richardii* Brongn. [*C. deltoidea* Benedict]**

Figure 96

Figure 96 *Ceratopteris richardii*

Sterile fronds ten to twenty inches long, about twelve inches wide; fertile fronds somewhat larger; annulus cells many. On mud or quiet water. Rare. Southern Florida, Louisiana; tropical America, Africa.

A pantropical species used in aquaria, *C. thalictroides* (L.) Brongn., has escaped from cultivation in Florida, Texas, and California and can be distinguished by its more divided (tripinnate to quadripinnate) sterile fronds.

CETERACH

Rhizome ascending, scaly with clathrate scales; leaves small, pinnatifid; veins free; sori elongate along veins, indusiate. Five species in warm to temperate climates, largely of Europe and Asia. (Fig. 97) .... *Ceterach dalhousiae* (Hooker) C.Chr.

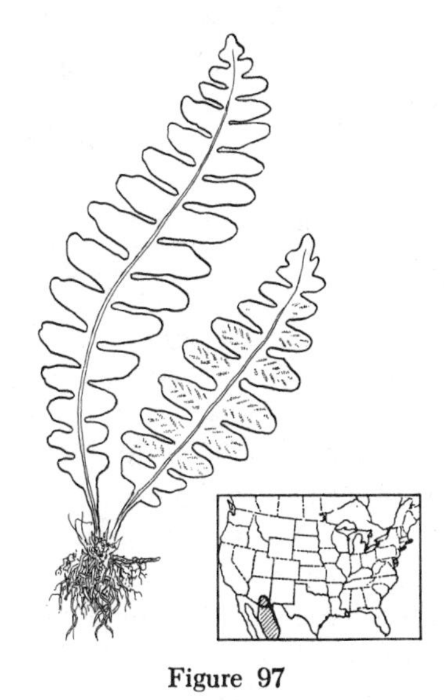

Figure 97

Figure 97 *Ceterach dalhousiae*

Blade naked, two to six inches long, three-fourths to one and three-fourths inch wide, pinnatifid with six to thirteen pairs of lobes, the lobes three-sixteenths to one-half inch wide; fronds spreading to form a rosette. Moist, rocky ravines. Rare. Southern Arizona; northern Mexico, Himalayas.

This species is sometimes placed in a separate genus, *Ceterachopsis,* since it is so distinct from the other species of *Ceterach,* which are very scaly.

CHEILANTHES

Lip Fern and Cloak Fern

Rhizome short- or long-creeping, occasionally ascending; scaly; fronds small to medium-sized, one to five times divided, often hairy, scaly or waxy on lower surface; veins free; sori near the margin, sometimes with the margin reflexed to protect the sori. Nearly two hundred species, largely of warm, dry, rocky regions.

The species included here are usually placed in two genera, *Cheilanthes* and *Notholaena*. *Cheilanthes* is alleged to have the sori protected by a reflexed, differentiated blade margin whereas *Notholaena* is said to have a flat, undifferentiated margin. Unfortunately, this character does not hold true. Many of the species generally placed in *Cheilanthes* do not have a modified margin at all, the margin being only slightly recurved, as in many species of *Notholaena*. One North American species *C. cooperae*, which is always placed in *Cheilanthes* and certainly has its relationships there, has a flat margin without the slightest sign of recurving. The species are, for the most part, readily distinguishable, the problem being which genus to place them in. Therefore, I consider it most appropriate to combine the two genera here. There are several groups of species within *Cheilanthes* that can be distinguished using characters other than the leaf margin, such as indument, anatomy, and form. This is already done by some botanists in a modest way by splitting off *C. siliquosa* and *C. californica* as the genus *Aspidotis*. Other splinters have been extracted in the tropics, such as *Mildella*, *Cheiloplecton*, and *Adiantopsis*. These are certainly natural groups of species, but they are so closely allied to *Cheilanthes* that it can be argued that they should all be maintained as one genus, *Cheilanthes*. A great deal of field work and laboratory study is necessary before a complete understanding of this group is possible.

1a **Stipe and rachis grooved on top; sori often small and discontinuous. 2**

1b **Stipe and rachis round to somewhat flattened on top but not grooved; sori nearly always continuous along the segment margins. 6**

2a **Blade naked or with a few sparse scales; glandular hairs lacking. 3**

2b **Blade with abundant short, glandular hairs. 5**

3a **Blade completely naked. 4**

3b **Blade with at least a few scales on the stipe and rachis. (Fig. 98). .... WRIGHT'S LIP-FERN, *Cheilanthes wrightii* Hooker**

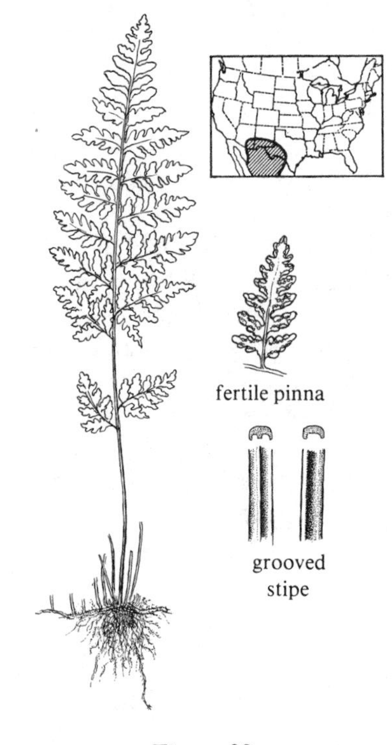

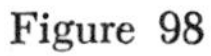
Figure 98

Figure 98 *Cheilanthes wrightii*

Rhizome slender, creeping, scales linear, orange; fronds mostly three to six inches tall; stipe chestnut brown, grooved, naked, one-third to one-half the frond length; sori small, discontinuous, protected only by the segment tips folded back. Igneous rocks and slopes. Frequent. Southwestern United States; northern Mexico.

Cheilanthes pringlei Davenp. is tripinnate to tripinnate-pinnatifid and has abundant scales on the stipe and rachis; its sori are protected by the recurved segment tips; limited to southern Arizona and northern Mexico.

4a Sori continuous along the margin; rhizome scales maroon or brown, concolorous. (Fig. 99). ...... INDIAN'S DREAM, *Cheilanthes siliquosa* Maxon [*Aspidotis densa* (Brack.) Lell.]

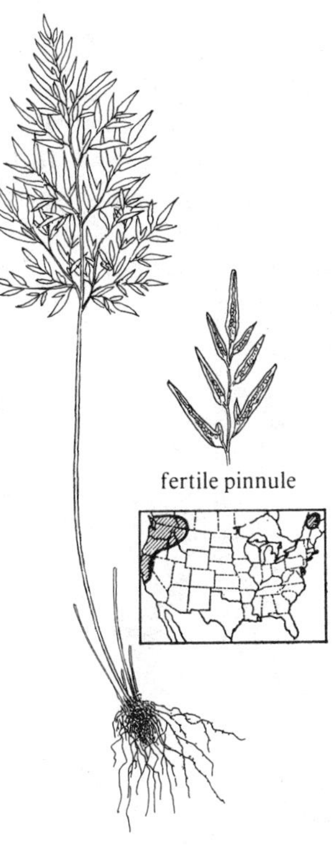

Figure 99

Figure 99 *Cheilanthes siliquosa*

Rhizome short-creeping, scales linear, shiny, dark maroon; fronds four to eight inches tall; stipe brown, grooved, one-half to three-fourths the frond length; blade triangular-oblong, bipinnate to tripinnate; pinnae close and overlapping; segments long and slender, four to eight times as long as wide; sori continuous along the segment margin, entire margin recurved to protect the sori. Serpentine rocks. Common. Western North America and the Gaspé Peninsula of eastern Canada.

Cheilanthes arizonica (Maxon) Mickel [*C. pyramidalis* var. *arizonica* Maxon] has a blade tripinnate to quadripinnate, the segments spreading, not strongly overlapping; the indusium margin has short, stubby hairs; the rhizome scales are dark reddish brown to nearly black; the under surface of the blade has numerous reddish-orange dots. It is limited to southernmost Arizona and northern Mexico.

4b Sori small, discontinuous; rhizome scales bicolorous, dark brown to black with a narrow pale margin. (Fig. 100). .... CALIFORNIA LACE-FERN, *Cheilanthes californica* (Hook.) Mett. [*Aspidotis californica* (Hooker) Nutt. ex Copel.]

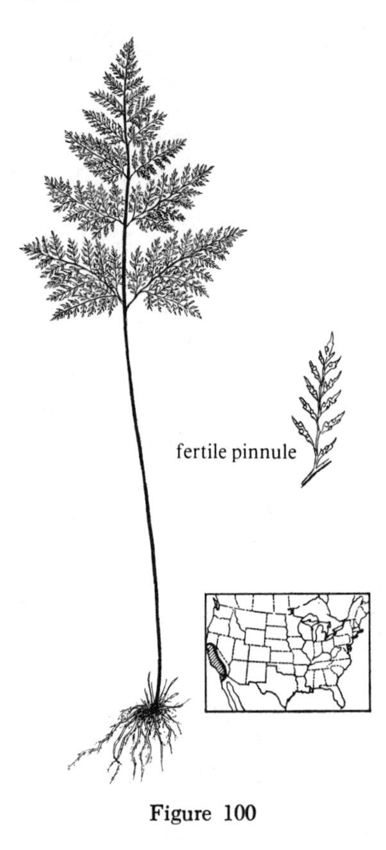

Figure 100

Figure 100 *Cheilanthes californica*

Rhizome short-creeping, scales dark brown to black; fronds mostly four to ten inches tall; stipes brown, grooved, one-third to one-half the front length; blade tripinnate to quadripinnate, triangular; pinnae mostly spreading, the segments sharp-pointed; sori small, round, discontinuous, protected by small recurved flap of the leaf margin. Granitic and serpentine rocks. Frequent. California; northern Mexico.

This species hybridizes with *C. siliquosa* to form a fertile hybrid; *C. carlotta-halliae* W. H. Wagner & Gilbert, on serpentine rocks in coastal California.

5a Blade broadly triangular or pentagonal; rhizome scales reddish black, straight; Texas. (Fig. 101). .... GLANDULAR LIP-FERN, *Cheilanthes kaulfussii* Kunze

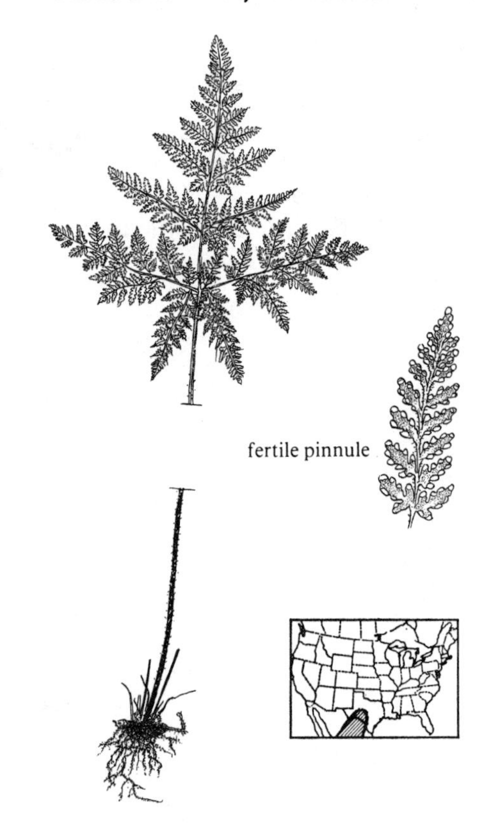

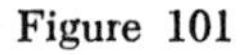
Figure 101

Figure 101 *Cheilanthes kaulfussii*

Rhizome stout, short-creeping to erect; scales reddish black, shining, straight; fronds four to sixteen inches tall; stipe slender, purplish brown, more than one-half the frond length; blade broadly triangular to pentagonal, the lowest pinnae with pinnules elongated downward, tripinnate to tripinnate-pinnatifid; all surfaces covered with glandular hairs; sori round, protected by slightly modified lobe margin. Rocky slopes. Rare. Texas; Middle and South America.

5b **Blade narrowly oblong; rhizome scales reddish orange, twisted; California. (Fig. 102). .... VISCID LIP-FERN, *Cheilanthes viscida* Davenp.**

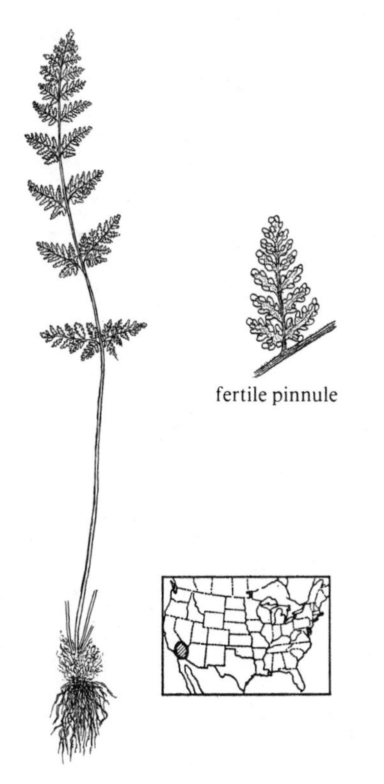

Figure 102

Figure 102 *Cheilanthes viscida*

Rhizome stout, short-creeping; scales reddish orange, twisted; fronds two to nine inches tall; stipes chestnut brown, one-third to one-half the frond length; blade oblong, widest about the middle, bipinnate-pinnatifid to tripinnate; sori protected by the recurved segment lobes. Frequent. Southern California; northern Mexico.

6a **(1b) Lower blade surface with white or yellow wax (rarely also with scales) or completely naked. (If lower surface is covered with scales, scrape some off to check for wax.) 7**

6b **Lower blade surface with hairs or scales or both, but without wax. 17**

7a **Indument lacking entirely; blade naked. 8**

7b **Indument of white or yellow wax glands, with or without scales. 9**

8a **Color of the minor rachises not running into the segment bases, but stopping abruptly at the segment base, the segments often falling off in old age; rachis slightly zig-zag; segments distant; blade tripinnate. (Fig. 103). .... SMALL-LEAVED CLOAK-FERN, *Cheilanthes parvifolia* (R. Tryon) Mickel [*Notholaena parvifolia* R. Tryon]**

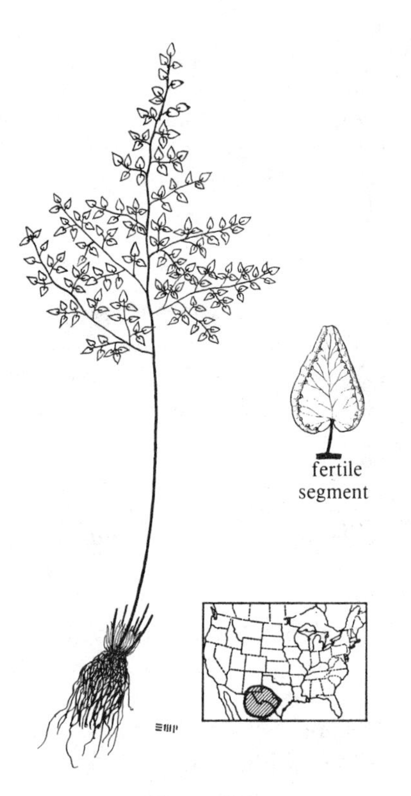

Figure 103

Figure 103 *Cheilanthes parvifolia*

Rhizome short-creeping; scales brown; fronds four to ten inches tall; stipe one-third to one-half the frond length; blade tripinnate to quadripinnate, triangular or broadly oval, naked; segments small, many, distant from one another, round or oblong, stalked, articulate; rachis zig-zags slightly. Limestone ledges and slopes. Common. Southwestern United States; northern Mexico.

Because of its naked blades and pale, bluish-green color, it is sometimes placed in the genus *Pellaea,* but it seems to be more closely allied to the finely dissected waxy-backed species of *Cheilanthes,* such as *C. dealbata* and *C. limitanea.*

8b Color of the minor rachises running into the segment bases and gradually tapering off; rachis straight; blade mostly bipinnate or rarely tripinnate at base; segments close to one another. (Fig. 104). .. .. JONES'S CLOAK-FERN, *Cheilanthes jonesii* (Maxon) Mickel [*Notholaena jonesii* Maxon]

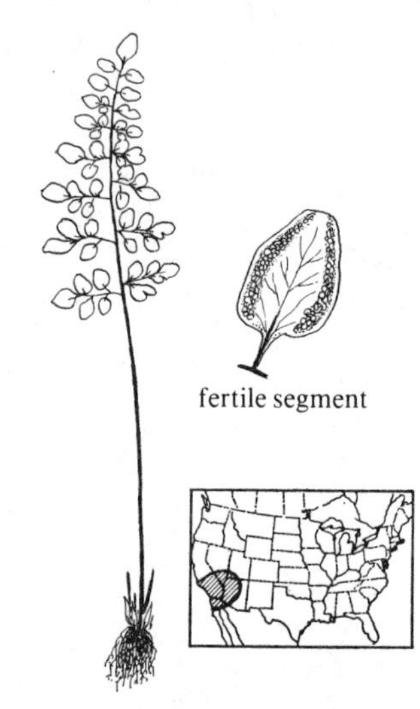

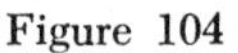

Figure 104

Figure 104 *Cheilanthes jonesii*

Rhizome short-creeping; scales brown; fronds two to six inches tall; stipe round, chestnut brown, naked, one-third to one-half the frond length; blade narrowly oblong, bipinnate (or rarely tripinnate at base); segments smooth-margined or slightly lobed, not articulate, the rachis color entering the segment base. Calcareous rock ledges. Frequent. Southwestern United States.

9a Blade indument of wax alone; rhizome scales rarely toothed. 10

9b Blade indument of wax and scales; rhizome scales toothed. 15

10a **All segments broadly attached to the minor rachises; rhizome scales dark (with or without a light margin, or dark with a very narrow pale, toothed margin in *C. deserti*). 11**

10b **Ultimate segments narrowly attached, some stalked; rhizome scales usually thin, brown, concolorous. 13**

11a **Blade broader than long; wax yellow. 12**

11b **Blade longer than broad; wax white. (Fig. 105). .... *Cheilanthes candida* Mart. & Gal. [*Notholaena candida* (Mart. & Gal.) Hooker]**

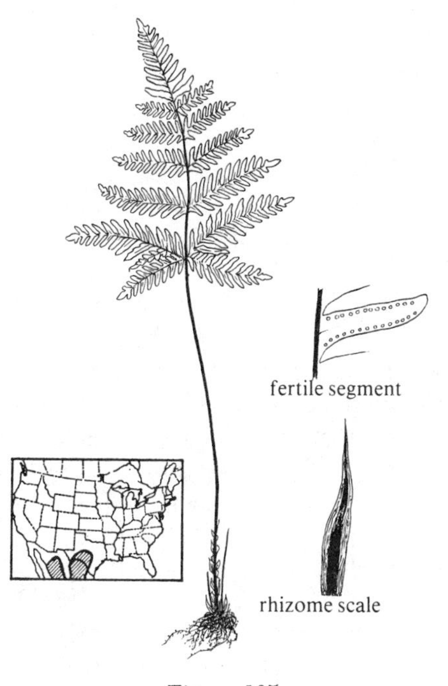

Figure 105

Figure 105 *Cheilanthes candida*

Rhizome short-creeping; scales bicolorous; fronds four to twelve inches tall; stipe thick, black, round, with scales at base; blade bipinnate-pinnatifid, the lowest pinnae with pinnules on basal side much longer, lower surface white waxy; segments wavy-margined to pinnatifid; blade pinnate-pinnatifid above the basal pinna pair. Calcareous rock ledges. Frequent. Southwestern Texas; northern Mexico.

The Texas specimens are of var. *copelandii* (C. C. Hall) Mickel with only one elongate exaggerated basal pinnule per pinna. It might be confused with *Pityrogramma triangularis,* but that species has sporangia along the veins, not just close to the margin.

Cheilanthes lemmonii (D. C. Eaton) Domin [*Notholaena lemmonii* D. C. Eaton], LEMMON'S CLOAK-FERN, is much longer than broad (three to four times), the lowest pinna pair similar to those above, not especially enlarged on the basal side. It is found frequently in southeastern Arizona and northern Mexico.

12a **Blade pinnatifid above the base. (Fig. 106). .... STAR CLOAK-FERN, *Cheilanthes standleyi* (Maxon) Mickel [*Notholaena standleyi* Maxon]**

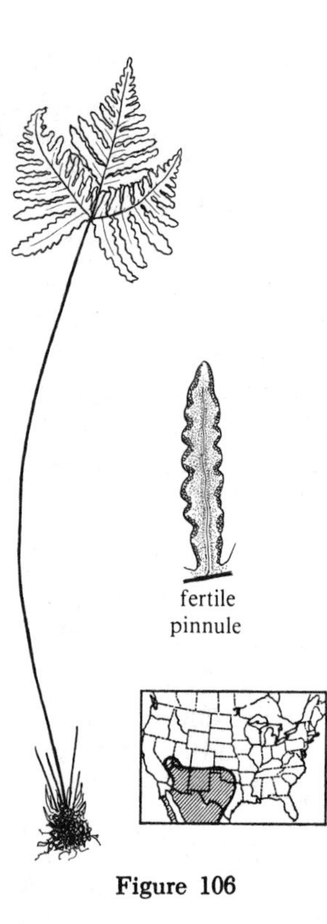

Figure 106

Figure 106 *Cheilanthes standleyi*

Rhizome short-creeping; scales with dark brown or black center and pale margin; fronds four to eight inches tall; stipe round, chestnut brown, much longer than the blade, more than three-fourths of the frond length; blade pentagonal, the basal pair of pinnae greatly elongated on the basal side, each basal pinna nearly as large as the rest of the blade above; pinnate-pinnatifid, pinnatifid above the base; lower surface covered with yellow wax. Calcareous or noncalcareous cliffs. Abundant. Southwestern United States; northern Mexico.

12b Blade pinnate-pinnatifid to bipinnate above the basal pinna pair. (Fig. 107). .... CALIFORNIA CLOAK-FERN, *Cheilanthes deserti* Mickel [*Notholaena californica* D. C. Eaton]

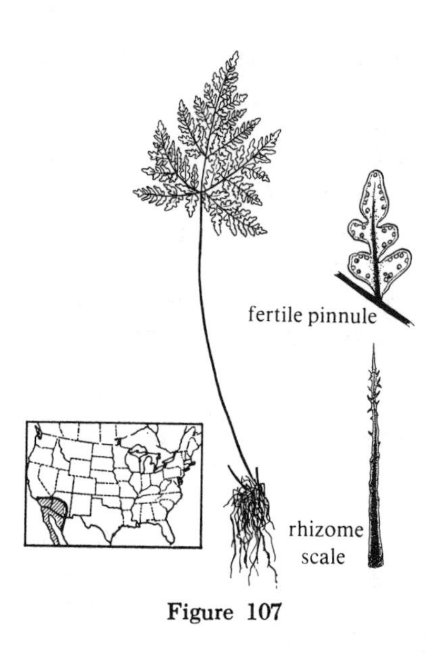

Figure 107

Figure 107 *Cheilanthes deserti*

Rhizome short-creeping; scales black-centered with pale margin; fronds one to five inches tall; stipe round, slender, chestnut brown, more than one-half the frond length; blade bipinnate, tripinnate at base; basal pinnae greatly enlarged on the basal side, upper surface glandular, lower surface yellow, or rarely white waxy. Noncalcareous rocks. Common. Southwestern United States; northwestern Mexico.

Two species have a narrower blade. *Cheilanthes neglecta* (Maxon) Mickel [*Notholaena neglecta* Maxon] is only sparsely glandular on the upper surface, the rhizome scales inconspicuously toothed or short-ciliate, and the stipe round and black. It occurs in northern Mexico and rarely gets into southern Arizona and Texas. GREGG'S CLOAK-FERN, *Cheilanthes greggii* (Mett.) Mickel [*Notholaena greggii* (Mett.) Maxon] is densely glandular on the upper surface, has rhizome scales inconspicuously toothed, fronds two to eight inches tall, stipe slightly grooved and brown, and occurs in northern Mexico and rarely the Big Bend of Texas.

13a **Rachis more or less straight.** **14**

13b **Rachis conspicuously zig-zag; rhizome scales conspicuous (three-eighths inch long). (Fig. 108).** ...... **LATTICEWORK CLOAK-FERN,** ***Cheilanthes cancellata*** **Mickel [*Notholaena fendleri* Kunze]**

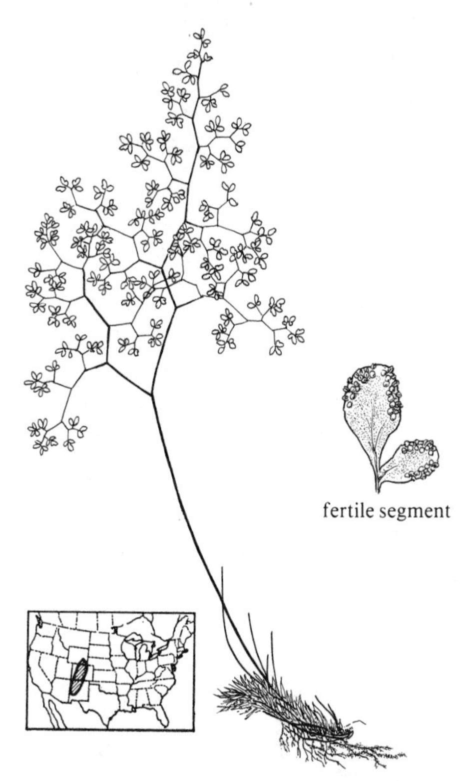

Figure 108

Figure 108 *Cheilanthes cancellata*

Rhizome short-creeping; scales chestnut brown; fronds four to twelve inches tall; stipe thick, round, chestnut brown, naked, about one-half the frond length; blade broadly triangular, five to six times pinnate; upper surface usually smooth, lower surface white waxy; rachis and pinna axes distinctly zig-zag. Dry cliffs. Common. Southwestern United States.

14a **Stipe one-thirty-second inch diameter; stipe and rachis purplish-black, with a slight whitish bloom; rhizome scales about one-fourth inch long and spreading. (Fig. 109).** .... ***Cheilanthes limitanea*** **(Maxon) Mickel [*Notholaena limitanea* Maxon]**

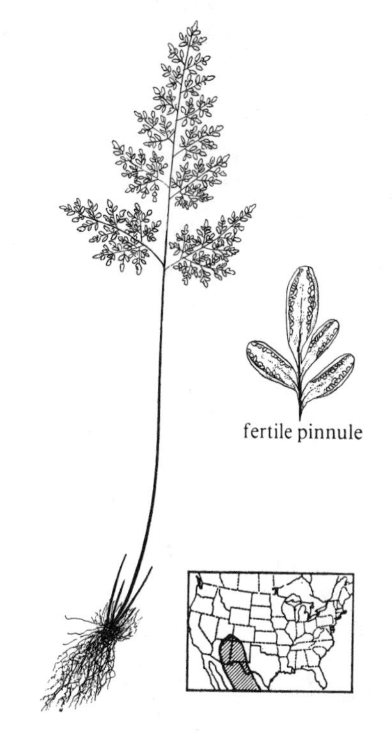

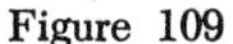

Figure 109 *Cheilanthes limitanea*

Rhizome short-creeping; scales chestnut brown; fronds four to ten inches tall; stipe round, brown, naked, about one-half the frond length; blade tripinnate to quadripinnate, triangular or oval; upper surface smooth, rarely slightly waxy, lower surface white waxy; segments small, crowded, oblong. Usually calcareous rocks. Frequent. Southwestern United States; northern Mexico.

The most common form is var. *limitanea*, but var. *mexicana* (Maxon) Mickel can be distinguished by its narrower fronds (oblong rather than triangular) and the basal pinnae are somewhat ascending.

14b **Stipe about one-sixty-fourth inch diameter; stipe and rachis chestnut brown; rhizome scales about three-sixteenths inch long and not spreading. (Fig. 110). .... POWDERY CLOAK-FERN, *Cheilanthes dealbata* Pursh [*Notholaena dealbata* (Pursh) Kunze]**

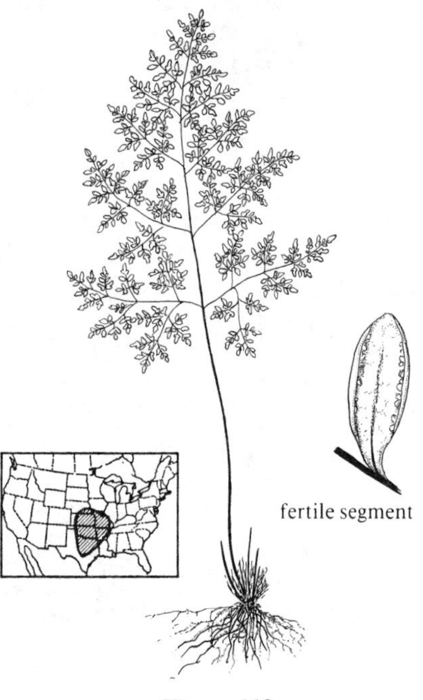

Figure 110

Figure 110 *Cheilanthes dealbata*

Rhizome short-creeping, slender; scales light brown; fronds two to six inches tall; stipe slender, round, naked, chestnut brown, at least one-half the frond length; blade triangular, sometimes broadly so, four to five times pinnate; upper surface naked or nearly so, lower surface white waxy (rarely yellow). Limestone cliffs. Common. Central United States.

15a **(9b) Blade scales lance-shaped, appressed, and abundant; rachises lacking glandular hairs, although wax glands present in *C. grayi*. 16**

15b **Blade scales slender, hair-like, sparse, spreading; rhizome scales brown to black-tipped, ciliate; rachises with minute, glandular hairs. (Fig. 111). .... NEALLEY'S CLOAK-FERN, *Cheilanthes nealleyi* Seaton ex Coulter [*Notholaena schaffneri* (Fourn.) Underw. ex Davenp.]**

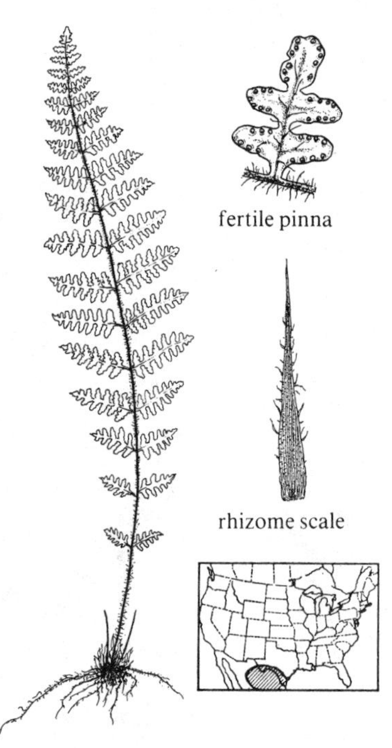

Figure 111

Figure 111 *Cheilanthes nealleyi*

Rhizome short-creeping; scales brown to black, ciliate; fronds two to ten inches tall; stipe one-fourth the frond length, round, brown to black, clothed with many slender scales and long, straight hairs; blade narrowly oblong, bipinnate-pinnatifid; upper surface sparsely glandular, lower surface white waxy, with long hairs on the midveins. Rocky ledges. Rare. Southern Texas; northern Mexico.

16a Scales densely covering the lower surface of the blade, the scales being long-ciliate or dissected; wax and sporangia obscured by the scales. (Fig. 112) .... ASCHENBORN'S CLOAK-FERN, *Cheilanthes aschenborniana* (Kl.) Mett. [*Notholaena aschenborniana* Kl.]

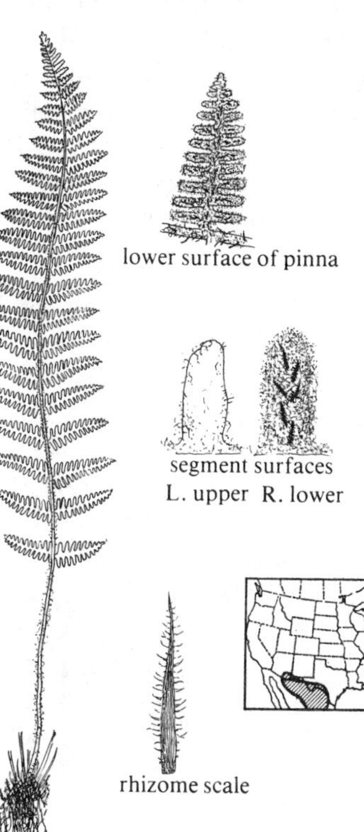

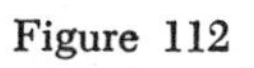

Figure 112 *Cheilanthes aschenborniana*

Rhizome short-creeping; scales black or dark brown, concolorous with long, hair-like teeth; fronds eight to eighteen inches tall; stipe one-fourth to one-third the frond length, thick, round, brown to black, clothed with many brown, ciliate scales; blade bipinnate, narrowly oblong; upper surface with dissected scales, appearing like stellate hairs, lower surface densely clothed with slender, long-ciliate scales, hiding the waxy surface. Limestone rocks and cliffs, higher elevation (4,500-9,000 feet). Frequent. Arizona and Texas; northern Mexico.

16b Scales essentially smooth-margined on the lower blade surface, linear-lance-shaped; wax evident without having to remove the scales; upper blade surface lightly waxy, lacking hairs. (Fig. 113). .... GRAY'S CLOAK-FERN, *Cheilanthes grayi* (Davenp.) Domin [*Notholaena grayi* Davenp.]

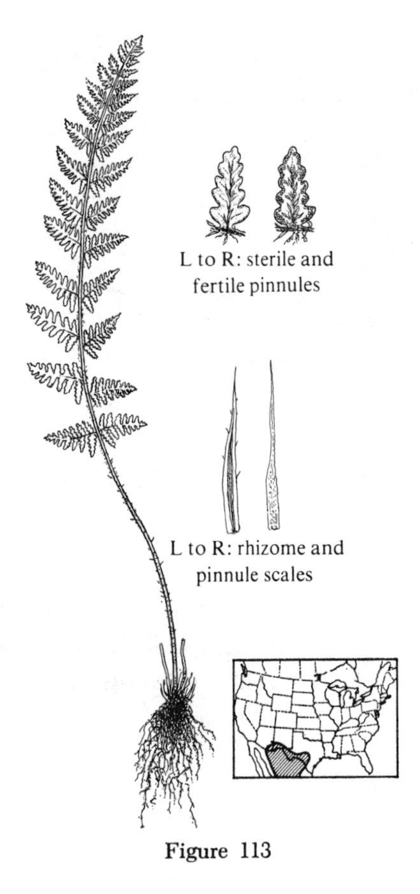

Figure 113

Figure 113 *Cheilanthes grayi*

Rhizome ascending; scales black to chestnut, with slightly toothed margin; fronds three to nine inches tall; stipe thick, pale brown, round, with a few scales and glands, about one-half the frond length; blade pinnate-pinnatifid to bipinnate-pinnatifid, very narrow; pinnae with lower facing pinnules longer than the upper facing ones; upper surface with a few wax glands, lower white waxy, and with reddish brown scales on rachises and midveins. On rocks (3,000-5,000 feet elevation). Frequent. Southwestern United States; northern Mexico.

Cheilanthes aliena (Maxon) Mickel [*Notholaena aliena* Maxon] is similar to *C. grayi* but has tortuous, delicate white hairs on the upper surface. It occurs in the Big Bend of Texas and northern Mexico. Rare.

17a (6b) Blade indument of hairs only. .. 18

17b Blade indument of scales or scales and hairs. 25

18a Blade hairs sparse, stiff, white, sharp-pointed on the leaf tissue, often orange on the rachises. 19

18b Blade hairs abundant. 20

19a Blade bipinnate to bipinnate-pinnatifid; stipe and rachis black; blade nearly naked on lower surface; upper side of rachis with abundant short hairs. (Fig. 114). .... ALABAMA LIP-FERN, *Cheilanthes alabamensis* (Buckl.) Kunze

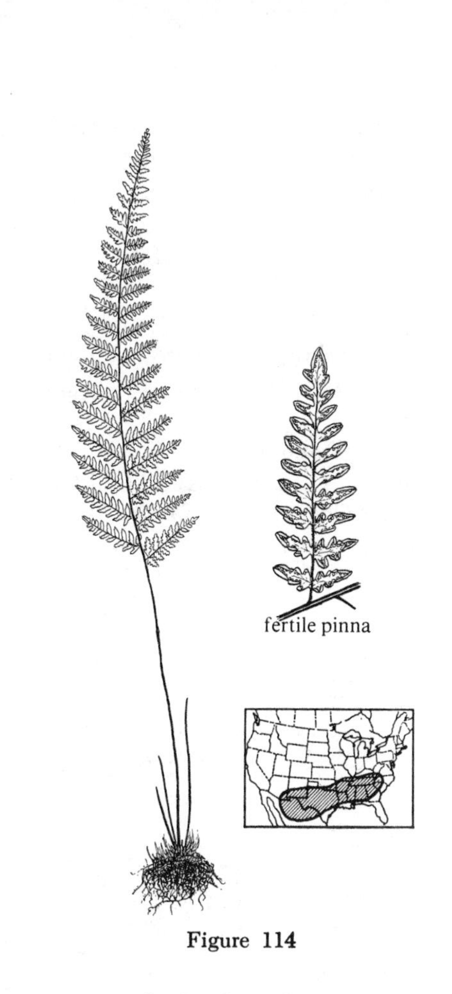

Figure 114

Figure 114 *Cheilanthes alabamensis*

Rhizome short-creeping; scales very slender, orange; fronds to twenty inches tall; stipe black, naked or with scales at base; blade narrowly oblong, bipinnate to bipinnate-pinnatifid; segments often pointed; upper and lower surfaces naked or with scattered short hairs beneath; indusium continuous. Calcareous rock slopes. Frequent to common. Southern United States; northern Mexico, Jamaica.

Cheilanthes aemula Maxon is similar but has a broadly triangular blade; rare in southern Texas, extending to southern Mexico.

Cheilanthes notholaenoides (Desv.) Maxon is rare in southern Texas but frequent in Mexico. It is bipinnate and has a margin not at all differentiated but slightly recurved; the rachis hairs are reddish and abundant.

19b Blade bipinnate-pinnatifid to tripinnate; stipe and rachis dark brown; small stiff white hairs on upper surface; rachis with short hairs on upper side and often with long hairs on lower side. (Fig. 115). .... SOUTHERN LIP-FERN, *Cheilanthes microphylla* Sw.

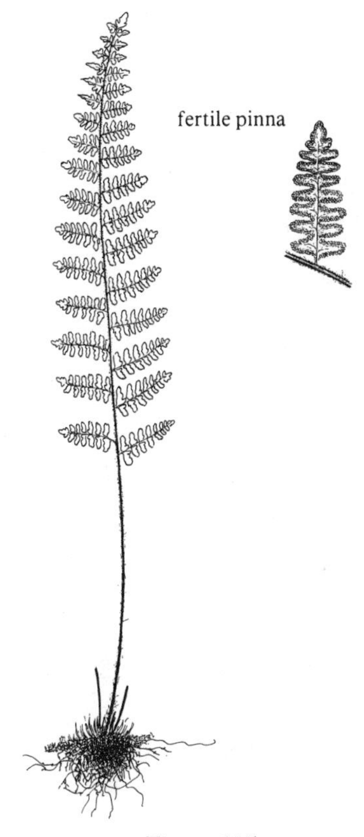

Figure 115

Figure 115 *Cheilanthes microphylla*

Rhizome slender, short-creeping; scales orange, slender; fronds to twenty inches tall; stipe black with many short, orange hairs; blade bipinnate-pinnatifid to nearly tripinnate; the segments blunt-tipped; sori and indusium continuous. Limestone outcrops. Uncommon. Peninsular Florida; tropical America.

20a Stipe and rachis dark; blade oval-lance-shaped to narrowly triangular. 21

20b Stipe and rachis yellow; blade broadly pentagonal. (Fig. 116). .... WHITE-FOOT LIP-FERN, *Cheilanthes leucopoda* Link

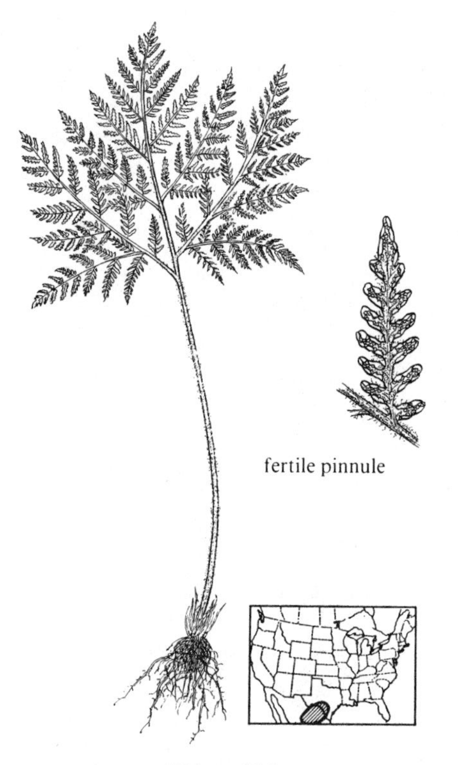

Figure 116

Figure 116 *Cheilanthes leucopoda*

Rhizome short-creeping; scales slender, light brown; fronds to fourteen inches tall, clumped; stipe yellow, with abundant hairs, about two-thirds the frond length; blade triangular to pentagonal, tripinnate to quadripinnate; lowest pinnae with long basal pinnules; upper surface with straight white hairs, the lower surface with straight rusty hairs; margin differentiated as indusium, continuous. Rocky slopes. Rare. Texas; northern Mexico.

21a Blade bipinnate or more divided; narrowly to broadly oblong. 22

21b Blade pinnate-pinnatifid; blade linear. (Fig. 117). .... GOLDEN CLOAK-FERN, *Cheilanthes bonariensis* (Willd.) Proctor [*Notholaena aurea* (Poir.) Desv.]

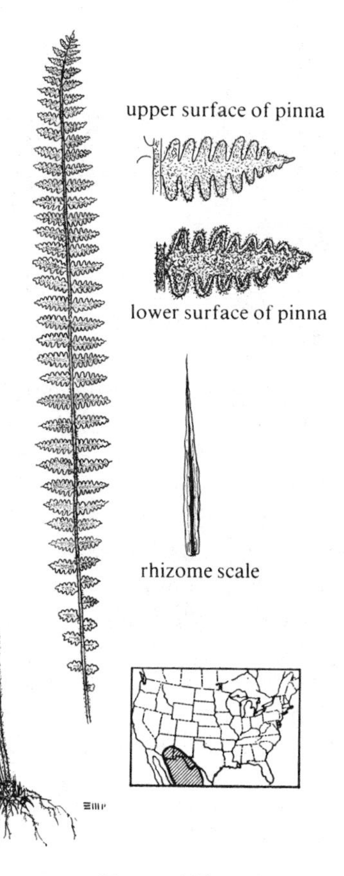

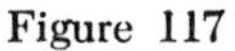

Figure 117

Figure 117 *Cheilanthes bonariensis*

Rhizome short-creeping; scales dark brown centered with pale margins; fronds eight to twenty-four inches tall; stipe one-sixth to one-fourth the frond length, round, dark brown or black, clothed with white hairs; blade very narrow, pinnate-pinnatifid; upper surface with few hairs, lower surface densely covered with long white to tan hairs. Among rocks. Common. Southwestern United States; tropical America.

22a Rhizome scales distinctly bicolorous; the rhizome very short-creeping; the fronds strongly clumped. 23

22b Rhizome scales concolorous or only slightly paler at margin; rhizome short- to long-creeping, the fronds loosely clumped to spaced apart. 24

23a Hairs of upper blade surface abundant, white; those of the lower surface white to light tan; hairs so fluffy and surpassing the segments as to appear like cotton. (Fig. 118). .... PARRY'S LIP-FERN, *Cheilanthes parryi* (D. C. Eaton) Domin [*Notholaena parryi* D. C. Eaton]

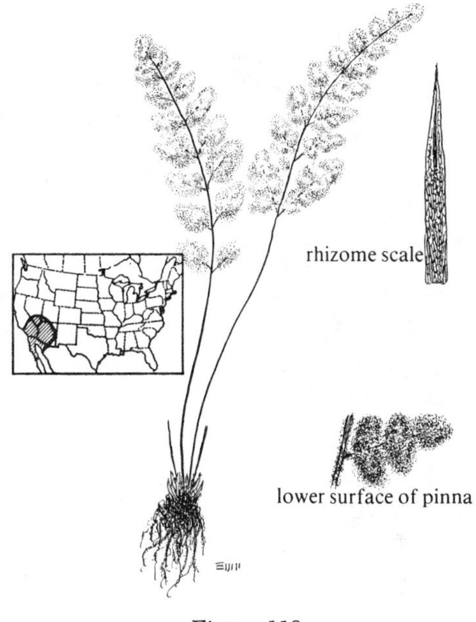

Figure 118

Figure 118 *Cheilanthes parryi*

Rhizome very short-creeping or ascending, stout; scales bicolorous with black center and orange margin; fronds three to six inches tall; stipe slender, round, chestnut brown, hairy with a few scales near the base, about one-half the frond length; blade narrowly triangular or oval, bipinnate-pinnatifid; upper and lower surfaces covered with long, curly, white to tan hairs. Among rocks. Common. Southwestern United States; northwestern Mexico.

23b Hairs of upper blade surface sparse, pale; those of the lower surface rusty; hairs not far surpassing the segments and not making the frond appear like a fluff of cotton. (Fig. 119). .... SLENDER LIP-FERN, *Cheilanthes feei* Moore

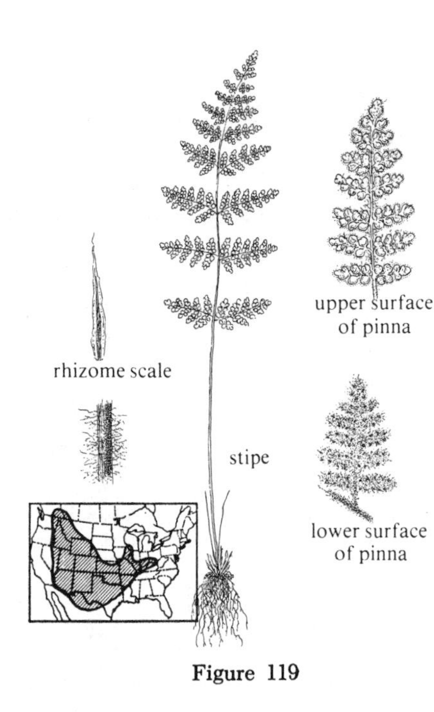

Figure 119 *Cheilanthes feei*

Rhizome very short-creeping or ascending; scales bicolorous with black centers and orange margin; fronds tufted, to ten inches tall; stipe slender, dark brown, more than one-half the frond length, hairy with a few scales at the base; blade narrowly oblong to oval, bipinnate-pinnatifid to tripinnate; upper surface with sparse, long, white hairs, lower surface densely hairy with tan or rusty hairs. Calcareous and noncalcareous cliffs. Frequent to common. Central and western North America.

24a Blade hairs reddish, straight or lax, not curled; rhizome scales orange to chestnut brown, toothed, lacking a kinky hair tip. (Fig. 120). .... HAIRY LIP-FERN, *Cheilanthes lanosa* (Michx.) D. C. Eaton

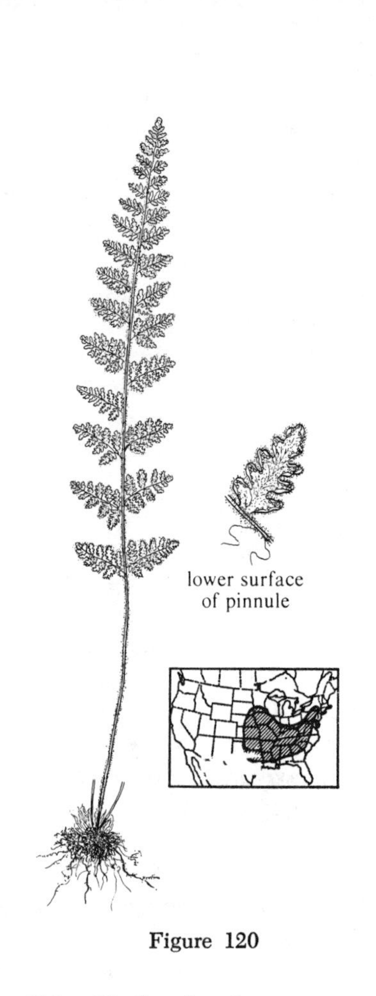

Figure 120 *Cheilanthes lanosa*

Rhizome slender, short-creeping; scales chestnut brown, toothed; fronds to sixteen inches tall, loosely clumped; stipe chestnut brown, with spreading hairs, one-fourth to one-third the frond length; blade narrowly oblong, bipinnate-pinnatifid to tripinnate; upper surface with sparse straight or lax reddish hairs; lower surface densely hairy with hairs similar to those above. Noncalcareous rocky slopes. Frequent. Eastern and southern United States.

Cheilanthes lendigera (Cav.) Sw. can be distinguished from *C. lanosa* in having the rhizome more widely creeping, the rhizome scales concolorous, the blade triangular and naked on the upper surface, and the indusium broad and pouch-like. It occurs among noncalcareous rocks in the southwestern states (Arizona to Texas) and south to South America; rare.

MRS. COOPER'S LIP-FERN, *Cheilanthes cooperae* D. C. Eaton, of California, has long glandular hairs, several cells long, the segment margin not at all differentiated from the rest of the leaf tissue, lying flat with small, round sori; rhizome scales not toothed; rare.

24b Blade hairs white, strongly curled, matted like wool; rhizome scales glossy maroon to nearly black, smooth-margined, shiny with a kinky hair tip. (Fig. 121). .... NEWBERRY'S CLOAK-FERN, *Cheilanthes newberryi* (D. C. Eaton) Domin [*Notholaena newberryi* D. C. Eaton]

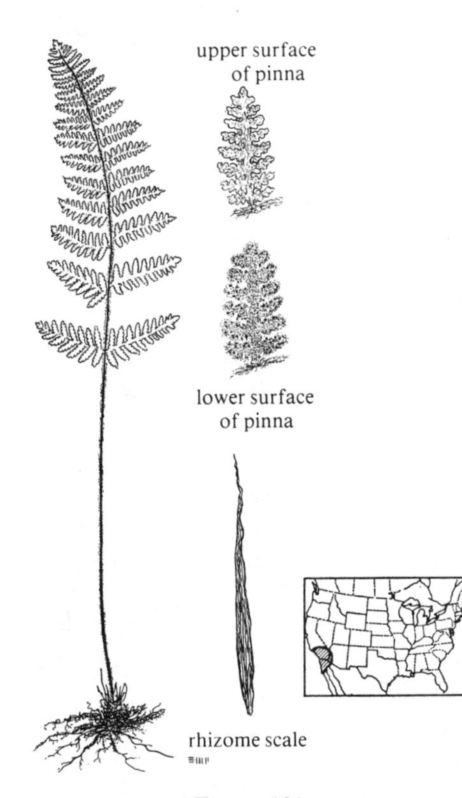

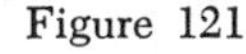

Figure 121

Figure 121 *Cheilanthes newberryi*

Rhizome short- to long-creeping; scales glossy maroon to nearly black, smooth-margined, with long kinky hair tips; fronds three to eight inches tall; stipe one-third to one-half the frond length, round, chestnut brown, clothed with long, fine, curled hairs; blade oblong, bipinnate to tripinnate; upper and lower surfaces clothed with fine, curled hairs, white above, tan below; segments tiny, round, bead-like. Among rocks. Common Southern California; Baja California.

25a (17b) Upper blade surface bearing conspicuous and abundant hairs. (Beware of hairs from below poking through between the segments to appear like hairs on the upper surface.) 26

25b Upper blade surface naked or with only a few finely-cut stellate scales. 30

26a Upper blade surface with lax white hairs; rhizome scales bicolorous or rarely concolorous. 27

26b Upper blade surface with short, stiff, stout, sharp, milk-white hairs; rhizome scales orange, concolorous. (Fig. 122). .. .. HISPID LIP-FERN, *Cheilanthes horridula* Maxon

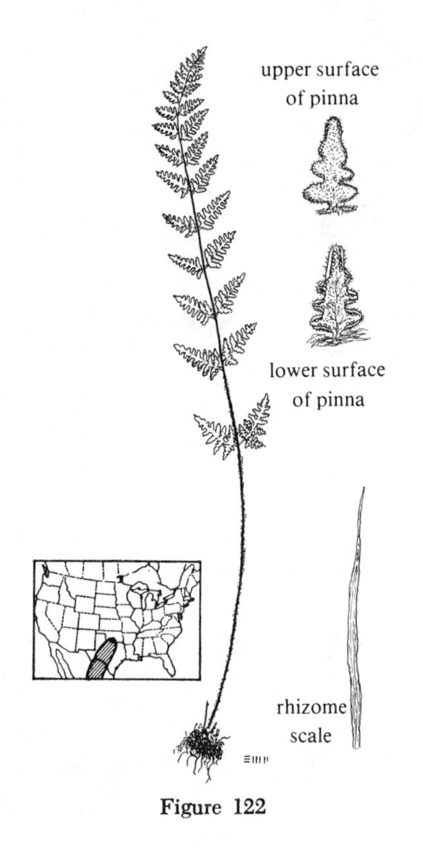

Figure 122 *Cheilanthes horridula*

Rhizome short-creeping, scales slender, reddish brown; fronds to twelve inches tall; stipes slender, one-third the frond length, dark brown to black, with narrow scales and bristly hairs; blade narrowly oblong, bipinnate to bipinnate-pinnatifid; upper and lower surfaces with short, stiff, sharp, white hairs; indusium continuous. Rocky crevices. Frequent. Southwestern United States; northern Mexico.

27a Scales (broadly oval or triangular-lance-shaped) covering most of the lower blade surface. 28

27b Scales not covering the whole lower blade surface; hairs the dominant covering. 29

28a Rhizome long-creeping, the fronds well spaced; rhizome scales oval, concolorous, tan to shiny maroon; scales of lower blade surface linear-lance-shaped to oval, ciliate at least in lower half of the scale; upper blade surface wooly. (Fig. 123). ...... LINDHEIMER'S LIP-FERN, FAIRY SWORDS, *Cheilanthes lindheimeri* Hooker

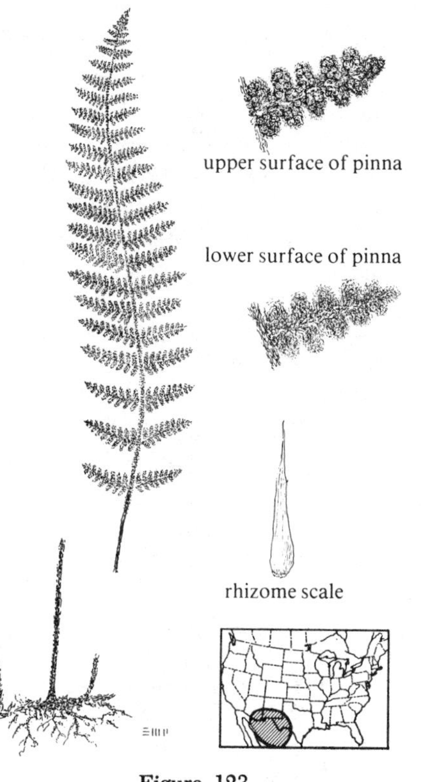

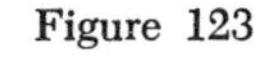
Figure 123

Figure 123 *Cheilanthes lindheimeri*

Rhizome long-creeping, slender; scales oval, brown, long-ciliate; fronds to fourteen inches tall, distant; stipe about one-half or more the frond length, dark brown, with scales and wooly hairs; blade oblong, tripinnate to quadripinnate; upper surface with white hairs, the lower surface chaffy with matted rusty hairs and/or rusty scales; segments beadlike; indusium continuous. Noncalcareous rocks. Common. Southwestern United States; northern Mexico.

28b Rhizome short-creeping, fronds clumped; rhizome scales bicolorous with black central stripe; blade scales triangular, not ciliate; upper surface loosely hairy. (Fig. 124). .... *Cheilanthes villosa* Davenp. ex Maxon

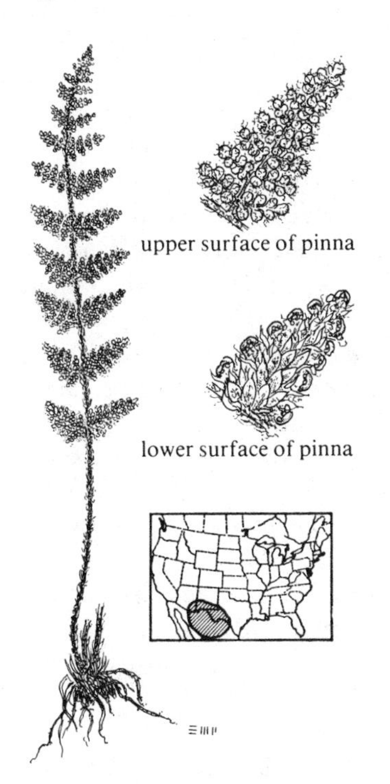

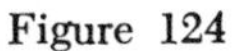
Figure 124

Figure 124 *Cheilanthes villosa*

Rhizome short-creeping to ascending, stout; rhizome scales narrow, bicolorous; fronds four to fourteen inches tall, clumped; stipe purplish brown, covered with both narrow and broad scales; blade narrowly oblong, tripinnate to quadripinnate; segments beadlike; sparsely hairy on upper surface, lower surface covered with triangular pale scales mixed with hairs; indusium continuous. Limestone or granitic rocks. Uncommon. Southwestern United States; northern Mexico.

29a Lower blade surface with essentially only hairs; scales linear, extremely narrow and looking much like hairs. (Fig. 125). .... WOOLY LIP-FERN, *Cheilanthes* tomentosa Link

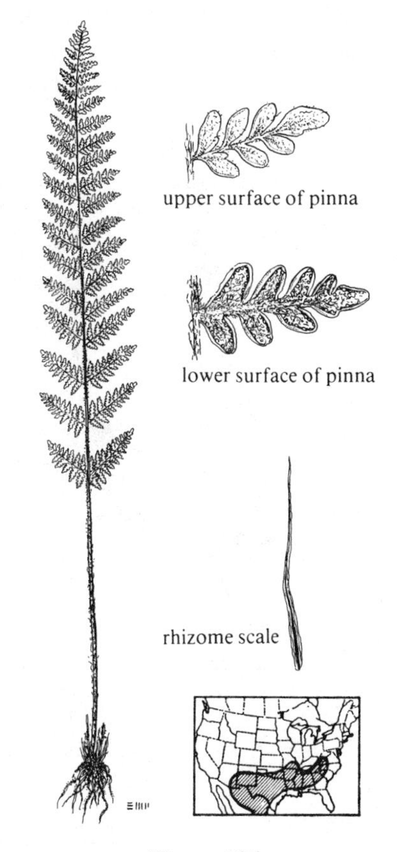

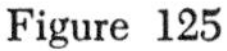
Figure 125

Figure 125 *Cheilanthes tomentosa*

Rhizome short-creeping, stout; scales brown or with black center, linear; fronds about twelve inches tall, up to twenty-four inches tall, clumped; stipe brown, one-third to one-half the frond length, covered with tan hairs and narrow scales; blade oblong to linear, bipinnate-pinnatifid to tripinnate; upper surface with white curly hairs, lower surface with dense mat of white, gray or brown hairs and very narrow scales; indusium continuous. Noncalcareous rocks. Common. Southern United States; northern Mexico.

29b Lower blade surface covered with hairs and many lance-shaped scales. (Fig. 126). .... EATON'S LIP-FERN, *Cheilanthes eatonii* Baker ex Hooker & Baker (incl. *C. castanea* Maxon)

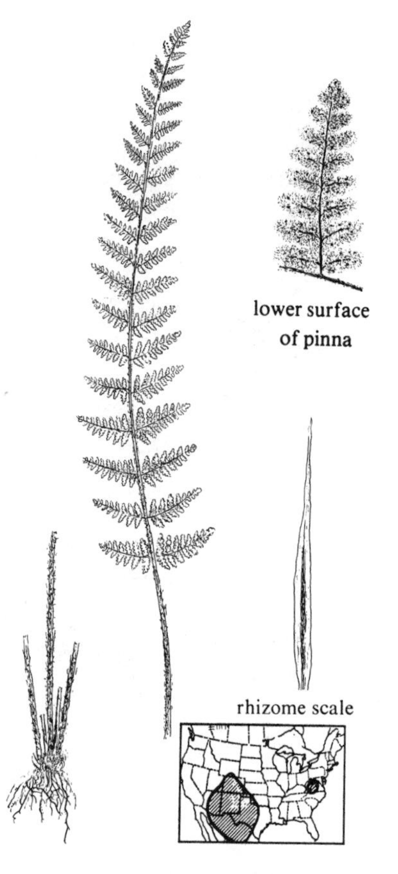

Figure 126

Figure 126 *Cheilanthes eatonii*

Rhizome short-creeping, stout; scales bicolorous, black or brown center; fronds to sixteen inches tall, clumped; stipe brown, one-third to one-half the frond length, with narrow scales and hairs more or less appressed; blade tripinnate to tripinnate-pinnatifid, oblong-lance-shaped or narrow lance-shaped; upper surface with white curly hairs, the lower surface with dense rusty hairs along with lance-shaped scales; indusium continuous. Calcareous and noncalcareous rocks. Common. Southwestern United States and Virginia; northern Mexico.

Cheilanthes castanea Maxon is sometimes held to be distinct from *C. eatonii* by its not having as dense a hair covering on the upper surface, even becomng naked with age, but this character is difficult to distinguish, and I believe it is better treated as a form of *C. eatonii.*

Cheilanthes × *parishii* Davenp. has long, sparse hairs that form a loose mat on the upper and lower srfaces, and the scales on the lower surface are coarsely toothed; southern California; rare. It is thought that this is a hybrid between *C. parryi* and *C. covillei.*

Cheilanthes fibrillosa Davenp. ex Underw. has rhizome scales that are linear-lance-shaped, maroon, shiny; lower blade surface densely wooly; rare; southern California.

30a (25b) Blade narrowly to broadly oblong, bipinnate or more divided. 31

30b Blade linear, pinnate or pinnate pinnatifid. (Fig. 127). .... WAVY CLOAK-FERN, *Cheilanthes sinuata* (Lag. ex Sw.) Domin [*Notholaena sinuata* (Lag. ex Sw.) Kaulf.]

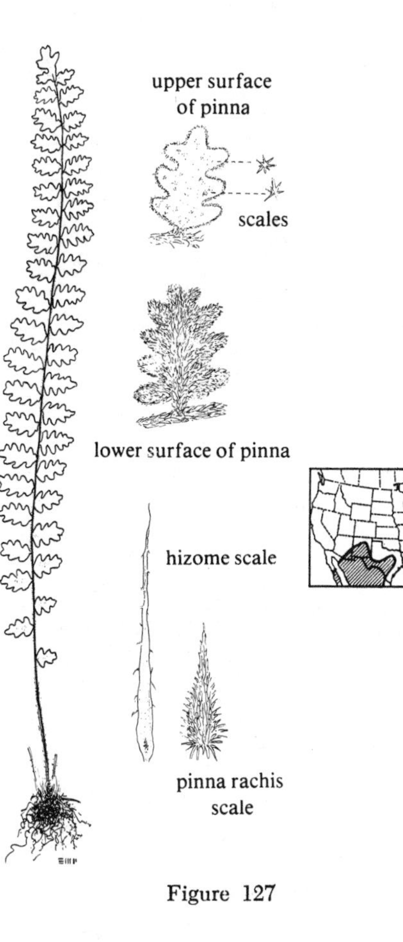

Figure 127 *Cheilanthes sinuata*

Rhizome stout, short-creeping; scales chestnut brown, linear, ciliate or toothed; fronds six to eighteen inches tall; stipe stout, round, only about one-eighth the frond length, clothed with white scales; blade very narrow, pinnate-pinnatifid; upper surface with a few white dissected scales, lower surface covered with brown (or white) ciliate scales; pinnae cut one-third to one-half way to the midvein, three to six pairs of lobes per pinna. Limestone rocks and slopes. Abundant. Southwestern United States; Middle and South America, Hispaniola.

Two common species closely related to *C. sinuata* are found in southwestern United States and northern Mexico, both with short pinnae less than one-half inch long. *Cheilanthes cochisensis* (Goodd.) Mickel [*Notholaena cochisensis* Goodd.] (Fig. 128) has rhizome scales smooth-margined instead of ciliate, and the blade scales on the underside are about one-sixteenth inch long; the pinnae have one to two pairs of shallow lobes. It is poisonous to livestock.

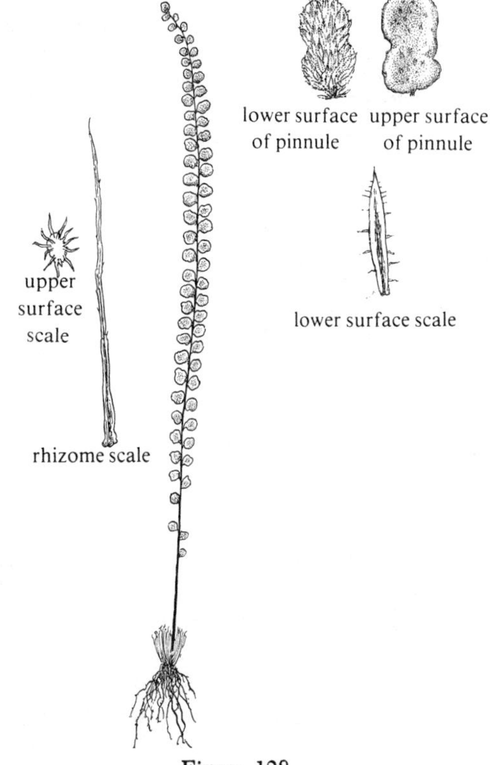

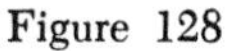
Figure 128

Figure 128 *Cheilanthes cochisensis*

Cheilanthes integerrima (Hooker) Mickel [*Notholaena integerrima* (Hooker) Hevly] (Fig. 129) has rhizome scales as in *C. sinuata,* but the blade scales on the lower surface are very small, only about one-thirty-second inch long; the pinnae have three pairs of shallow lobes. It may have originated as a hybrid between *C. sinuata* and *C. cochisensis.*

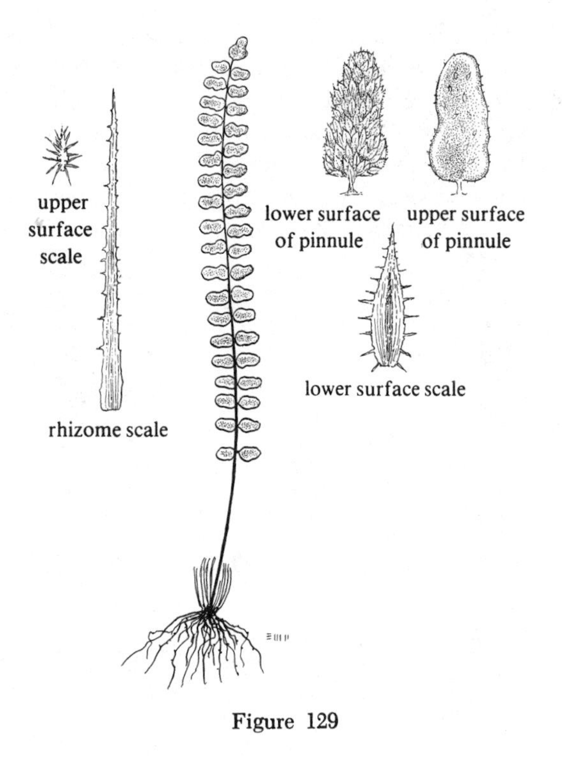

Figure 129 *Cheilanthes integerrima*

31a Scales of lower surface ciliate, at least at the scale base; scales broadly overlapping, completely covering the segments. 32

31b Scales of lower surface not ciliate, barely overlapping, often not covering the segment. (Fig. 130). .... FENDLER'S LIP-FERN, *Cheilanthes fendleri* Hook.

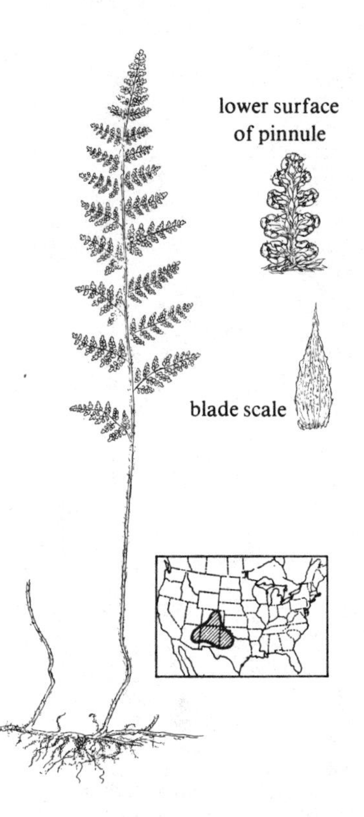

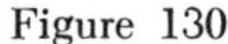
Figure 130

Figure 130 *Cheilanthes fendleri*

Rhizome slender, wide-creeping; scales smooth-margined, tan or with shiny darkened patches (not thickened central streak); fronds to twelve inches tall, spaced apart; stipe chestnut brown, scaly, one-half the frond length; blade tripinnate, narrowly oval; upper surface naked, lower surface covered with white to red-brown, lance-shaped scales, no hairs; segments round; indusium continuous. Among rocks. Common. Southwestern United States; northern Mexico.

32a Rhizome scales linear or narrowly lance-shaped, blackish, often with a tendency to become bicolorous. 33

32b **Rhizome scales broadly lance-shaped, pale reddish-brown or partially shiny darkened, but not in a central streak. (Fig. 131). .... BEADED LIP-FERN, *Cheilanthes wootonii* Maxon**

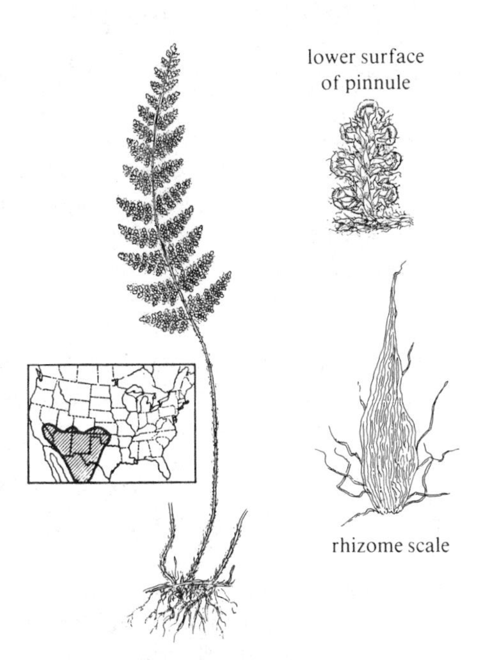

Figure 131

Figure 131 *Cheilanthes wootonii*

Rhizome slender, wide-creeping; scales oval to lance-shaped, toothed, pale brown or with dark brown center; fronds four to twelve inches tall, distant (one-fourth inch or more apart on rhizome); stipe slender, chestnut brown, about one-half the frond length, scaly; blade oblong, tripinnate to quadripinnate; blade scales long-ciliate, covering the lower surface, upper surface naked; segments minute, bead-like; indusium continuous and recurvd but not strongly differentiated. Noncalcareous rocks. Uncommon. Southwestern United States; northern Mexico.

33a **Fronds small, many, close; rhizome stout, much-branched; blade scales one-sixteenth inch or more long. 34**

33b **Fronds few, large; rhizome slender, wide-creeping, sparsely branched; blade scales about one-thirty-second inch long, covering the lower surface but not far exceeding it, not conspicuous when viewed from the top of the frond. (Fig. 132). .... CLEVELAND'S LIP-FERN, *Cheilanthes clevelandii* D. C. Eaton**

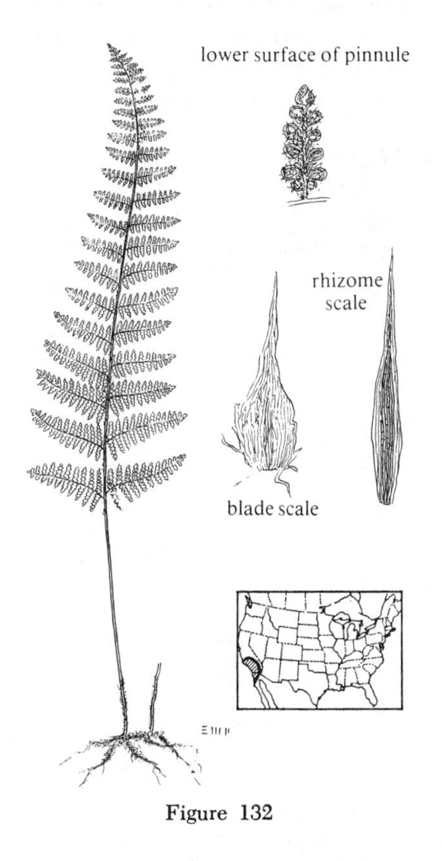

Figure 132

Figure 132 *Cheilanthes clevelandii*

Rhizome long-creeping, slender; scales glossy brown, darker in the center, linear; fronds four to sixteen inches tall, spaced apart; stipe brown, densely covered with brown linear scales to nearly naked, one-third to one-half the frond length; blade tripinnate to quadripinnate, lance-shaped or narrowly triangular, upper surface naked, lower surface covered with lance-shaped, reddish brown ciliate scales; segments round, bead-like; indusium continuous. Among rocks. Common. Southern California; Baja California.

34a Scales of lower blade surface bright red-brown, linear or lance-shaped, generally not exceeding the segments; many long hairs, entangled, forming a mat on the lower surface; upper surface often with a few white, stellate scales. 35

34b Scales of lower blade surface white to pale reddish brown, large, much exceeding the segments; upper blade surface naked. (Fig. 133). ...... COVILLE'S LIP-FERN, *Cheilanthes covillei* Maxon

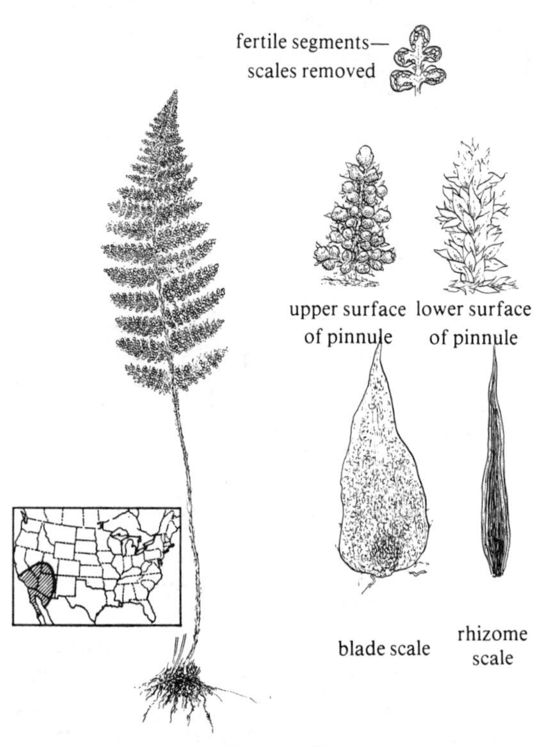

Figure 133

Figure 133 *Cheilanthes covillei*

Rhizome short-creeping; scales brown, somewhat darker in center, linear to lance-shaped; fronds four to twelve inches tall, clumped; stipe brown to purplish, clothed with pale lance-shaped scales, one-third to one-half the frond length; blade tripinnate, oblong to triangular; segments round, bead-like, upper surface naked, lower surface covered with broad, triangular white to pale reddish brown nonciliate scales; indusium continuous. Among rocks. Common. Southwestern United States; northwestern Mexico.

35a Blade bipinnate; rhizome scales concolorous (or slightly bicolorous only with age); lower surface with very narrow scales mixed with a mat of hairs. (Fig. 134). .... LACE FERN, *Cheilanthes gracillima* D. C. Eaton

Rhizome short-creeping; scales linear, light brown or orange, concolorous, older scales with dark center streak; fronds to ten inches tall, tufted; stipe dark brown, usually more than one-half the frond length, nearly naked or with appressed scales that are linear with ciliate bases; blade bipinnate, oblong; upper surface with a few stellate scales, lower surface with dense cinnamon hairs and very narrow scales; segments oblong; indusium continuous. Igneous rocks. Abundant. Northwestern United States.

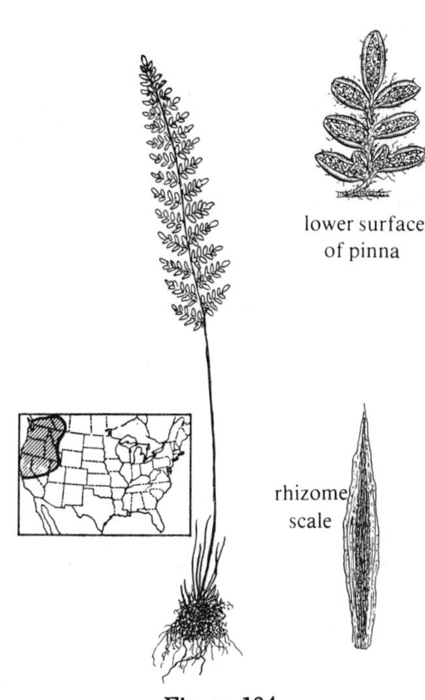

Figure 134 *Cheilanthes gracillima*

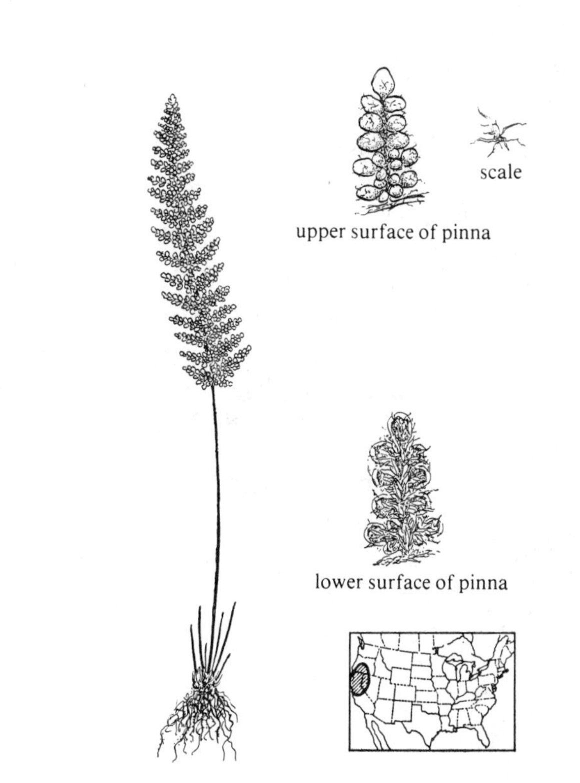

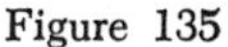
Figure 135

35b Blade tripinnate; rhizome scales distinctly bicolorous; lower surface with abundant, lance-shaped scales covering a mat of hairs. (Fig. 135). ...... COASTAL LIP-FERN, *Cheilanthes intertexta* (Maxon) Maxon

Figure 135 *Cheilanthes intertexta*

Rhizome short-creeping; scales linear, maroon, bicolorous; fronds to eleven inches tall, clumped; stipe brown, often with scattered scales, usually more than one-half the frond length; blade tripinnate, oval-triangular or oblong; upper surface with a few white stellate scales, lower surface densely clothed with cinnamon-colored, lance-shaped scales and hairs; segments bead-like; indusium continuous. Among rocks. Frequent. Western United States.

Some botanists have allied this species with *C. covillei,* with which I can see little similarity. On the other hand, it appears to be very closely related to *C. gracillima* and is thought to hybridize with it.

CRYPTOGRAMMA

Rock-Brake, Cliff-Brake

Rhizome short-creeping or ascending, scaly; fronds small, naked, pinnate to tripinnate, somewhat to strongly dimorphic; sori marginal, protected by the recurved margin. Four species of north temperate regions.

**1a Rhizome stout, ascending; fronds numerous, clustered; blades thick, opaque. (Fig. 136). ....
PARSLEY FERN, AMERICAN ROCK-BRAKE, *Cryptogramma acrostichoides* R. Brown [*C. crispa* var. *acrostichoides* (R. Brown) Clarke]**

Figure 136 *Cryptogramma acrostichoides*

Rhizome stout, ascending; fronds clustered, three to six inches tall, one to one-and-one-half inches wide, bipinnate to tripinnate; fertile fronds taller with more slender segments. Evergreen. Open rocky barrens. Common. Northern North America.

Whether this should be considered a variety of the Eurasian *C. crispa* is a matter of interpretation.

**1b Rhizome quite slender, creeping; fronds few, spaced along the rhizome; blade thin, papery when dry. (Fig. 137). ....
SLENDER CLIFF-BRAKE, *Cryptogramma stelleri* (Gmel.) Prantl**

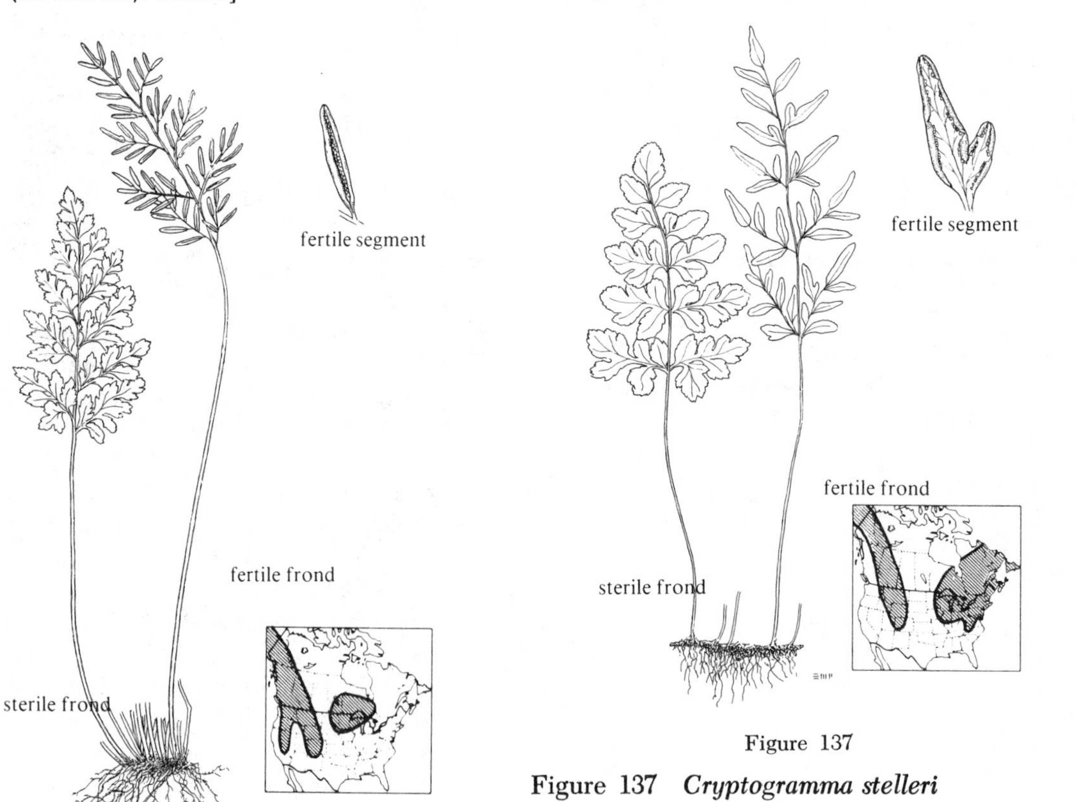

Figure 136

Figure 137

Figure 137 *Cryptogramma stelleri*

Rhizome creeping; sterile fronds three to six inches tall, one-and-one-half to two inches wide, pinnate-pinnatifid, thin-textured; fertile fronds taller than the sterile, bipinnate, the segments more slender than the sterile. Deciduous. Cool, moist, shaded, limestone or other calcareous rock ledges. Frequent. Northern North America; Europe, Asia.

CTENITIS

Rhizome short-creeping or ascending and trunk-like, scaly; fronds large, with characteristic small, jointed hairs; sori round; indusium kidney-shaped. One-hundred-fifty species of tropical regions.

1a **Blade bipinnate-pinnatifid to tripinnate; rhizome scales red-brown; conspicuous, about one inch long; rhizome erect, trunk-like. (Fig. 138). .... FLORIDA TREE FERN, *Ctenitis sloanei* (Poepp.) Morton [*C. ampla* of authors]**

fertile pinnule

stipe base

Figure 138

Figure 138 *Ctenitis sloanei*

Rhizome erect, like a short trunk, clothed with reddish-brown scales; fronds clumped to form a vase-like crown, thirty to sixty inches tall, ten to twenty inches wide; stipe nearly one-half the frond length; rachis and blade underside scaly and hairy. Moist woods. Uncommon. Southern Florida; tropical America.

1b **Blade pinnate-pinnatifid; rhizome scales brown, inconspicuous; rhizome short-creeping. (Fig. 139). .... *Ctenitis submarginalis* (Langsd. & Fisch.) Copel.**

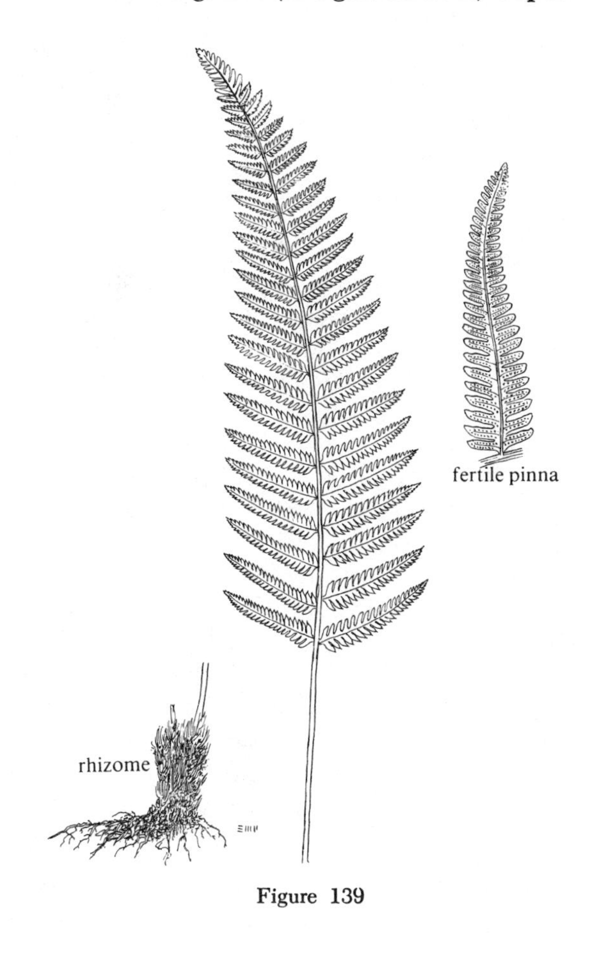

Figure 139

Figure 139 *Ctenitis submarginalis*

Rhizome short-creeping, scales brown; fronds pinnate-pinnatifid, not clumped vase-like, twenty-four to thirty-six inches long, eight to twelve inches wide; stipe about one-fourth of frond length; rachis and blade underside hairy. Wet woods. Rare. Southern Florida; tropical America.

This species closely resembles a large *Thelypteris,* but can be distinguished from members of that genus by the jointed hairs of *Ctenitis* as opposed to the needle-like hairs of *Thelypteris.*

CYRTOMIUM

Holly Fern

Rhizome stout, ascending, with large tan scales; fronds medium-sized, pinnate, leathery; veins netted; sori medial, round; indusium round, umbrella-like. Twelve species of eastern Asia; naturalized in the southern United States but hardy in protected areas north to New York.

Closely related to *Polystichum* and *Phanerophlebia* with similar umbrella-shaped indusia and leathery blade texture but distinguished from them by its netted venation. (Fig. 140) .... JAPANESE HOLLY-FERN, *Cyrtomium falcatum* (L.f.) Presl

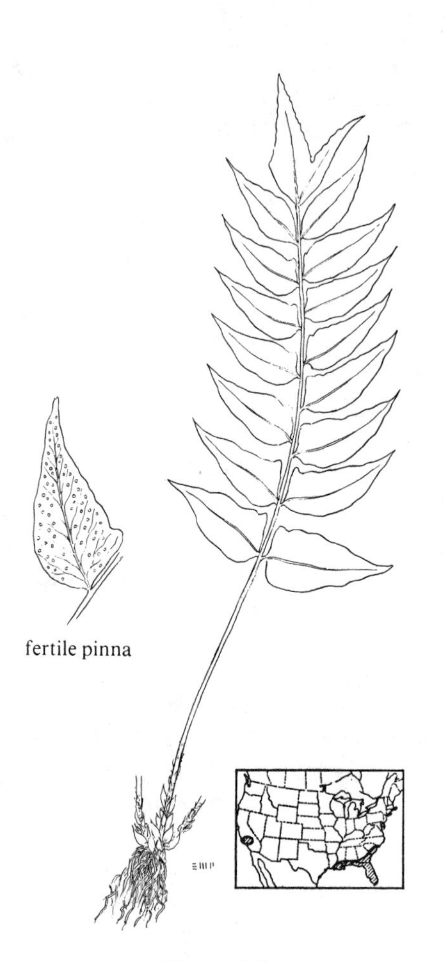

Figure 140

Figure 140 *Cyrtomium falcatum*

Rhizome erect, stout; fronds few, forming a vase-like crown, ten to twenty inches tall, three to five inches wide, leathery, glossy dark green; pinnae few (four to ten, rarely more), three-fourths to one inch wide, usually auricled, often with coarsely toothed margin like a holly. Escaped or naturalized in scattered areas in the Southeast north to South Carolina, and in California; hardy north to New York; native of eastern Asia.

Another species, *C. fortunei* J. Smith, is also escaped and naturalized north to South Carolina, though it is hardy farther north. It can be distinguished by its greater number of one-half inch-wide pinnae (nine to twenty pairs), which are paler green, more dull in appearance, have fewer teeth on the margin and no auricles on the pinnae.

CYSTOPTERIS

Fragile Fern

Rhizome creeping, scaly; fronds small to medium-sized, thin-textured; veins free; sori medial, round; indusium attached laterally, covering the sorus like a hood. Ten species of temperate regions.

1a Blade triangular (basal pinnae the longest); veins ending in the teeth; rhizome short- or long-creeping. 2

1b Blade oval or lance-shaped, the basal pinnae not the longest; veins ending in notches in the teeth; rhizome short-creeping. (Fig. 141). .... FRAGILE FERN, BRITTLE FERN, *Cystopteris fragilis* (L.) Bernh.

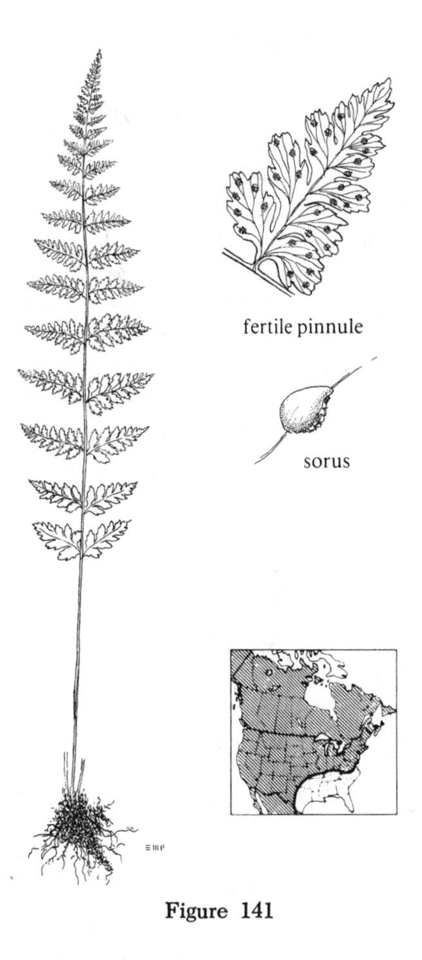

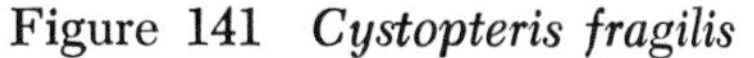

Figure 141 *Cystopteris fragilis*

Rhizome short-creeping, its tip not projecting noticeably beyond the stipe bases; fronds clumped; stipe one-third to one-half the frond length; blade bipinnate-pinnatifid; indusium lacking glands. Moist woods and cliffs. Abundant. Most of North America; most regions of the world.

Most specimens in Arizona and New Mexico are larger and have longer, more attenuate pinnae, and are called *C. fragilis* var. *tenuifolia* (Clute) Broun.

Cystopteris protrusa (Weath.) Blasdell is distinguished by its rhizome, which tends to be slightly more wide-creeping, and the rhizome tip extends slightly beyond the stipe bases. It occurs usually in woods, along streams, and is most common in the southeastern United States.

2a **Blade broadly triangular, lacking bulblets; pinnae about ten pairs; rhizome long-creeping and cord-like; basal pinnae with downward-pointing pinnules longer than those pointing upward. (Fig. 142). .... MOUNTAIN FRAGILE-FERN, *Cystopteris montana* (Lam.) Bernh.**

Figure 142

Figure 142 *Cystopteris montana*

Rhizome very long and slender; fronds one to two inches distant; stipe one-half to three-fourths the frond length; blade tripinnate-pinnatifid at base; basal pinnae nearly as large as rest of the blade above; basal pinnae with downward-pointing pinnules longer than those pointing upward; indusium glandular. Wet woods, rocks, and meadows. Rare. Northern North America; Europe, Asia.

2b **Blade narrowly triangular, bearing bulblets on rachis and pinna rachises; pinnae over twenty pairs; rhizome short-creeping; basal pinnae with downward-pointing pinnules equal to or shorter than those pointing upward. (Fig. 143). .... BULBLET BLADDER-FERN, *Cystopteris bulbifera* (L.) Bernh.**

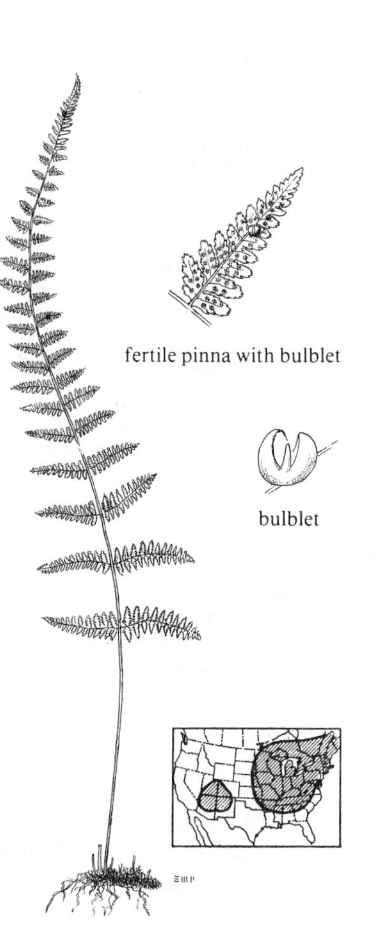

Figure 143

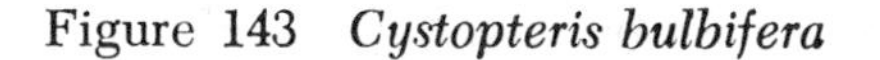
Figure 143 *Cystopteris bulbifera*

Rhizome short-creeping; fronds clumped; stipe one-sixth to one-third the frond length; blade bipinnate-pinnatifid; basal pinnae not especially different from other pinnae; indusium glandular. Usually on or among wet limestone rocks. Common. Central and eastern North America.

This distinct species hybridizes with *C. fragilis* to form *C. laurentiana* (Weath.) Blasdell and with *C. protrusa* to form *C. tennesseensis Shaver*. Both hybrids have at least a few scaly bulblets, and the blades that are broadest at the base. *Cystopteris tennesseensis* is most abundant in central Tennessee, the Ozarks, and Kansas, whereas *C. laurentiana* occurs mainly in the northern Great Lakes region. Both hybrids are fertile.

DENNSTAEDTIA

Rhizome long- or short-creeping, hairy; fronds medium-sized to large; sori marginal in cup-like structures made of an inner and outer indusium. Seventy species of tropical wet forests, but our most common species is of moist temperate woodlands. (Fig. 144) ....
HAY-SCENTED FERN, *Dennstaedtia punctilobula* (Michx.) Moore

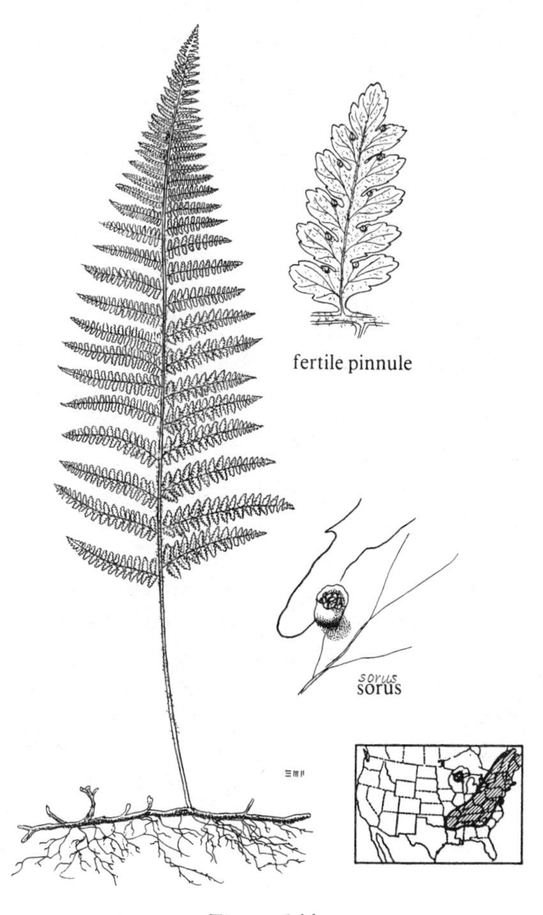

Figure 144

Figure 144 *Dennstaedtia punctilobula*

Rhizome long-creeping; fronds fifteen to thirty inches tall, bipinnate to bipinnate-pinnatifid, oval-oblong in outline, yellow-green, quite hairy, thin-textured; stipe one-fourth to one-third of the frond length; sori cylindrical. Open, sandy meadows or forest clearings; tends to be quite weedy and aggressive. Abundant. Northeastern North America.

Two tropical species have been found rarely in southern United States. Both have fronds three to five feet tall, bipinnate-pinnatifid to tripinnate, of firm texture and nearly naked. *Dennstaedtia bipinnata* (Cav.) Maxon [*D. adiantoides* (Humb. & Bonpl. ex Willd.) Moore], which is common in tropical America, becomes established in peninsular Florida. It is leathery, has wedge-shaped segments, and cylindrical sori. *Dennstaedtia globulifera* (Poir.) Hieron. of Middle and South America has been found in southern Texas. It has broader segments, firm texture, and globular sori.

DICRANOPTERIS

Forking Fern

Rhizome long-creeping, slender, clothed with bristle-like hairs; frond forking, the tip of the central rachis often becoming dormant and the two pinnatifid pinnae in turn forking; veins free, forking two to three times between the midveins of the segment and the segment margin; blade with a whitish bloom on the under surface, naked; sori round, medial, without an indusium. Ten species of tropical regions. (Fig. 145) .... FORKING FERN, *Dicranopteris flexuosa* (Schrad.) Underw.

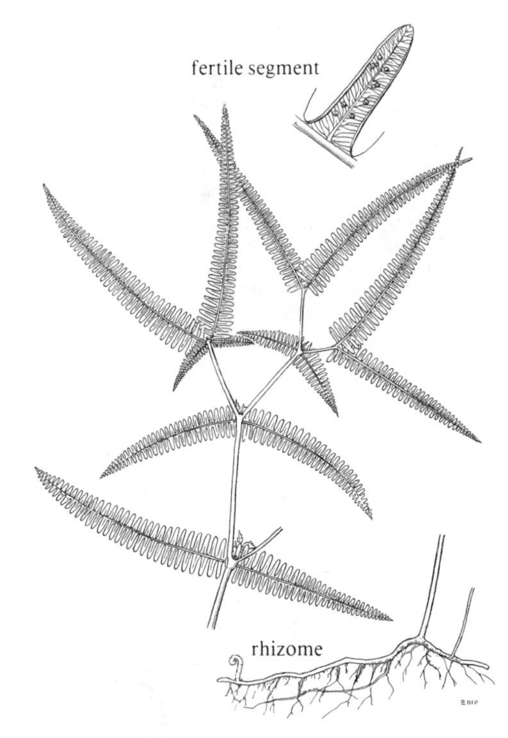

Figure 145

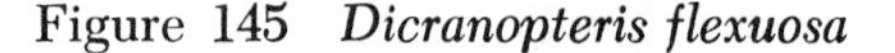

Figure 145 *Dicranopteris flexuosa*

Fronds two to four feet tall; pinnae five to six inches long, three-fourths to one inch wide; basal segment of each pinna enlarged and pinnatifid, making accessory pinnae. Moist open woods. Very rare and sporadic; not well established in North America—only three localities known and it is extinct in two of them. Alabama and Florida; tropical America.

DIPLAZIUM

Rhizome ascending or creeping, scaly; roots coarse and wiry; fronds medium-sized to large, pinnate to tripinnate; veins free; sori elongate along the veins with at least some sori back to back on the same vein; indusium present. Four hundred species of the tropics.

1a Blade pinnate-pinnatifid; fronds eight to twenty inches tall. (Fig. 146). .... *Diplazium lonchophyllum* Kunze

Figure 146

Figure 146 *Diplazium lonchophyllum*

Rhizome stout, ascending; fronds clumped, ten to twenty inches long, four to eight inches wide, triangular; pinnae short-stalked, the lobes finely toothed; texture firm; blade naked. Moist wooded slopes. Very rare. Southern Louisiana; tropical America.

Diplazium japonicum (Thunb.) Bedd. [*Athyrium japonicum* Thunb], native of eastern Asia, has been found rarely escaped in Florida. It can be distinguished by its short, twisted hairs and narrow scales on the blade.

1b Blade bipinnate to tripinnate; fronds thirty to sixty inches tall. (Fig. 147). .... VEGETABLE FERN, *Diplazium esculentum* (Retz.) Sw.

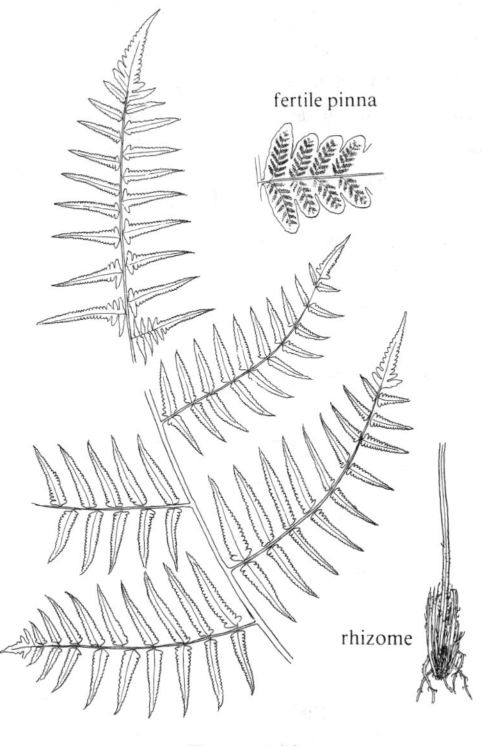

Figure 147

Figure 147 *Diplazium esculentum*

Rhizome stout, ascending; fronds large, thirty to sixty inches tall; ten to thirty inches wide, broadly oval to triangular; the segments shallowly lobed, naked. Moist woods. Scattered escape records in southernmost Florida; native of eastern Asia and the South Pacific where it is commonly eaten.

DRYOPTERIS

Wood Fern, Shield Fern

Rhizome short-creeping or ascending, scaly; fronds closely set, crown-forming, medium to large size, of firm or leathery texture; hairs lacking on all parts except for very short glands in a few species; blades pinnate-pinnatifid to tripinnate; veins free; sori round, on the lower surface, protected by a kidney-shaped indusium. One hundred fifty species of temperate regions.

Hybridization is frequent, resulting in numerous sterile hybrids (see list); a few fertile hybrids function as distinct species (see Fig. 148).

Figure 148 Relationships of some wood ferns. Each letter represents a set of chromosomes: A = assimilis set, I =intermedia set, etc. "semicristata" is still hypothetical. (From W. H. Wagner, Jr., Evolution of *Dryopteris* in relation to the Appalachians, *In* Holt, P. C. (ed.), The Distributional History of the Biota of the Southern Appalachians. Part II. Flora. Research Division Monograph 2. Virginia Polytechnic Institute and State University, Blacksburg, Va. (1971). With permission of the author.)

1a **Scales sparse or lacking on underside of frond; fronds twelve to forty-four inches tall; pinnae not overlapping; fronds not aromatic. 2**

1b **Scales abundant on underside of frond; fronds two to twelve inches tall; pinnae overlapping or nearly so; fronds aromatic. (Fig. 149). .... FRAGRANT WOOD-FERN, *Dryopteris fragrans* (L.) Schott**

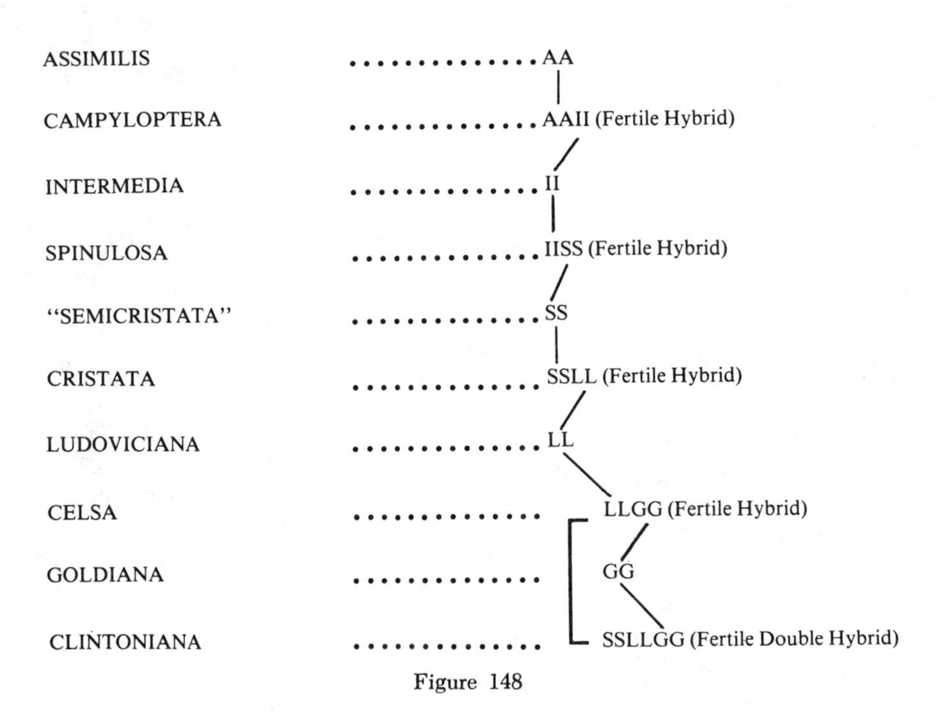

Figure 148

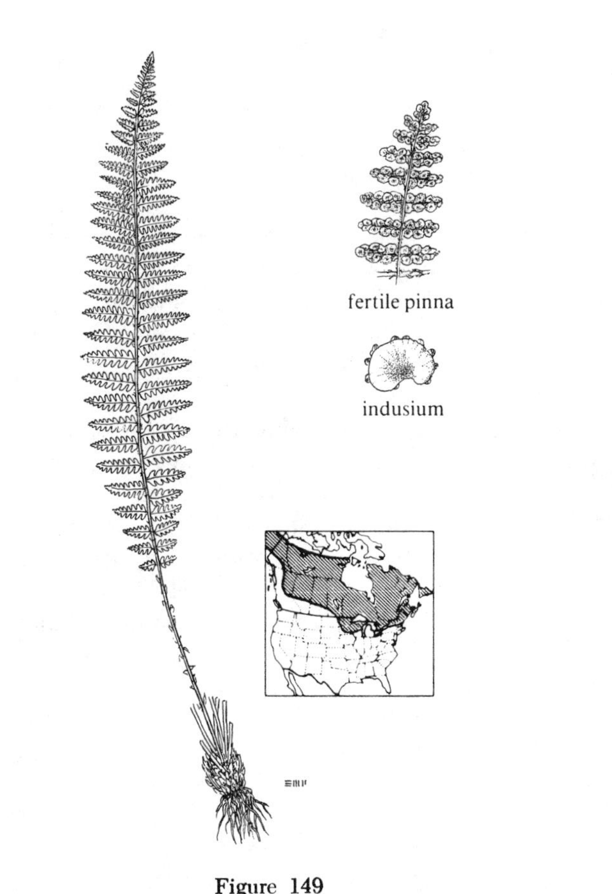

Figure 149

Figure 149 *Dryopteris fragrans*

Stipe one-fourth the frond length, with abundant light brown scales; fronds four to twelve inches tall, one to two and one-half inches wide, the underside nearly covered with scales and glandular hairs; blade pinnate-pinnatifid. Rock crevices. Frequent. Arctic and subarctic regions of North America; Europe, Asia.

2a Sori medial or near the midvein. **3**

2b Sori nearly marginal. (Fig. 150). .... **MARGINAL SHIELD-FERN, *Dryopteris marginalis* (L.) A. Gray**

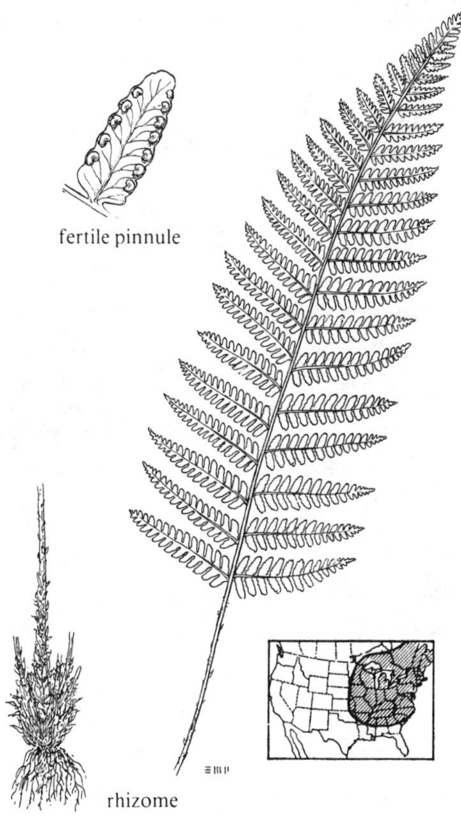

Figure 150

Figure 150 *Dryopteris marginalis*

Stipe one-fourth to one-third the frond length, with many pale brown scales, especially at base; frond eighteen to twenty-four inches long, six to ten inches wide; blade bipinnate to bipinnate-pinnatifid, leathery; sori located near margin. Evergreen. Rocky wooded slopes. Common. Northeastern North America.

3a Blade pinnate-pinnatifid. **4**

3b Blade bipinnate to tripinnate. **10**

4a **Fertile fronds or parts of fronds noticeably narrower or more erect than the sterile.** **5**

4b **Fertile and sterile fronds or parts of fronds not distinctly different.** **6**

5a **Fertile fronds bearing sori on most pinnae; fertile pinnae not much narrower than the sterile; fertile fronds more erect than the sterile; rhizome ascending; pinnae of fertile fronds turned horizontally like open venetian blinds; northern. (Fig. 151).** .... **CRESTED SHIELD-FERN, *Dryopteris cristata* (L.) A. Gray**

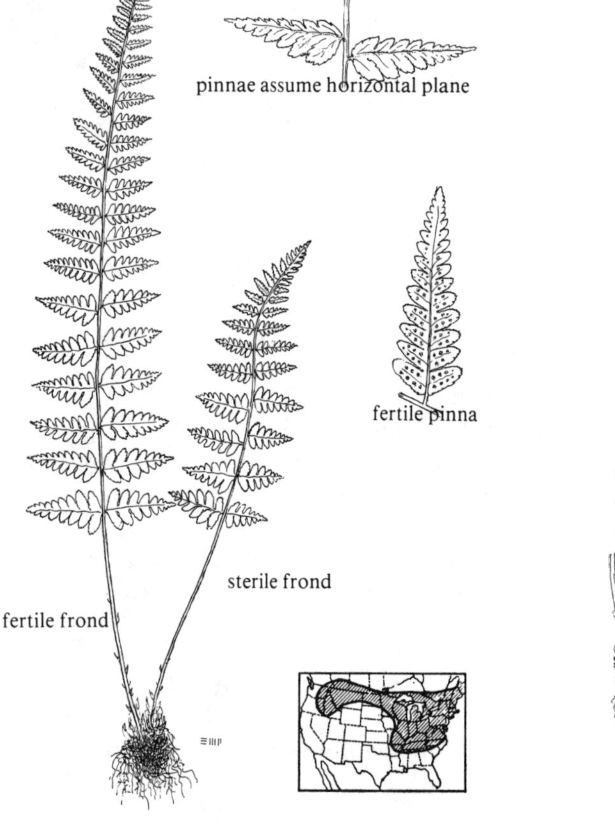

Figure 151

Figure 151 *Dryopteris cristata*

Rhizome and stipe scales light brown, stipe one-fourth to one-third the frond length; blade pinnate-pinnatifid, narrowly oblong, pinnae broadest near the stipe, more or less triangular, turned at right angles to the plane of the frond when fertile, like open venetian blinds, of firm texture; small sterile leaves evergreen; sori medial. Marshes, bogs, and swamps. Common. Northern and eastern North America; Europe.

5b **Fertile fronds bearing sori only in the upper half; fertile pinnae much narrower than the sterile, not turned horizontally; rhizome short-creeping, horizontal; southern coastal states. (Fig. 152).** .... **SOUTHERN SHIELD-FERN, *Dryopteris ludoviciana* (Kunze) Small**

fertile pinna

rhizome

Figure 152

Figure 152 *Dryopteris ludoviciana*

Rhizome, stipe and rachis with tan scales; stipe one-fourth the frond length; blade pinnate-pinnatifid, lustrous dark green, leathery, evergreen; fertile pinnae noticeably contracted; sori medial. Wet woods, swamps, shaded limestone outcrops, and margins of cypress swamps. Frequent. Southeastern coastal United States.

6a **Rhizome scales dark, at least in the center. 7**

6b **Rhizome scales tan. 9**

7a **Sori medial; blade tip narrowing gradually. 8**

7b **Sori nearer the midvein than the margin; blade tip narrowing abruptly. (Fig. 153). .... GOLDIE'S WOOD-FERN, *Dryopteris goldiana* (Hooker) A. Gray**

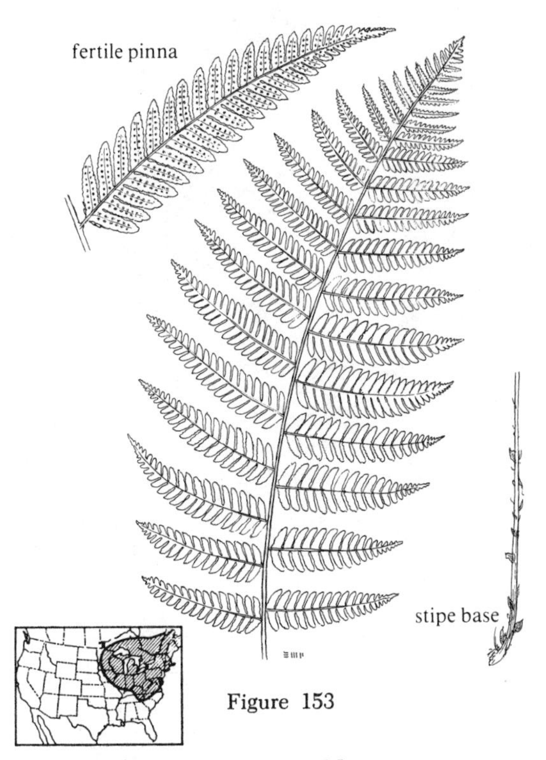

Figure 153

Figure 153 *Dryopteris goldiana*

Rhizome and stipe scales tan with dark brown centers or dark throughout; fronds three to four feet tall, one foot wide; stipe one-third the frond length; blade oblong-triangular, pinnate-pinnatifid, tapering abruptly at the apex, of firm texture; sori nearer the midvein than the margin. Cool, moist woods. Frequent. Northeastern North America.

8a **Rhizome scales dark only in the center; basal pinnae strongly triangular, their basal pinnules the longest. (Fig. 154). .... CLINTON'S WOOD-FERN, *Dryopteris clintoniana* (D. C. Eaton) Dowell**

Figure 154

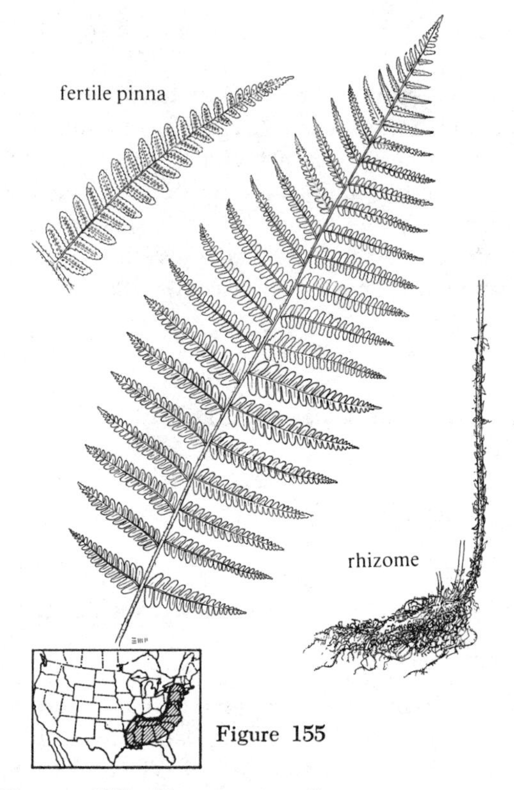

Figure 155

Figure 154 *Dryopteris clintoniana*

Rhizome and stipe scales light brown with darker center; fronds three to four feet tall, six to eight inches wide; stipe about one-third the frond length; blade pinnate-pinnatifid, narrowly oblong, of firm texture; sori medial. Swamps and wet woods. Frequent. Northeastern North America.

This species apparently originated as a hybrid between *D. goldiana* and *D. cristata.* It is now a fertile hexaploid and functions as a distinct species.

8b Rhizome scales completely dark; basal pinnae narrow, not strongly triangular, their basal pinnules not the longest. (Fig. 155). .... LOG FERN, *Dryopteris celsa* (Palmer) Small

Figure 155 *Dryopteris celsa*

Rhizome and stipe scales generally dark, especially in the center; fronds three to four feet tall, eight to twelve inches wide; stipe about one-third the frond length; blade pinnate-pinnatifid, oblong, of firm texture; evergreen, at least in the southern part of its range. Swamps and wet woods. Rare. Eastern United States, though mainly southern.

This is thought to be the fertile tetraploid hybrid between *D. goldiana* and *D. ludoviciana.*

9a Pinnules with many fine, spreading teeth; pinnules tending to taper toward the tip; far western. (Fig. 156). .... WESTERN SHIELD-FERN, COASTAL WOOD-FERN, *Dryopteris arguta* (Kaulf.) Watt

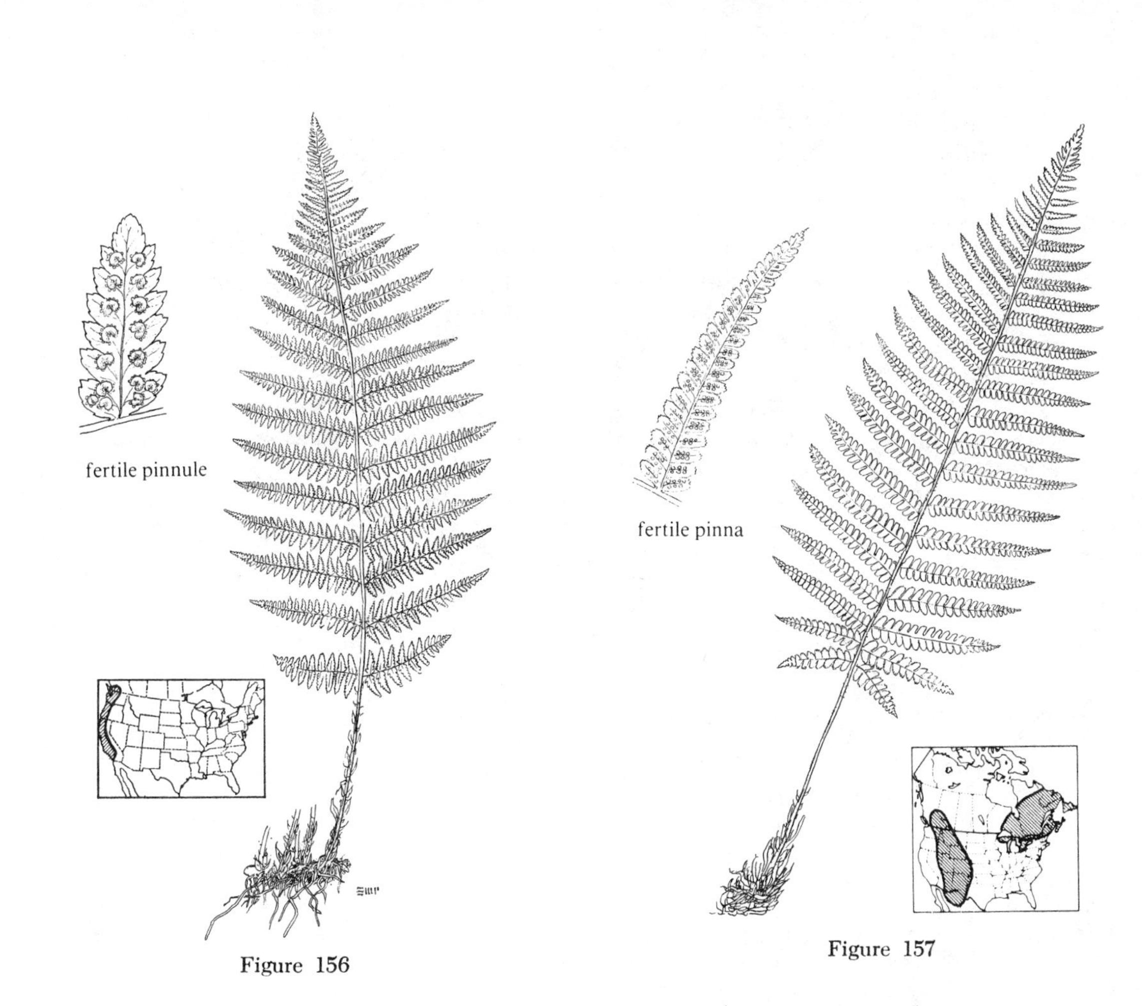

Figure 156

Figure 157

Figure 156 *Dryopteris arguta*

Rhizome and stipe scales light brown; fronds one to three feet tall, six to twelve inches wide; stipe short, one-fourth to one-third the frond length; blade pinnate-pinnatifid; the pinnules tapering toward the tip, with fine spreading teeth; sori medial. Evergreen. Sparsely wooded slopes. Frequent. Western North America.

9b Pinnules with few incurved teeth; pinnules essentially parallel-sided, blunt-tipped; western and northeastern. (Fig. 157). .... MALE FERN, *Dryopteris filix-mas* (L.) Schott

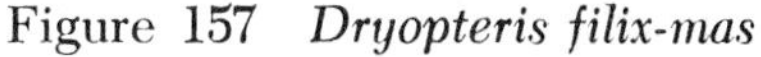

Figure 157 *Dryopteris filix-mas*

Rhizome and stipe scales light brown; fronds two to four feet tall, six to twelve inches wide; stipe short, less than one-fourth the frond length; blade pinnate-pinnatifid; pinnules parallel-sided, blunt-tipped with few inconspicuous incurved teeth; sori nearer the midvein than the margin; indusium often with a glandular margin. Evergreen. Cool, moist, rocky woods. Rare. Western and northeastern North America; Europe, Asia.

10a (3b) Lowest downward pointing pinnule of the lowest pinna usually longer than the pinnule next to it; no glands on

the blade; blade yellowish-green, turning yellow-brown in winter. 11

10b Lowest downward pointing pinnule of the lowest pinna usually shorter than or equal to the next; blade with tiny glands, especially visible on young indusia and upper rachis; evergreen. (Fig. 158). .... GLANDULAR WOOD-FERN, FANCY FERN, EVERGREEN WOOD-FERN, *Dryopteris intermedia* (Muhl. ex Willd.) A. Gray. [*D. spinulosa* var. *intermedia* (Muhl.) Underw.]

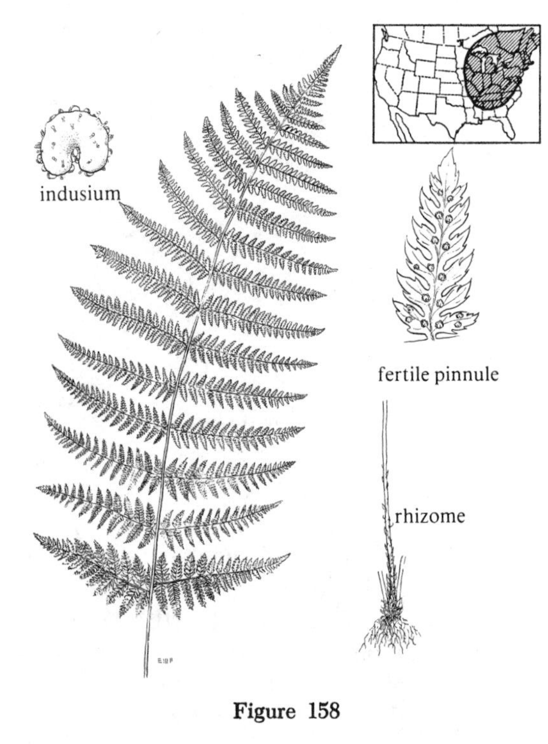

Figure 158

Figure 158 *Dryopteris intermedia*

Rhizome and stipe scales light brown; fronds eighteen to thirty-six inches tall, six to ten inches wide; stipe one-fourth to one-third the frond length; blade bipinnate-pinnatifid to tripinnate, oval to narrowly triangular; pinna outline essentially parallel-sided, tapering only gradually to the tip; the upward and downward facing pinnules on most pinnae about the same length; thin texture; evergreen, often with a slightly bluish cast; tiny glands all over the blade; sori medial. Moist shaded woods and rocky slopes. Abundant. Eastern North America.

11a Blade oval or oval-triangular; downward pointing pinnules of basal pinnae two to three times longer than the upward pointing pinnules on the same pinna; all pinnae broadest next to the rachis. (Fig. 159). .... SPINULOSE SHIELD-FERN, *Dryopteris spinulosa* (O. F. Muell.) Watt

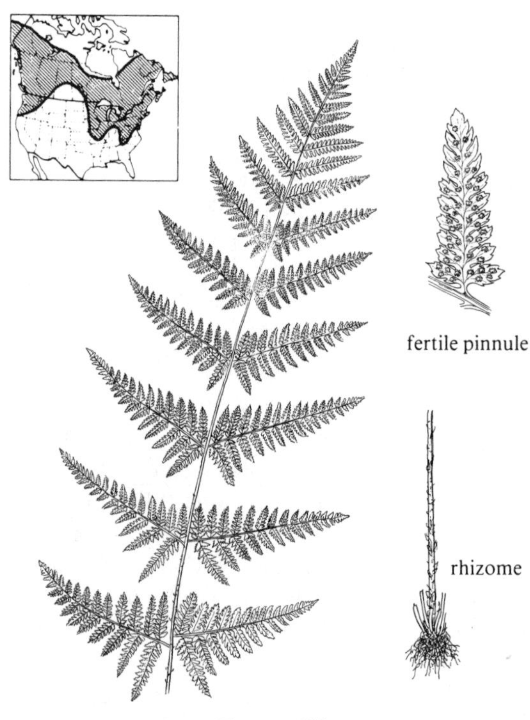

Figure 159

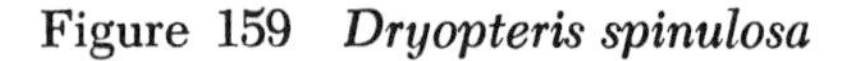

Figure 159 *Dryopteris spinulosa*

Rhizome and stipe scales light brown; fronds one and one-half to two and one-half feet tall;

stipe one-fourth to one-third the frond length; blade bipinnate-pinnatifid, narrowly oval to triangular; pinna outline narrowly triangular, the basal pinnules of each pinna, especially the lower ones, noticeably longer than the next; downward facing pinnules of lower pinnae longer than those facing upward; texture thin to medium firm; green to yellowish green; glands lacking; sori medial. Swamps and wet woods. Common. Northern North America.

This species is thought to have originated as a hybrid between *D. intermedia* and one other species not yet known (*D. "semicristata"*). It is a fertile tetraploid.

Dryopteris patula (Sw.) Underw. is limited to cliffs in southeastern Arizona; widespread in Mexico and other parts of Latin America.

Dryopteris cinnamomea (Cav.) C.Chr. of Mexico, has been reported once from southern Texas.

11b Blade broadly triangular or pentagonal; downward pointing pinnules of basal pinnae three to five times longer than the upward pointing pinnules of the same pinna; other pinnae with upward pointing and downward pointing pinnules about the same length, the pinna outline essentially parallel-sided most of its length. (Fig. 160). .... MOUNTAIN WOOD-FERN, SPREADING WOOD-FERN, *Dryopteris campyloptera* Clarkson [*D. spinulosa* var. *americana* (Fisch.) Fernald]

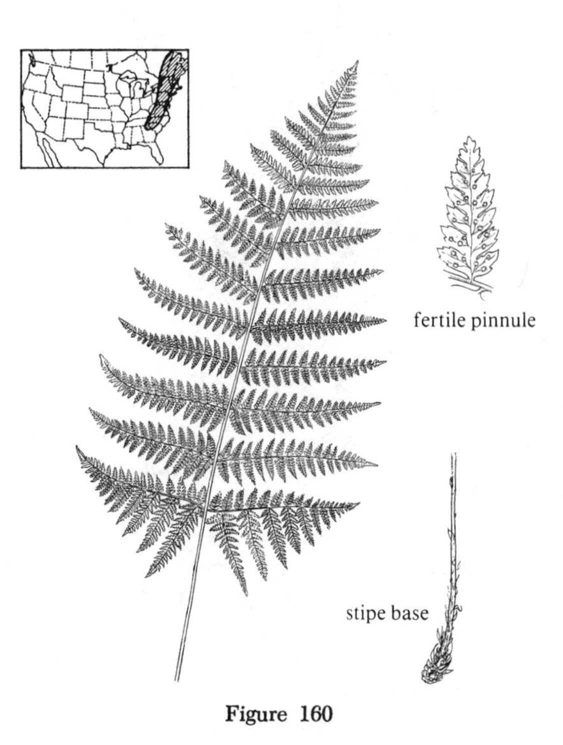

Figure 160 *Dryopteris campyloptera*

Rhizome and stipe scales light brown; fronds two to three feet tall, ten to twelve inches wide; stipe about one-half the frond length; blade tripinnate-pinnatifid, thin textured, broadly triangular to pentagonal, naked or with some glands; sori medial. Moist deciduous woods. Frequent to common. Northern New England and mountain tops of the Appalachian Mountains.

This apparently is a fertile hybrid between *D. expansa* and *D. intermedia*. *Dryopteris expansa* (Presl) Fraser-Jenkens and Jermy (*D. assimilis* S. Walker) is similar to *D. campyloptera* and occurs in the Northwest, the Lake Superior area, and Europe. It differs in its broader, more oval frond form and more delicate texture.

Many sterile hybrids are known in this genus. Some of the more common have been given binomial names. The following are known where their parents occur in close proximity.

D. campyloptera × *intermedia* (3n)
D. campyloptera × *marginalis* (3n)
D. celsa × *cristata* (4n)
D. celsa × *goldiana* (3n)
D. celsa × *intermedia* [*D.* × *separabilis* (Palmer) Small] (3n)
D. celsa × *ludoviciana* [*D.* × *australis* (Wherry) Small] (3n)
D. celsa × *marginalis* [*D.* × *leedsii* Wherry] (3n)
D. clintoniana × *cristata* (5n)
D. clintoniana × *goldiana* (4n)
D. clintoniana × *intermedia* [*D.* × *dowellii* (Farw.) Wherry] (4n)
D. clintoniana × *marginalis* [*D.* × *burgessii* Boivin ?] (4n)
D. clintoniana × *spinulosa* [*D.* × *benedictii* Wherry] (5n)
D. cristata × *intermedia* [*D.* × *boottii* (Tuckerm.) Underw.] (3n)
D. cristata × *marginalis* [*D.* × *slossonae* Wherry] (3n)
D. cristata × *spinulosa* [*D.* × *uliginosa* (A. Braun) Druce] (4n)
D. expansa × *intermedia* (2n)
D. expansa × *marginalis* (2n)
D. filix-mas × *marginalis* (3n)
D. fragrans × *marginalis* [*D.* × *algonquinensis* Britton] (2n)
D. goldiana × *intermedia* (2n)
D. goldiana × *marginalis* [*D.* × *neowherryi* W. H. Wagner] (2n)
D. intermedia × *marginalis* (2n)
D. intermedia × *spinulosa* [*D.* × *triploidea* Wherry] (3n)
D. marginalis × *spinulosa* [*D.* × *pittsfordensis* Slosson] (3n)

EQUISETUM

Horsetail, Scouring-Rush

Subterranean rhizome with jointed, grooved aerial stems; leaves small, whorled, fused to form a sheath at the joints; aerial stems generally with a large central canal and smaller vallecular canals (under the grooves) and carinal canals (under the ridges); spores produced in terminal cones, the sporangia borne on umbrella-like structures that make up the cone. Fifteen species, largely of temperate regions.

1a **Plants growing in dry to wet places, only rarely in the water; stems with central cavity usually less than three-fourths the diameter of the stem; vallecular canals present; stem ridges usually distinct. 2**

1b **Plants growing in water or ditches; stems with central cavity about four-fifths of the stem diameter; vallecular canals generally lacking (except near the base); stem ridges nearly lacking. (Fig. 161). .... WATER HORSETAIL, PIPES, *Equisetum fluviatile* L.**

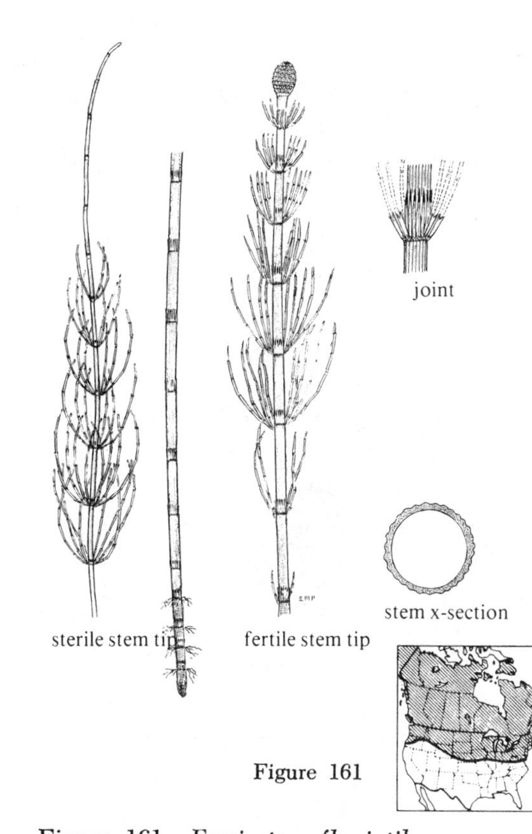

Figure 161 *Equisetum fluviatile*

Aerial stems to three feet tall, branched or unbranched (first aerial stems of the year unbranched; branches appear later); sheath teeth twelve to twenty, dark, narrow, appressed; branches with first internode shorter than or equal to the subtending stem sheath of the main stem. Stems annual. Shallow ponds, wet ditches. Common. Northern North America; Europe.

Equisetum × *litorale* Kuehl., the SHORE HORSETAIL, is the hybrid between *E. fluviatile* and *E. arvense*. It is often mistaken for the former but has vallecular canals, fewer teeth (eight to fourteen), first branch internodes longer than the subtending stem sheath, cones with abortive spores in unopened sporangia.

2a **Vegetative stems regularly branched (fertile stems unbranched in *E. arvense* and *E. telmateia*). 3**

2b **Stems usually unbranched, vegetative and fertile stems alike. 7**

3a **Branches usually unbranched; sheath teeth white to brown or black. 4**

3b **Branches rebranched; sheath teeth reddish brown. (Fig. 162). .... WOODLAND HORSETAIL, *Equisetum sylvaticum* L.**

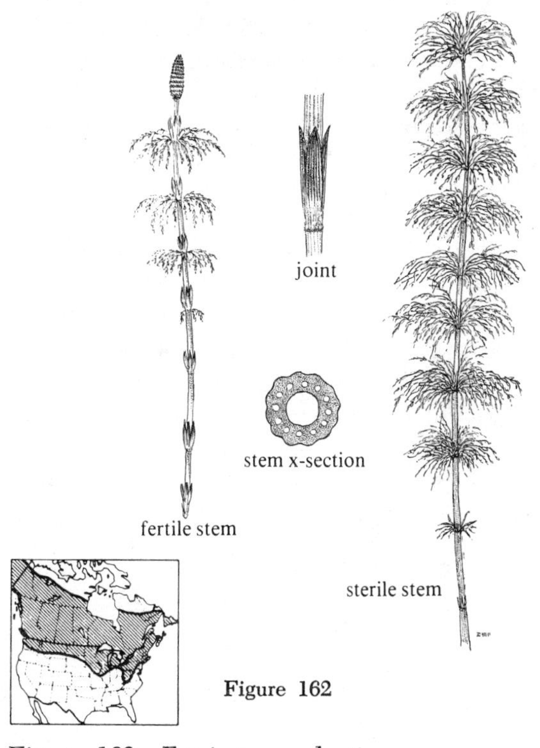

Figure 162 *Equisetum sylvaticum*

Aerial stems ten to twenty inches tall; branches slender, rebranched, giving the plants a lacy appearance; branches often drooping; sheath teeth reddish brown, eight to sixteen; cones on normal stems, though branching and green color may be developing as the cone withers; central cavity about one-half the stem diameter; vallecular canals large. Stems annual. Moist woodlands. Common. Northern North America; Europe.

4a **Plants small, usually less than eighteen inches (rarely to thirty inches) tall; stem ridges fewer than seventeen; widespread. 5**

4b **Plants large, twelve to seventy-two inches tall; stem ridges sixteen to forty; limited to the Pacific coast. (Fig. 163). .... GIANT HORSETAIL, *Equisetum telmateia* Ehrh.**

Aerial stems twelve to seventy-two inches tall; fertile and vegetative stems dimorphic; sterile stems regularly branched; sheaths one-half to one and one-half inches long, pale with brown tips, teeth sixteen to forty, persistent; first internode of branches much longer than subtending stem sheath; central canal to three-fourths the stem diameter; vallecular canals large; fertile stems white to tan, unbranched, appearing before the sterile; cone one and one-half to three inches long, blunt. Stems annual. Moist woods, ditches. Frequent. Northwestern North America; Europe, Asia, northern Africa.

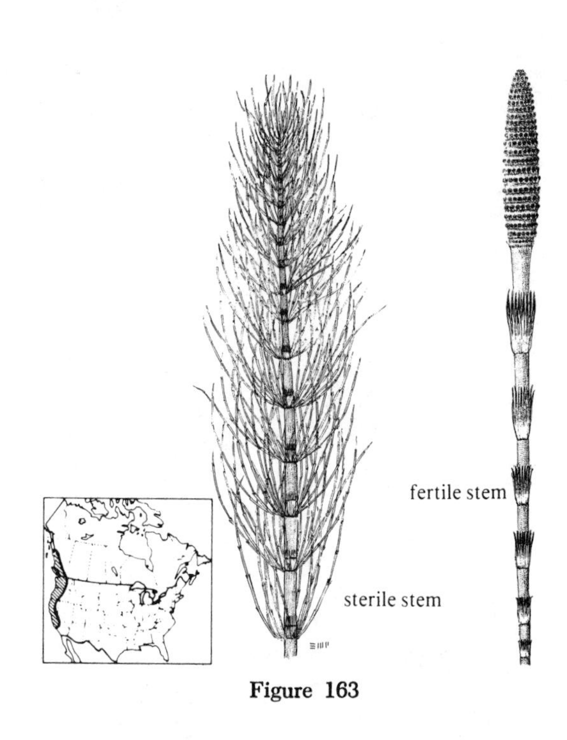

Figure 163

Figure 163 *Equisetum telmateia*

Equisetum ramosissum Desf. of the Eastern Hemisphere has recently been discovered in North Carolina and Florida. It can be distinguished by its branching evergreen stems and withered sheath teeth.

5a **Sheath teeth bicolorous; fertile stems living through the growing season, either equal to the vegetative stems or at least producing green branches after spores are released. 6**

5b **Sheath teeth dark, concolorous; fertile stems ephemeral, dying down soon after shedding spores, never producing green branches. (Fig. 164). .... FIELD HORSETAIL, *Equisetum arvense* L.**

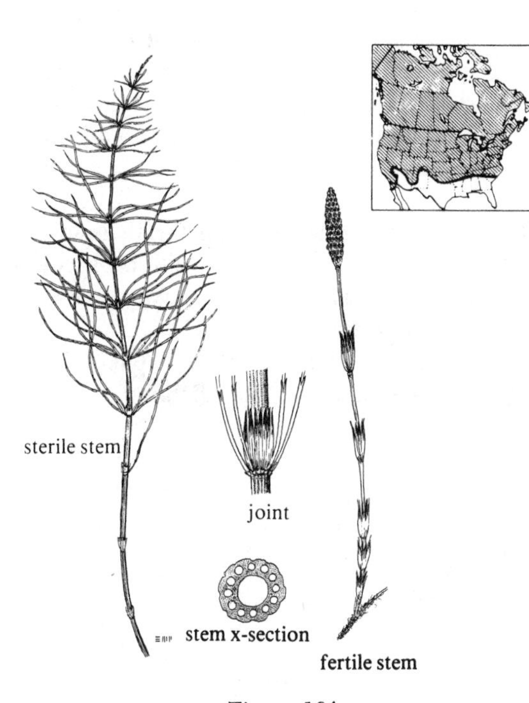

Figure 164

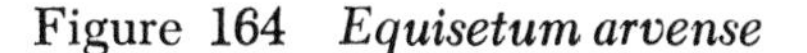

Figure 164 *Equisetum arvense*

Aerial stems usually less than eighteen inches (rarely to thirty inches) tall; fertile and sterile stems dimorphic; sterile stems branched, branches not rebranched; sheath teeth six to twelve, pale brown, persistent; first branch internode longer than the subtending stem sheath; branches with tubercles on the ridges, the main stem smooth; branch sheath teeth longer than wide; central canal one-half the diameter of the stem; vallecular canals large; fertile stems pink or tan, appearing before the sterile and fade quickly after spores are shed. Stems annual. Open meadows, lake shores, railroad embankments. Abundant. Widespread in North America; Europe, Asia.

This is our most common and variable species of horsetail. It is often an aggressive weed and difficult to control.

6a Branches three-sided, spreading; fertile and sterile stems dimorphic. (Fig. 165). .... MEADOW HORSETAIL, *Equisetum pratense* Ehrh.

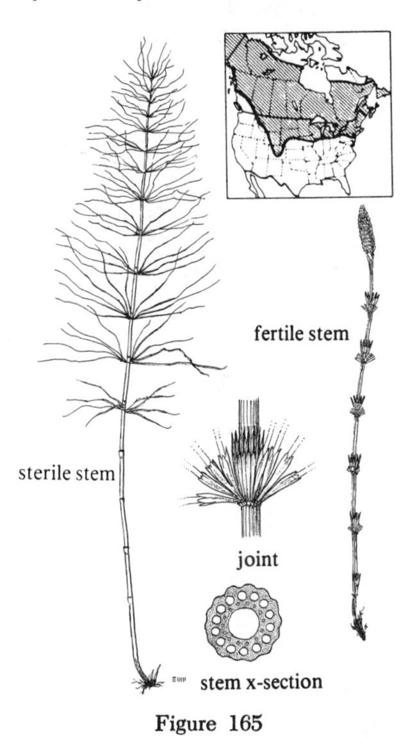

Figure 165

Figure 165 *Equisetum pratense*

Stems dimorphic, the fertile unbranched at first, later may become branched and similar to the sterile stems as the cone shrivels; sterile stems to twenty inches tall, slender, only three thirty-seconds inch diameter; sheaths of sterile stems green, slightly longer than broad; sheath teeth eight to sixteen, with dark center and colorless margin, persistent; branches three-angled, unbranched, with teeth broad as long; first branch internode equal to the subtending sheath or longer only on upper whorls; central canal one-third to one-half the stem diameter; vallecular canals well developed. Stems annual. Moist woods. Frequent. Circumboreal.

Equisetum pratense is similar to *E. ar-*

vense but has tubercles on the ridges of the main stem (branch ridges smooth), bicolorous teeth, and branch teeth broad as long.

6b Branches five- to seven-sided, ascending; fertile and sterile stems alike. (Fig. 166). .... MARSH HORSETAIL, *Equisetum palustre* L.

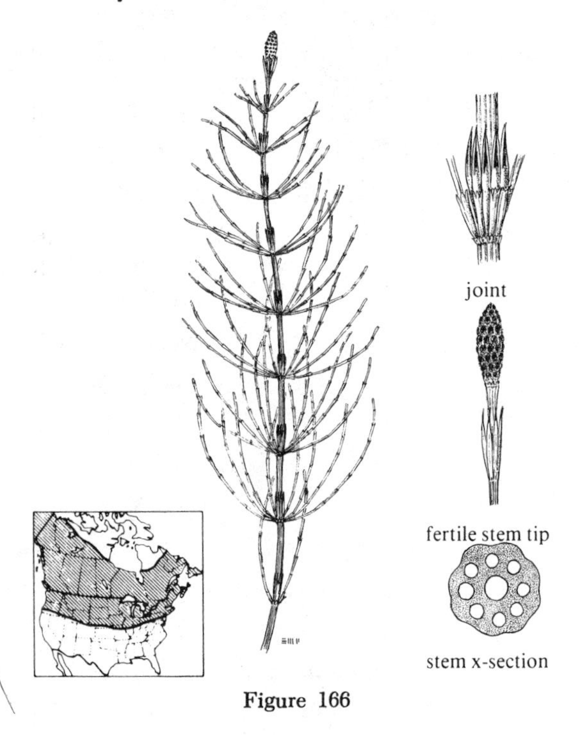

Figure 166

Figure 166 *Equisetum palustre*

Fertile and sterile stems alike, to twenty-four inches tall; sheath longer than wide, teeth eight to twelve, dark with white margins, persistent; ridges of main stem lacking tubercles; branches five- to seven-angled, first internode only one-half as long as the subtending stem sheath; central canal one-sixth to one-third the stem diameter; vallecular canals nearly as large as the central canal. Stems annual. Wet places. Frequent. Circumboreal.

7a Stems ten to forty inches tall (rarely only six inches tall in *E. variegatum*), erect, straight, one-sixteenth to one-fourth inch thick; central canal present; vallecular canals five or more. 8

7b Stems three to six inches tall, low and flexuous, less than one-thirty-seconds inch thick; central canal lacking; vallecular canals three. (Fig. 167). .... DWARF SCOURING-RUSH, *Equisetum scirpoides* Michx.

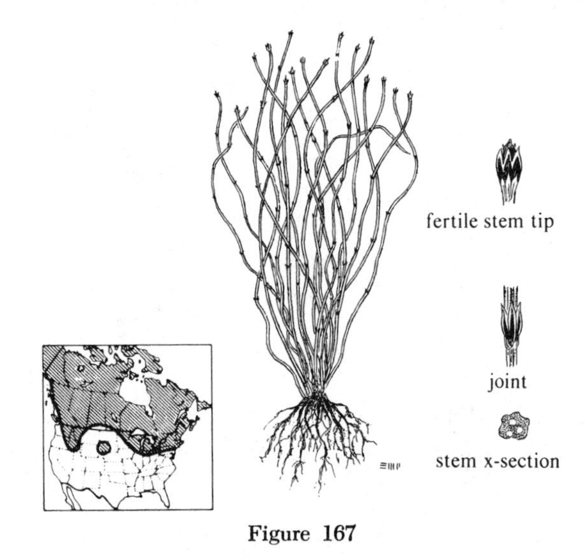

Figure 167

Figure 167 *Equisetum scirpoides*

Stems all alike, to six inches tall, prostrate to ascending, usually twisted; stems with three ridges, each ridge deeply grooved so may appear as six ridges; sheath darkened above; teeth three, black-centered with white margins; central canal lacking; vallecular canals conspicuous; cone tiny, pointed at tip. Stems evergreen. Sandy, moist woods. Frequent. Circumboreal.

8a Stems with fourteen to forty ridges; sheath teeth deciduous; ridges rounded with single row of tubercles or cross-bands. 9

8b Stems with five to twelve ridges; sheath teeth persistent; vallecular canals few, large (as large as central canal. (Fig. 168). .... VARIEGATED SCOURING-RUSH, *Equisetum variegatum* Schleich.

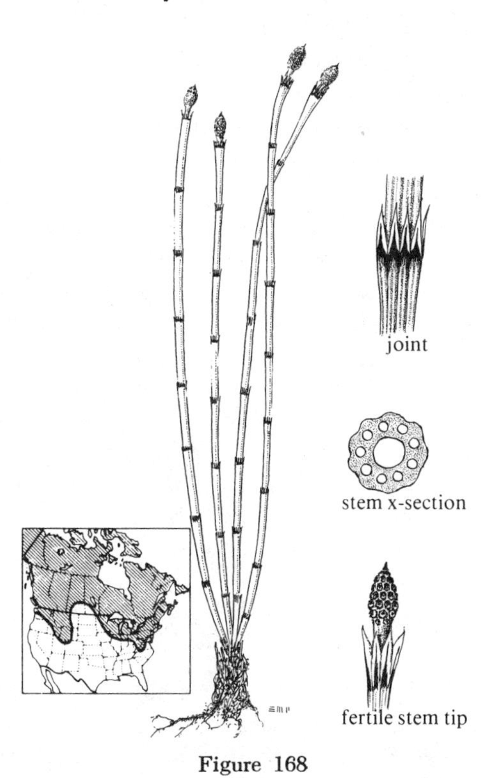

Figure 168

Figure 168 *Equisetum variegatum*

Stems all alike, to fourteen inches tall; ridges five to ten, grooved, with two rows of tubercles; sheath with broad black band at tip; teeth five to ten, black-centered with white margin, persistent; central canal about one-third the stem diameter; vallecular canals conspicuous. Stems evergreen. Sandy shores, wet stream banks, ditches. Frequent. Circumboreal.

Most widespread is var. *variegatum*. Another variety, var. *alaskanum* A. A. Eaton, has more robust habit, sheath teeth eight to twelve, black and incurved. It is limited to southern Alaska coast, but intergradients with var. *variegatum* are frequent.

Equisetum × *trachyodon* A. Braun is the hybrid between *E. hyemale* and *E. variegatum.*

Equisetum × *nelsonii* (A. A. Eaton) Schaffner is the hybrid between *E. laevigatum* and *E. variegatum.* It resembles the latter species most closely but has partially annual stems.

9a Stems smooth, annual, cones blunt-tipped; sheaths widened upwards with one narrow dark band at top. (Fig. 169). .... SMOOTH SCOURING-RUSH, *Equisetum laevigatum* A. Braun

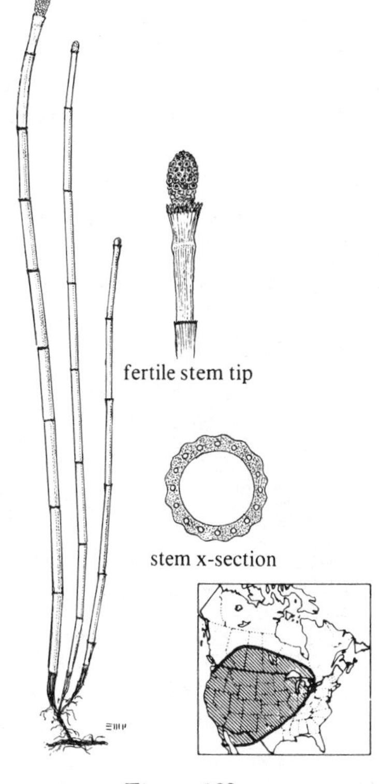

Figure 169

Figure 169 *Equisetum laevigatum*

Stems all alike, usually annual, to three feet tall; ridges eighteen to twenty-four but roundish, not harsh; sheaths flare out slightly at the tip, twice as long as wide; green with a narrow black band at the tip; teeth dark-centered with white margin but usually quickly deciduous; central canal three-fourths the stem diameter; vallecular canals conspicuous; cone blunt, rarely sharp-tipped. Open fields, sand pits, lake shores. Common. Western and midwestern North America, south to northern Mexico.

Equisetum × *ferrissii* Clute is a frequent hybrid of *E. laevigatum* and *E. hyemale.* It has partially evergreen stems, and the sheath has a black basal band, at least in the sheaths of the lower half of the stem.

9b Stems rough, with prominent ridges, evergreen; cones with sharp terminal point; sheath nearly cylindrical, mostly gray with a black band at the base and the tip. (Fig. 170). .... SCOURING-RUSH, *Equisetum hyemale* L.

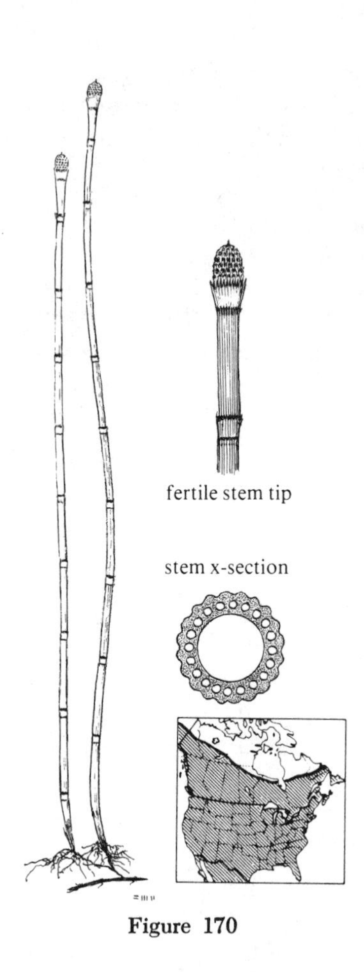

Figure 170

Figure 170 *Equisetum hyemale*

Stems all alike, evergreen, unbranched except after injury, to three and one-half feet tall; ridges eighteen to forty, rough; sheath cylindrical, as long as broad; black band above and below the sheath; teeth narrow, dark brown with white margin, deciduous or persistent; central canal two-thirds the stem diameter; vallecular canals large; cone sharp-pointed. Wide range of damp situations. Abundant. Most of North America, south to Guatemala; Europe, Asia.

The American material belongs to var. *affine* (Engelm.) A. A. Eaton.

GRAMMITIS

Rhizome short-creeping, scaly; fronds clustered, very small, hairy, pinnatifid; veins free; sori round; indusium lacking; spores green. Four hundred species of tropical and subtropical regions. (Fig. 171) ...... *Grammitis nimbata* (Jenm.) Proctor

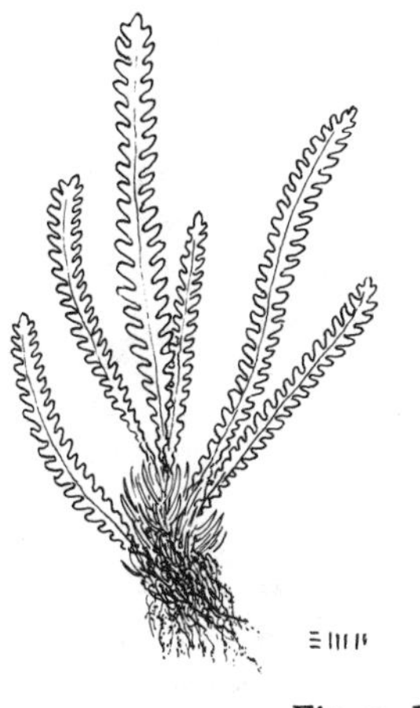

fertile frond tip

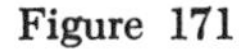

Figure 171 *Grammitis nimbata*

Fronds one to two inches tall, one-half to three-eighths inch wide, with scattered hairs. It closely resembles a miniature, delicate *Polypodium* but with hairs on the blade. Wet rocks under waterfalls. Extremely rare in the United States. Western North Carolina; West Indies.

GYMNOCARPIUM

Oak Fern

Rhizome long-creeping, slender, scaly; fronds distant, small, triangular, bipinnate to tripinnate; sori medial to near the margin, round; indusium lacking. Three or four species of temperate or boreal regions.

1a Rachis and blade glandless or nearly so; only the basal pinna pair stalked; margins not recurved; first pinnule on lower side of basal pinna about half the length of the main rachis or longer. (Fig. 172). .... OAK FERN, *Gymnocarpium dryopteris* (L.) Newm.

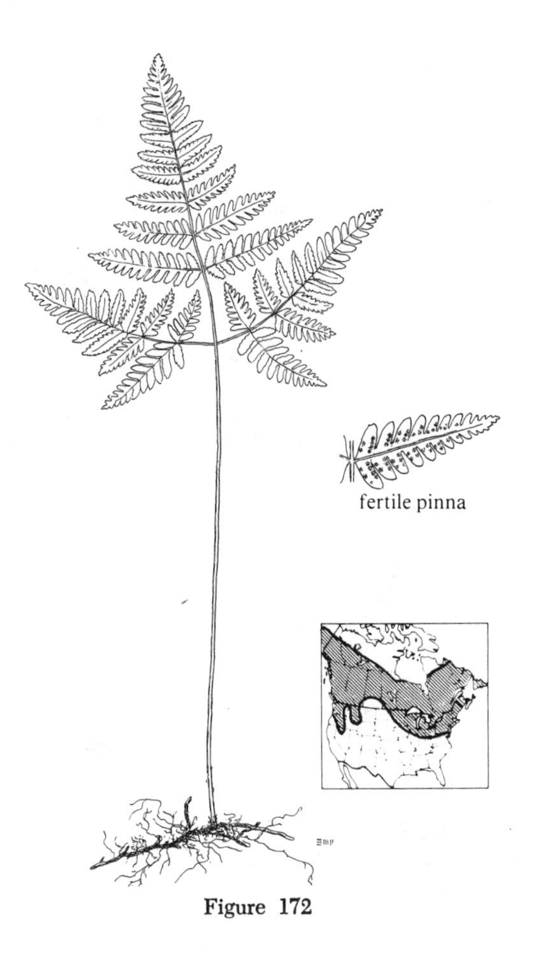

Figure 172 *Gymnocarpium dryopteris*

Fronds nine to eighteen inches tall, five to ten inches broad. Wet woods. Common. Northern North America; Europe, Asia.

1b Rachis and blade densely glandular; lowest and next pinna pair stalked; mar-

gins often recurved; first pinnule on lower side of basal pinna about one-fourth the length of the main rachis or shorter. (Fig. 173). .... LIMESTONE OAK-FERN, ROBERT'S OAK-FERN, *Gymnocarpium robertianum* (Hoffm.) Newm.

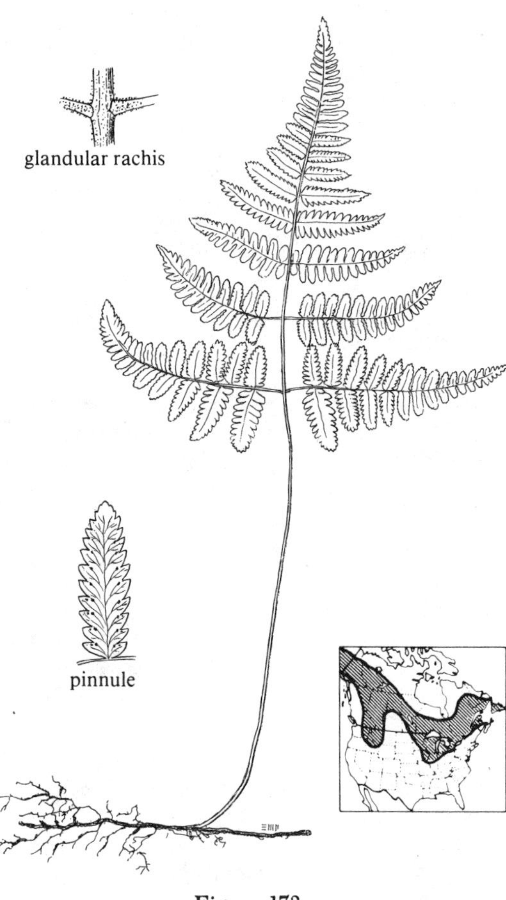

Figure 173

Figure 173 *Gymnocarpium robertianum*

Fronds nine to eighteen inches tall, four to eight inches broad. Shaded wooded limestone rocks and calcareous swamps. Rare. Northern North America; Europe, Asia.

The widespread hybrid between these two species, called *G.* × *heterosporum* W. H. Wagner, is know from Pennsylvania, the Upper Great Lakes region, Pacific Northwest, and Alaska.

HYMENOPHYLLUM

Filmy Fern

Rhizome thread-like, long-creeping, hairy; fronds small, pinnate to tripinnate, very thin, translucent, only one cell thick; sori in marginal, bivalvate cups. Two hundred fifty species, largely of tropical regions.

1a **Margins toothed; southeastern United States. (Fig. 174). .... TUNBRIDGE FILMY-FERN, *Hymenophyllum tunbrigense* (L.) J. E. Smith**

Fronds one to two inches tall, about one-half inch wide, bipinnate to bipinnate-pinnatifid, naked; margin toothed. Moist shaded rocks. Very rare. Western North and South Carolina; tropical America, Europe, Asia.

1b **Margins smooth, not toothed; Pacific Northwest. .... WRIGHT'S FILMY-FERN, *Hymenophyllum wrightii* v.d.B. [*Mecodium wrightii* (v.d.B.) Copeland]**

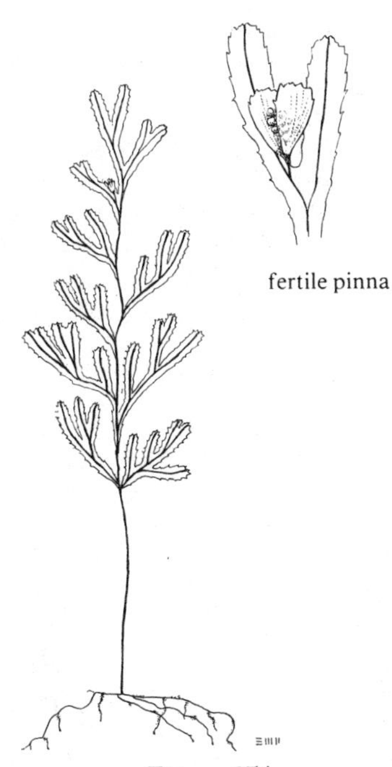

Figure 174

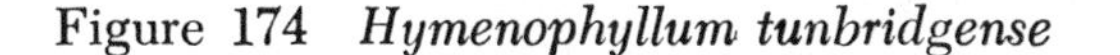

Figure 174 *Hymenophyllum tunbridgense*

Fronds one to two inches tall, pinnate to pinnate-pinnatifid, naked, margin smooth. Wet shaded rocks or tree bases. Very rare. Coastal British Columbia; eastern Asia.

Gametophytes of this species have been found in southern Alaska without the sporophytes. Apparently the gametophytes are more hardy than the sporophytes.

Gametophytes and very young sporophytes of a distinct but unknown species of filmy fern have been found in western North Carolina. Because of the distinctive stellate hairs, it can be placed in the section *Sphaerocionium,* other members of which are not known from the United States.

HYPOLEPIS

Bramble Fern

Rhizome long-creeping, hairy; fronds spaced apart, large, forming a thicket, tripinnate to quadripinnate; sorus short, near margin, protected by a recurved tooth. Fifty species of tropical and south temperate regions. (Fig. 175) .... BRAMBLE FERN, *Hypolepis repens* (L.) Presl

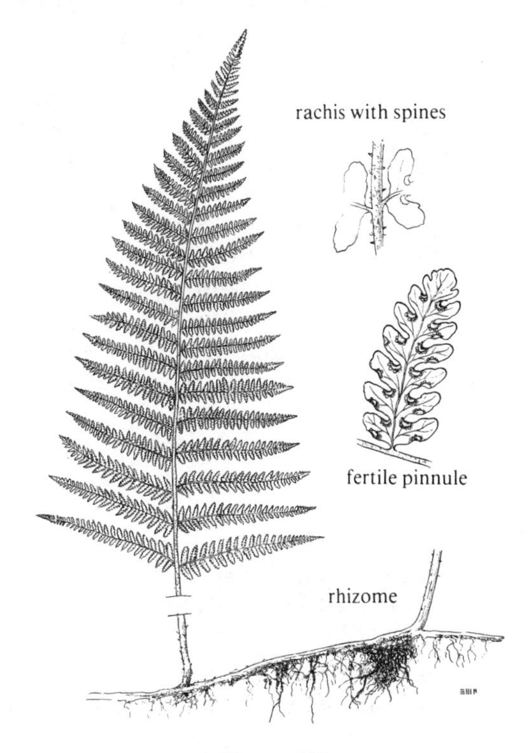

Figure 175

Figure 175 *Hypolepis repens*

Fronds three to eight feet tall, triangular, tripinnate to quadripinnate; stipe and rachis with small prickles; blade sparsely hairy. Moist open woods and wet seepage meadows. Rare. Central Florida; tropical America.

ISOETES

Quillwort

Plants appearing as grass- or onion-like tufts in or by water; stem erect, corm-like, subterranean, bearing fleshy roots in grooves between the stem's two or three lobes; leaves borne on top of the stem, long and slender, grass-like, with a single vein (occasionally with additional vein-like thickenings parallel to the vein, called peripheral strands) and internal air canals and possessing a tiny, flap-like ligule at the base; sporangia embedded in top side of leaf bases, at least partially covered by a membrane, called the velum; heterosporous, the microspores and megaspores borne in different sporangia on the same plant. Sixty species, largely of temperate regions.

This is an extremely difficult genus, greatly needing modern study. Several species exhibit considerable variation, but to what extent the differences are ecologically or genetically induced is not understood. The classification of *Isoetes* is based largely on megaspore ornamentation and requires magnification of about thirty times.

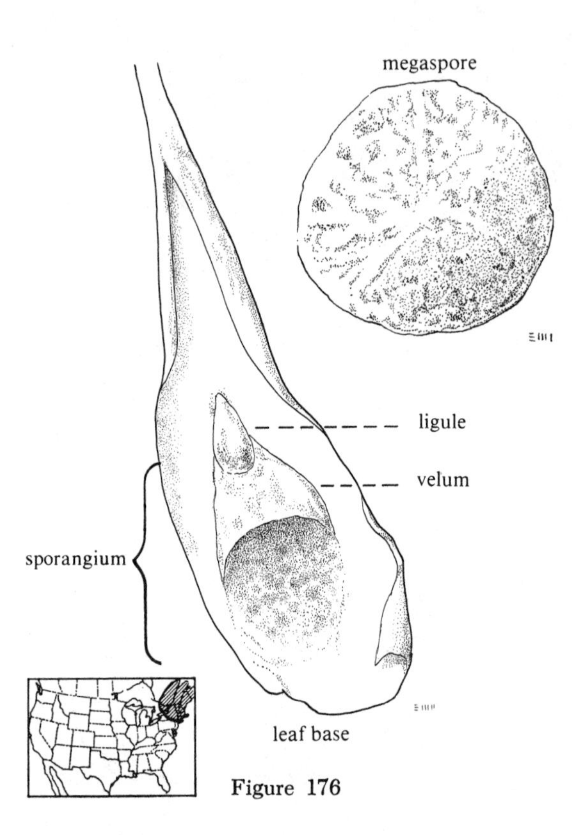

Figure 176 *Isoetes tuckermanii*

Stem two- or three-lobed; leaves to seven inches long; sporangia round or oblong, one-third covered by the velum; megaspores white, four hundred fifty to six hundred fifty microns diameter, with crests (ridges) on upper faces, reticulate on the lower surface. Submersed. Frequent. Northeastern North America.

1a **Megaspores with reticulate ridges, at least on the lower surface. 2**

1b **Megaspores with spines, tubercles, crests or smooth, not regularly reticulate. 4**

2a **Megaspores reticulate throughout. 3**

2b **Megaspores reticulate only on the outer (lower) surface, the three small upper faces crested. (Fig. 176). .... TUCKERMAN'S QUILLWORT, *Isoetes tuckermanii* A. Braun**

3a **Sporangia spotted. (Fig. 177). .... EATON'S QUILLWORT, *Isoetes eatonii* Dodge**

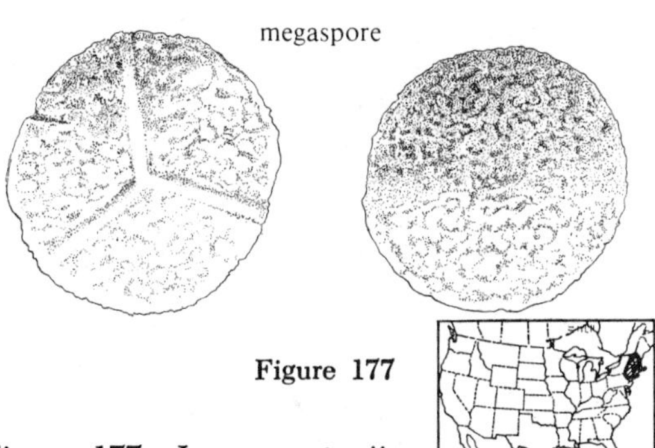

Figure 177

Figure 177 *Isoetes eatonii*

Stem two-lobed; leaves to twenty-four inches long, with peripheral strands; sporangia oblong, spotted, one-sixth to one-fourth covered by the velum; megaspores white, three hundred seventy-five to five hundred twenty-five microns diameter, with narrow reticulate ridges. Amphibious. Frequent. Northeastern North America.

Isoetes louisianensis Thieret, of streams in eastern Louisiana, differs in having the sporangium one-third to one-half covered by the velum, the megaspores five hundred to six hundred twenty-five microns diameter, with irregularly netted ridges. Rare.

Isoetes foveolata A. A. Eaton, of the Northeast, differs from *I. eatonii* in lacking any perpheral strands. Rare.

3b **Sporangia not spotted. (Fig. 178). .... ENGELMANN'S QUILLWORT, *Isoetes engelmannii* A. Braun**

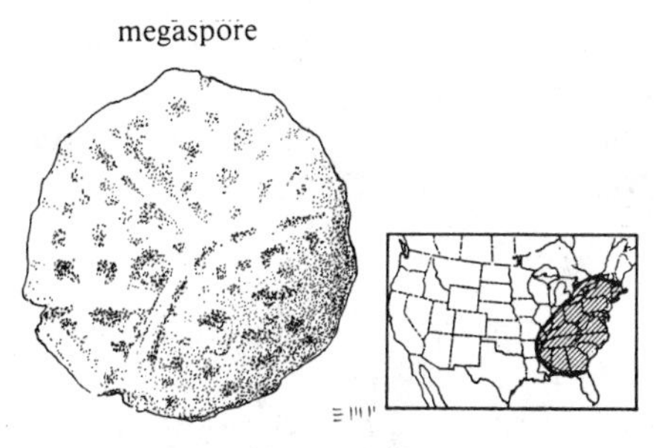

Figure 178

Figure 178 *Isoetes engelmannii*

Stem two-lobed; leaves to nineteen inches long; sporangia oblong, not spotted, one-sixth covered by velum; megaspores white, four hundred to six hundred microns diameter, with network of narrow ridges. Submersed. Frequent. Eastern United States.

4a **Megaspores with tubercles, crests or smooth; projections not sharp-pointed. 5**

4b **Megaspores with sharp-pointed spines. (Fig. 179). .... SPINY-SPORED QUILLWORT, *Isoetes echinospora* Dur. [*I. braunii* Dur., *I. flettii* Pfeiffer, *I. maritima* Underw., *I. muricata* Dur.]**

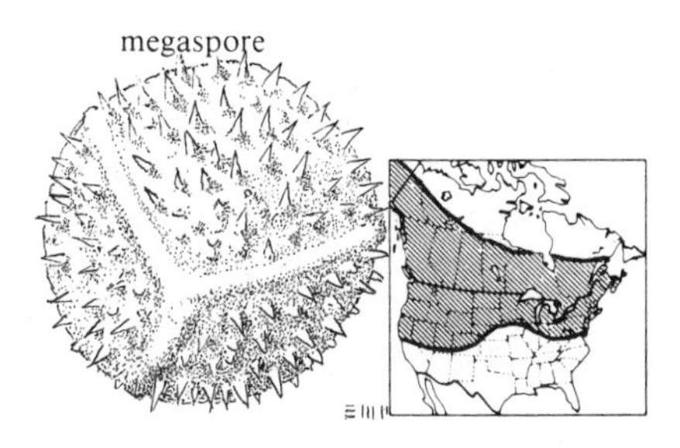

Figure 179

Figure 179 *Isoetes echinospora*

Stem two-lobed; leaves to eight inches long; sporangia often spotted, one-half to three-fourths covered by the velum; megaspores white, four hundred fifty to five hundred fifty microns diameter, with sharp spines. Usually submersed in cold, clear lakes. Frequent. Most of North America; Europe, Asia.

5a **Plants east of the Rocky Mountains. 6**

5b **Plants of the Rocky Mountains or farther west. 11**

6a Megaspores white; leaves two to sixteen inches long. 7

6b Megaspores blackish when wet; leaves one to three inches long. (Fig. 180). .... BLACK-SPORED QUILLWORT, *Isoetes melanospora* Engelm.

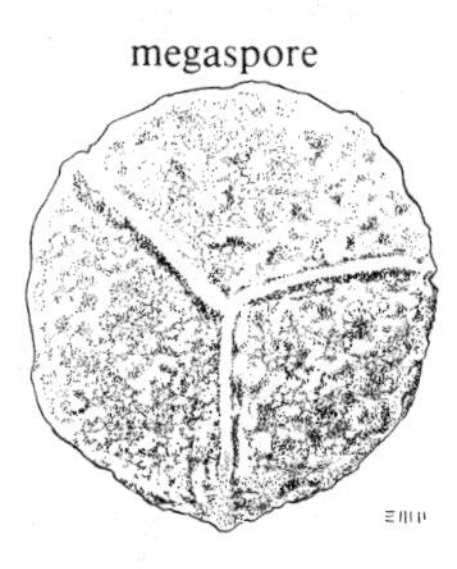

Figure 180

Figure 180 *Isoetes melanospora*

Stem two-lobed; leaves one to three inches long; sporangia round, completely covered by velum; megaspores dark gray to black when wet, four hundred to four hundred eighty microns diameter, with bumps that are sometimes elongated into short ridges. Shallow ponds and depressions. Frequent. Northern Georgia.

A newly discovered *Isoetes* from the granitic outcrops of northern Georgia has long, narrow outgrowths of the corm, forming a pseudorhizome. Whether this represents a variation of *I. melanospora* or a distinct species is not yet known.

7a Megaspores with tubercles. 8

7b Megaspores with crests or ridges. 10

8a Leaf bases green to light brown. 9

8b Leaf bases black. (Fig. 181). .... BLACK-FOOTED QUILLWORT, *Isoetes melanopoda* Gay & Dur.

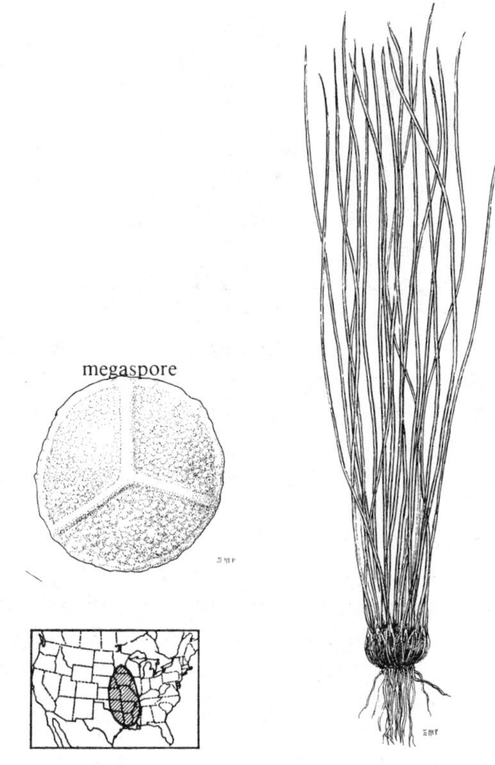

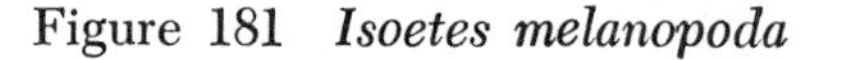

Figure 181

Figure 181 *Isoetes melanopoda*

Stem two-lobed; leaves four to sixteen inches long; sporangia oblong, spotted, up to one-half covered by the velum; megaspores granular to nearly smooth, white, two hundred fifty to four hundred fifty microns diameter. Depressions and ponds. Frequent. Eastern and central United States.

9a Megaspore tubercles crowded. (Fig. 182). ...... RIVERBANK QUILLWORT, *Isoetes riparia* Engelm. ex A. Braun (*I. saccharata* Engelm.)

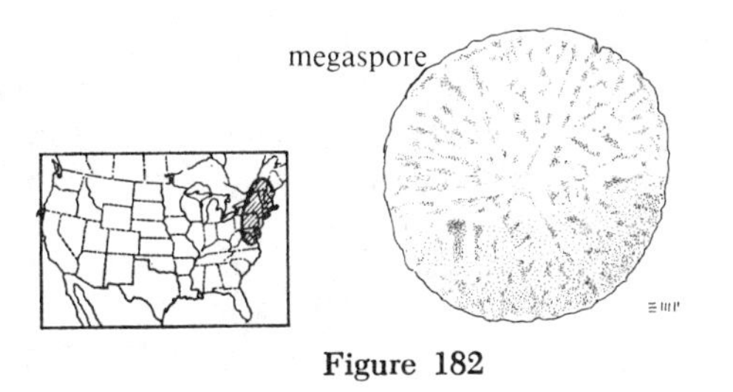

Figure 182

Figure 182 *Isoetes riparia*

Stem two-lobed; leaves three to twelve inches long; sporangia oblong, spotted or unspotted, one-fourth to one-third covered by the velum; megaspores white, with crowded tubercles or rough crests, four hundred fifty to six hundred fifty microns diameter. Banks of ponds and rivers. Frequent. Northeastern North America.

Isoetes butleri Engelm., of Tennessee and Arkansas to Kansas, has leaves three to six inches long, sporangia marked with brown lines, megaspores three hundred sixty to six hundred fifty microns diameter, with numerous tubercles. Frequent.

9b Megaspore tubercles not crowded. (Fig. 183). .... *Isoetes flaccida* Shuttlew. ex A. Braun

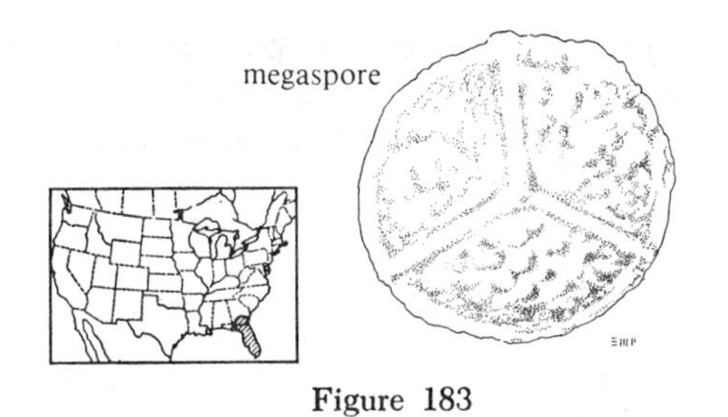

Figure 183

Figure 183 *Isoetes flaccida*

Stem two-lobed; leaves four to sixteen inches long; sporangia oblong, unspotted, completely covered by the velum; megaspores tuberculate on top, with short ridges on bottom surface, white, three hundred to five hundred microns diameter. Streams and pond margins. Rare. Southeastern United States.

10a Megaspores large, five hundred fifty to eight hundred microns diameter, without tubercles. (Fig. 184). .... LARGE-SPORED QUILLWORT, *Isoetes macrospora* Dur.

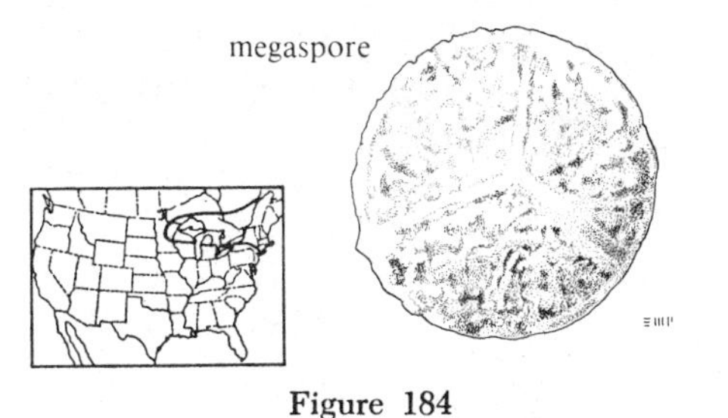

Figure 184

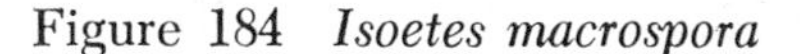

Figure 184 *Isoetes macrospora*

Stem two-lobed; leaves three to eight inches long; sporangium oblong, unspotted, one-third to one-half covered by the velum; megaspores white, five hundred fifty to eight hundred microns diameter, with crests. Submersed. Frequent. Northeastern North America.

10b Megaspores smaller, four hundred fifty to six hundred fifty microns diameter, at least partly tuberculate. .... RIVERBANK QUILLWORT, *Isoetes riparia* Engelm. ex A. Braun (see above)

Isoetes lithophila Pfeiffer, known only from Texas, has leaves four to five inches long, sporangium completely covered by the velum, megaspores two hundred ninety to three hundred sixty microns diameter, that are gray when dry, brown when wet. Rare.

Isoetes virginica Pfeiffer, of Virginia, has leaves six to twelve inches long that are brownish at base, sporangia generally round, about one-fourth covered by the velum, megaspores white, four hundred forty to four hundred eighty microns diameter, with short, discontinuous ridges. Rare.

11a (5b) Megaspores distinctly ornamented; leaves one to eleven inches long. 12

11b Megaspores smooth or with minute bumps; leaves one to two and one-half inches long. (Fig. 185). .... ORCUTT'S QUILLWORT, *Isoetes orcuttii* A. A. Eaton

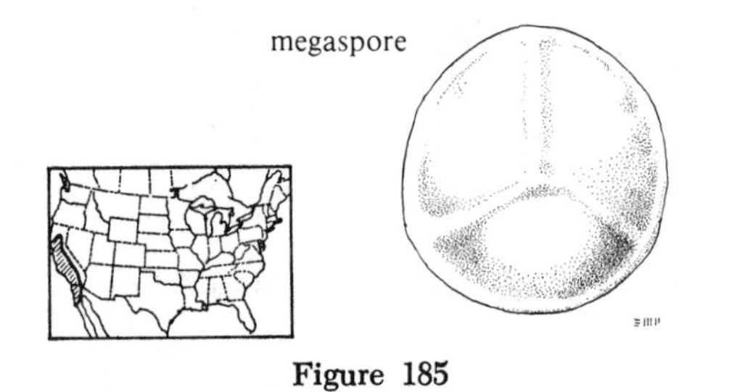

Figure 185

Figure 185 *Isoetes orcuttii*

Stem three-lobed; leaves one to two and one-half inches long; sporangia round to slightly oblong, completely covered by the velum; megaspores gray (dry) or brown (wet), two hundred to three hundred sixty (rarely to four hundred eighty) microns diameter, smooth or with minute, indistinct bumps. Amphibious. Frequent. California; Baja California.

12a Megaspores with tubercles, or with some of the tubercles elongated laterally to form slight crests. 13

12b Megaspores crested. (Fig. 186). .... WESTERN QUILLWORT, *Isoetes occidentalis* Hend. [*I. paupercula* (Engelm.) A. A. Eaton ex Maxon, *I. piperi* A. A. Eaton]

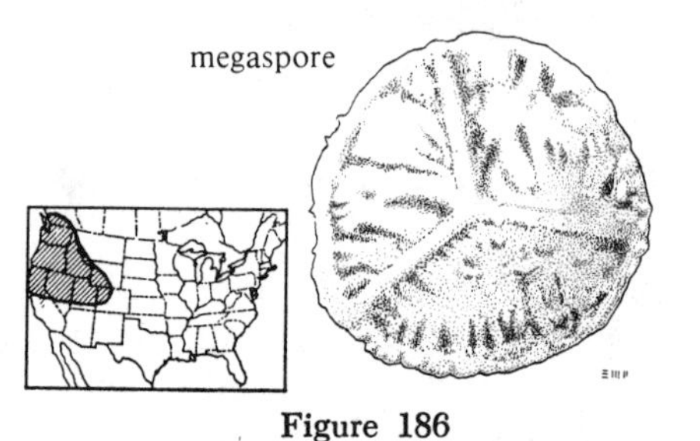

Figure 186

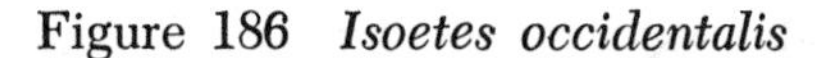

Figure 186 *Isoetes occidentalis*

Stem two-lobed; leaves two to eight inches long, stiff; sporangia nearly round, one-fourth to one-third covered by the velum; megaspores cream-colored, four hundred to five hundred microns diameter, with sharp crests, rarely tuberculate to smooth. Submersed. Frequent. Northwestern North America.

13a Leaves with peripheral strands present; plants terrestrial or amphibious. 14

13b Leaves lacking peripheral strands; plants submersed. (Fig. 187). .... BOLANDER'S QUILLWORT, *Isoetes bolanderi* Engelm.

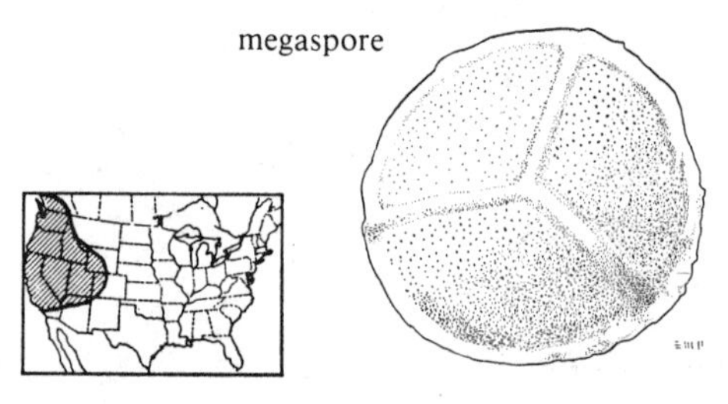

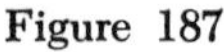

Figure 187

Figure 187 *Isoetes bolanderi*

Stem two-lobed; leaves two to six inches long, soft; sporangia oblong, one-third covered by the velum; megaspores white, three hundred to four hundred fifty microns diameter, with tubercles and short, low ridges. Submersed. Frequent. Western North America.

14a Velum completely covering the sporangium; stem three-lobed; peripheral strands three; plants terrestrial. (Fig. 188). .... NUTTALL'S QUILLWORT, *Isoetes nuttallii* A. Braun ex Engelm.

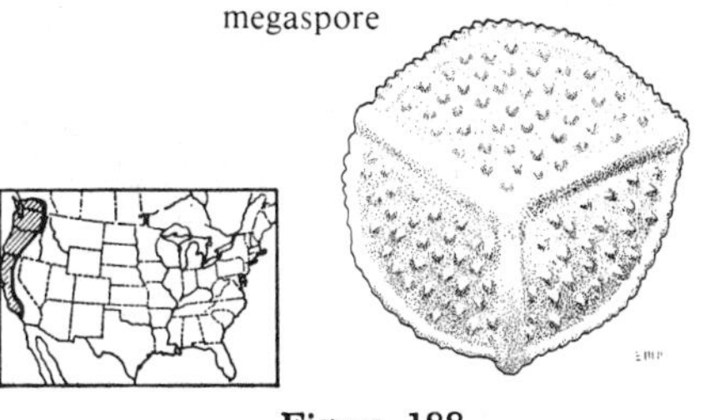

Figure 188

Figure 188 *Isoetes nuttallii*

Stem three-lobed; leaves three to seven inches long; sporangia oblong, completely covered by the velum; megaspores white, four hundred to five hundred microns diameter, densely covered with short papillae. Terrestrial to partially submersed. Frequent. Northwestern North America.

14b Velum one-third covering the sporangium; stem two-lobed; peripheral strands four to twelve; plants amphibious. (Fig. 189). .... HOWELL'S QUILLWORT, *Isoetes howellii* Engelm.

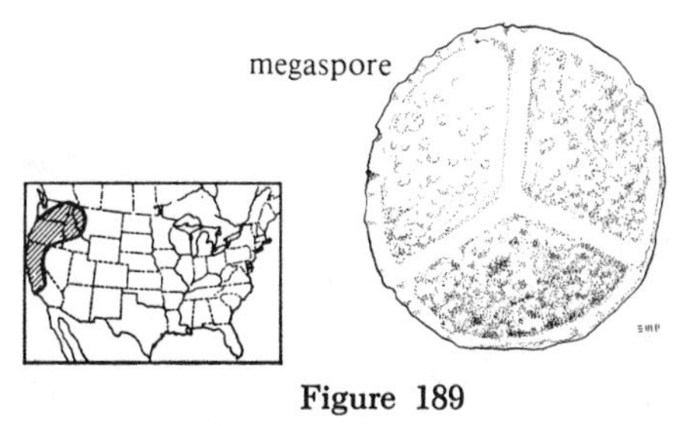

Figure 189

Figure 189 *Isoetes howellii*

Stem two-lobed; leaves two to ten inches long; sporangia round to oblong, often spotted, one-third covered by the velum; megaspores white, four hundred twenty to five hundred twenty microns diameter, with some tubercles and short ridges. Amphibious. Frequent. Northwestern North America.

LOMARIOPSIS

Climbing Holly-Fern

Rooted in the ground but the rhizome vine-like and climbing trees; fronds pinnate, medium-sized, dimorphic; sterile fronds with pinnae coarsely-toothed, holly-like, when juvenile, or finely toothed when mature; fertile fronds also pinnate, pinnae narrower than the sterile, covered with sporangia in no distinct sori. Forty species of tropical regions. (Fig. 190) .... CLIMBING HOLLY-FERN, *Lomariopsis kunzeana* (Underw.) Holttum [*Stenochlaena kunzeana* Underw.]

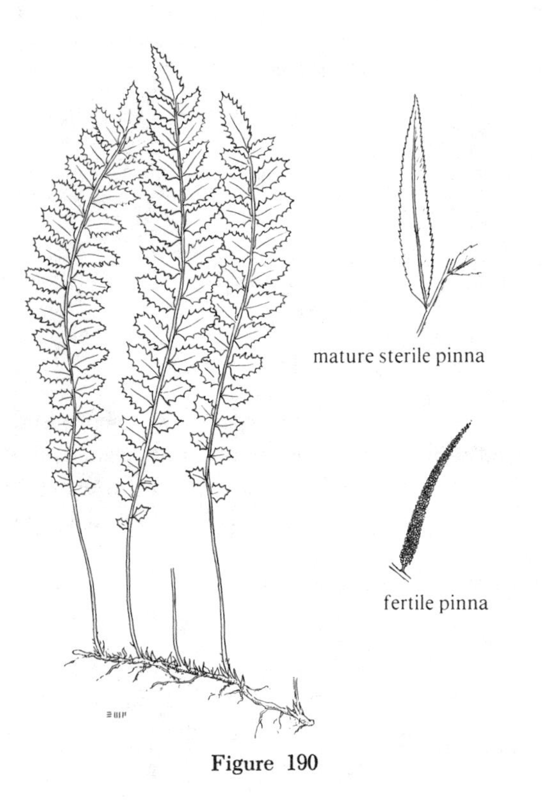

Figure 190

Figure 190 *Lomariopsis kunzeana*

Fronds eight to sixteen inches long, two to four inches wide; sterile pinnae shiny, oblong; fertile pinnae only about one-eighth inch wide. Moist woods. Frequent. Southernmost Florida; West Indies.

LYCOPODIUM

Clubmoss

Leaves needle- or scale-like, with a single vein; arranged spirally, often appearing four-ranked; sporangia borne in axils of either vegetative leaves or specialized leaves of a cone; homosporous. Four hundred species of both temperate and tropical regions.

1a **Sporangia borne in axils of vegetative leaves; no cone formed. 2**

1b **Sporangia borne in axils of specialized leaves which form a cone. 4**

2a **Plants terrestrial, erect; northern. 3**

2b **Plants epiphytic, pendant; southern Florida, very rare. (Fig. 191). .... HANGING CLUBMOSS, *Lycopodium dichotomum* Jacq.**

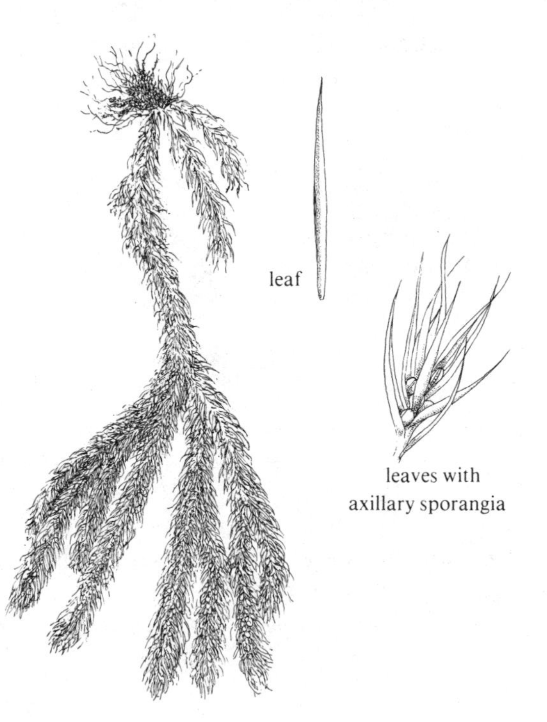

Figure 191

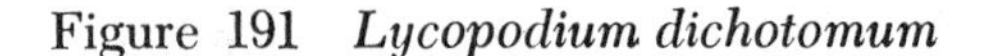

Figure 191 *Lycopodium dichotomum*

Epiphyte, hanging, forking zero to two times; sporangia not in cones; leaves three-eighths to three-fourths inch long, many and overlapping; stems six to twenty-four inches long. Florida

(Collier Co.), but not found in recent years; West Indies, Middle and South America.

3a Leaves coarsely toothed to barely toothed, usually broadest above the middle, or linear, spreading to reflexed, one-fourth to five-eighths inch long; leaves in alternating bands of shorter and longer leaves; base of plant short-creeping; gemma lobes broad; stomata lacking on top of leaves. (Fig. 192). .... SHINING CLUBMOSS, *Lycopodium lucidulum* Michx.

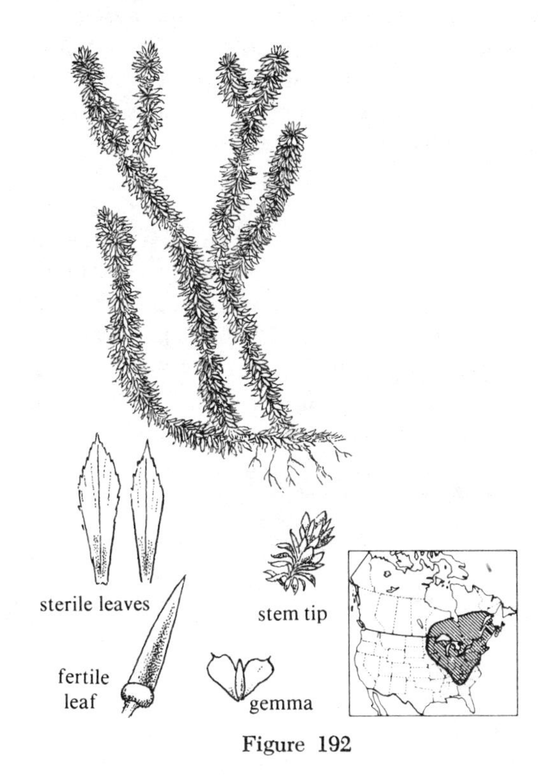

Figure 192

Figure 192 *Lycopodium lucidulum*

Stems all erect to form a clump about six to eight inches tall, the older parts of the stems lying down and rooting; leaves one-fourth to five-eighths inch long, broadened and toothed above the middle, spreading or rarely turned down; distinct zones of longer and shorter leaves; sporangia borne in the axils of vegetative leaves; cones absent; gemmae formed among the upper leaves; gemma lobes broad. Evergreeen. Rich, moist woods. Common. Easern North America.

This species hybridizes with *L. selago* and *L. porophilum.*

3b Leaves smooth-margined, linear, appressed or spreading, one-eighth to three-eighths inch long, rarely longer; leaves all the same size or in slightly alternating bands of different lengths; stem bases more or less erect, rarely short-creeping before ascending; gemma lobes broad or narrow, tapering to a sharp point; stomata on both leaf surfaces. (Fig. 193). .... FIR CLUBMOSS, *Lycopodium selago* L.

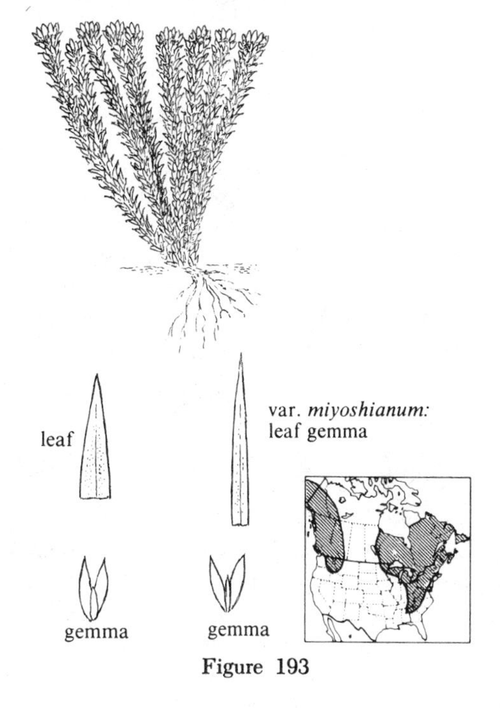

Figure 193

Figure 193 *Lycopodium selago*

Stems erect to form a clump usually four to six inches tall, rarely taller, the older parts only slightly lying down; leaves linear, more or less equal width throughout their length, smooth-margined, one-fourth to one-half inch long, appressed to spreading; no distinct zones of different sized leaves; sporangia borne in axils of vegetative leaves; cones absent; gemmae formed among the leaves. Evergreen. Rocky ledges to acid bogs. Northern North America; northern Europe, Asia. Frequent.

Lycopodium selago has several varieties in North America which probably represent a complex of separate species, hybrids and environmental forms. Var. *appressum* Desv. has the leaves held close to the stem and has narrow-lobed gemmae. It occurs in exposed boreal and alpine habitats, south on the alpine zones of the mountains of New England and the southern Appalachians. Var. *selago,* with slightly appressed to spreading leaves and broad-lobed gemmae, occurs in Europe and in boreal areas of North America. Var. *miyoshianum* Makino has long, narrow, spreading to reflexed leaves and narrow gemma lobes, and occurs only in the Pacific Northwest and eastern Asia. Var. *patens* (Beauv.) Desv., with spreading leaves, probably represents environmental forms of the other varieties as well as hybrids between *L. selago* and *L. lucidulum,* which are commonly found with *L. selago. Lycopodium porophilum* is often erroneously placed here.

The ROCK CLUBMOSS, *Lycopodium porophilum* Lloyd & Underw., closely resembles *L. selago* in having spreading leaves and narrow gemma lobes. It can be distinguished from that species by its stem bases being slightly more creeping, its variation in leaf length zones, and its leaves being somewhat wider in the middle. It occurs exclusively on acid, rocky ledges (usually sandstone) from Minnesota to Pennsylvania, south to Alabama. It frequently hybridizes with *L. lucidulum.*

Hybrids in the *L. selago* complex are recognized by their abortive spores.

4a **Erect stems unbranched, tipped with a single cone; sterile stems never ascending. 5**

4b **Erect stems branched, bearing one or more cones, or often sterile. 9**

5a **Leaves of prostrate stems spreading, relatively narrow (about one thirty-second inch wide), toothed or not; leaves of the erect stems many and overlapping, spirally arranged. 6**

5b **Leaves of prostrate stems lying close to the ground, giving the stems a flattened appearance, leaves relatively broad (one-sixteenth inch wide), not toothed; leaves of erect stems few, not overlapping, whorled. (Fig. 194). .... SLENDER CLUBMOSS, *Lycopodium carolinianum* L.**

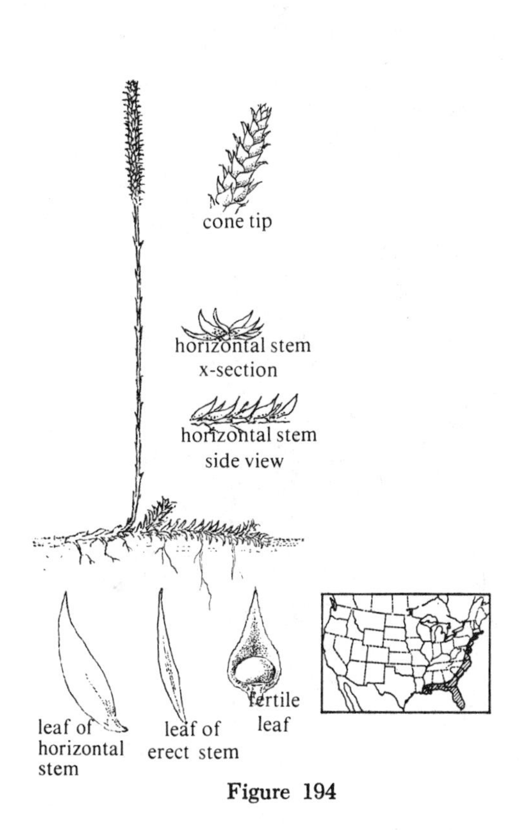

Figure 194 *Lycopodium carolinianum*

Vegetative stems creeping, with erect fertile branches; rhizome leaves spreading mostly laterally to give a flattened appearance; leaves one-sixteenth inch wide, smooth-margined; fertile branches mostly six to ten inches tall, with whorled, non-overlapping leaves; cones one to five inches long, one-eighth inch thick. Deciduous. Acid bogs and wet meadows. Frequent. Eastern and southern coastal states.

6a **Erect stems generally five to twelve inches tall; leaves of prostrate stems toothed. 7**

6b **Plant small, erect stems usually one to three inches tall; leaves of prostrate stems not toothed. (Fig. 195). .... BOG CLUBMOSS, *Lycopodium inundatum* L.**

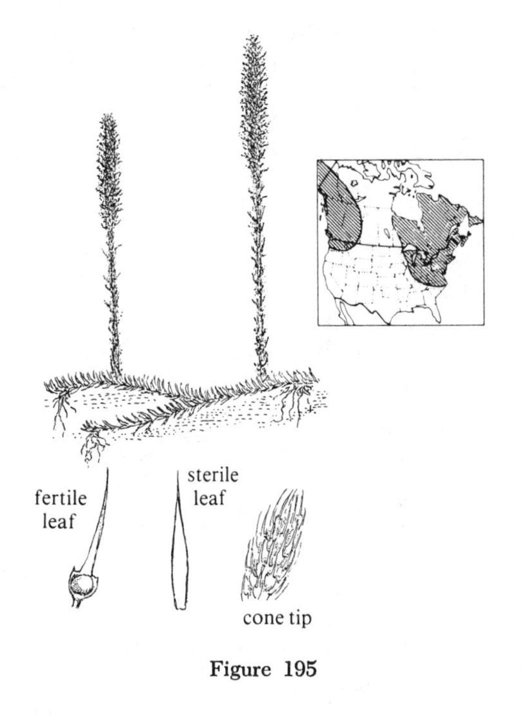

Figure 195 *Lycopodium inundatum*

Vegetative stems creeping, erect fertile branches usually only one or two, mostly one and one-half to three inches tall. Cone one-eighth to three-sixteenths inch thick, one inch long. Leaves in many spiral rows, one thirty-second inch wide, entire. Deciduous. Acid bogs and wet meadows. Eastern and southeastern United States; Europe, Asia.

The bog clubmosses (*L. carolinianum* through *L. prostratum*) often present difficulties in identification. Hybridization is known to occur, confusing the species delimitations.

7a **Leaves of erect stems spreading, strongly toothed; cone three-eighths to three-fourths inch in diameter. 8**

7b **Leaves of erect stem appressed, slightly toothed to entire; cone slender, one-eighth to one-fourth inch in diameter, hardly wider than the peduncle. (Fig. 196). ....**

APPRESSED CLUBMOSS, *Lycopodium appressum* **(Chapm.) Lloyd & Underw.**

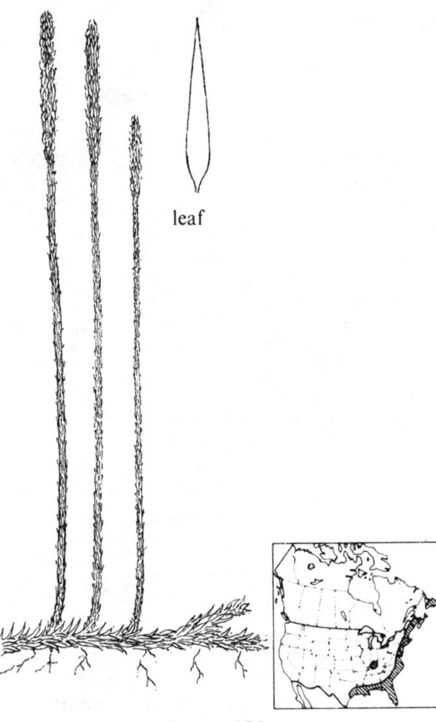

Figure 196

Figure 196 *Lycopodium appressum*

Vegetative stems long-creeping, rooting throughout; rhizome leaves in many spiral rows, leaves spreading, toothed; erect fertile branches four to nine inches tall, its leaves appressed to the stem, not toothed. Cone one-eighth to one-fourth inch thick, one and one-half inches long. Deciduous. Acid bogs and wet meadows. Common. Eastern and southern coastal states and Kentucky; West Indies.

8a Prostrate stems somewhat arching, rooting only where the stem touches the ground; leaves of erect stem spreading to slightly ascending. (Fig. 197) .... FOXTAIL CLUBMOSS, *Lycopodium alopecuroides* L.

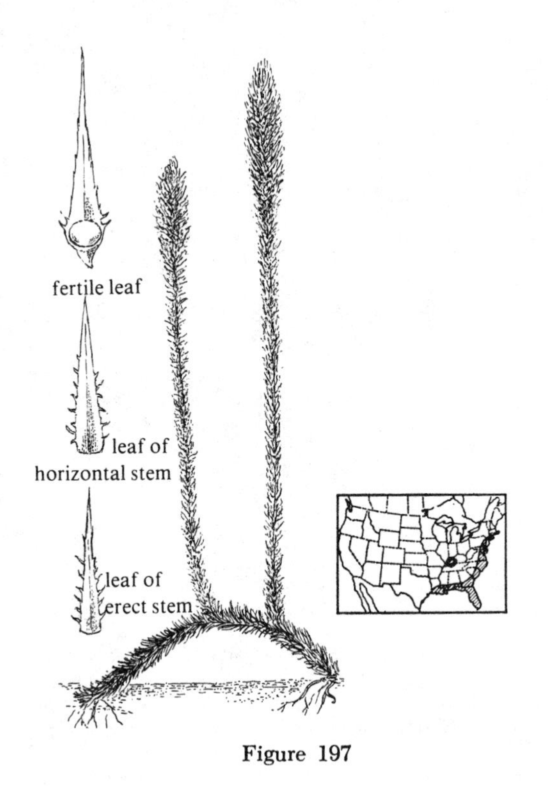

Figure 197

Figure 197 *Lycopodium alopecuroides*

Vegetative stems long-creeping, arching between the few rooted places; leaves in many rows, strongly toothed throughout; erect fertile branches several, six to ten inches tall, its leaves spreading to give a bushy effect; cone one-sixth to three-fourths inch thick, one to two inches long. Deciduous. Acid bogs and wet meadows. Common. Eastern and southern coastal states and Kentucky; West Indies, South America.

8b Prostrate stems not arching, rooting throughout; leaves of erect stems somewhat ascending. (Fig. 198) .... SOUTHERN CLUBMOSS, *Lycopodium prostratum* Harper

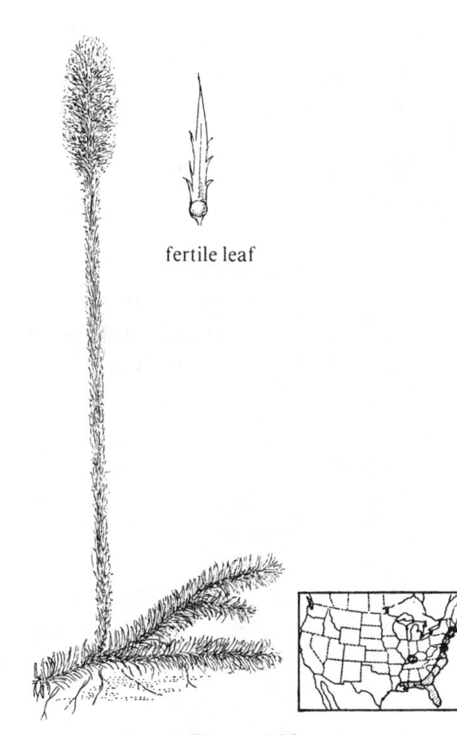

Figure 198

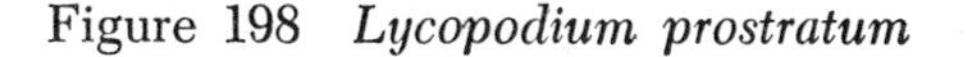

Figure 198 *Lycopodium prostratum*

Vegetative stems long-creeping, rooting throughout; leaves in many spiral rows, leaves strongly toothed throughout; erect fertile branches five to ten inches tall, leaves somewhat ascending; cone one-half to three-fourths inch thick, one to two inches long. Deciduous. Acid bogs and wet meadows. Common. Eastern and southern coastal states and Kentucky.

9a (**4b**) **Leaves radially arranged, all equal. 10**

9b Leaves in four rows on slightly to strongly flattened aerial stems. 14

10a Plants less than twelve inches tall; cones erect, one to one and one-half inches long; northern. 11

10b Plants twelve to thirty inches tall; cones one-fourth and one-half inch long, pendant; Gulf Coast. (Fig. 199) .... NODDING CLUBMOSS, ***Lycopodium cernuum*** **L.**

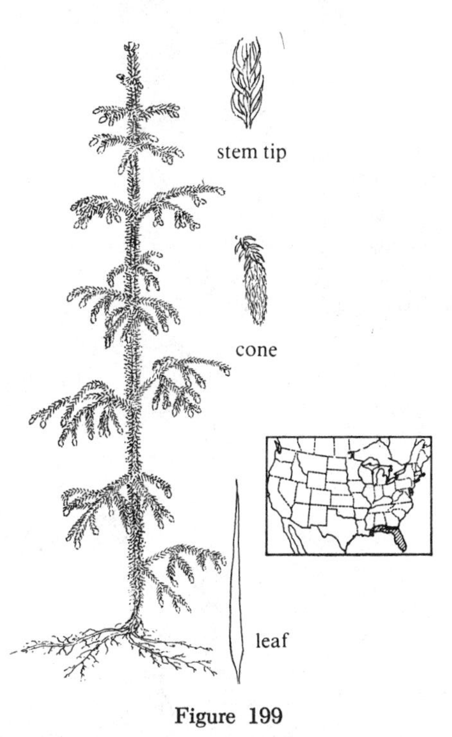

Figure 199

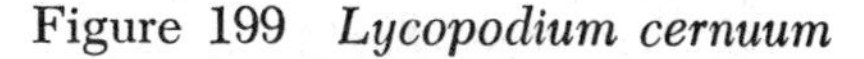

Figure 199 *Lycopodium cernuum*

Tree-like, to thirty inches tall, much branched; leaves spreading and up-curved, one-eighth to one-fourth inch long. Cones many, sessile, drooping from branch tips, one-fourth to one-half inch long. Stems occasionally send out branches along ground that root and send up new aerial shoots. Moist open areas along ditches, stream banks, and meadows. Rare. Florida to Mississippi; tropical regions of the world.

11a Leaves lacking a hair tip; cones sessile. 12

11b Leaves tipped with a long hair (though lacking in many western specimens); cone long-pedunculate. (Fig. 200) .... GROUND PINE, *Lycopodium clavatum* L.

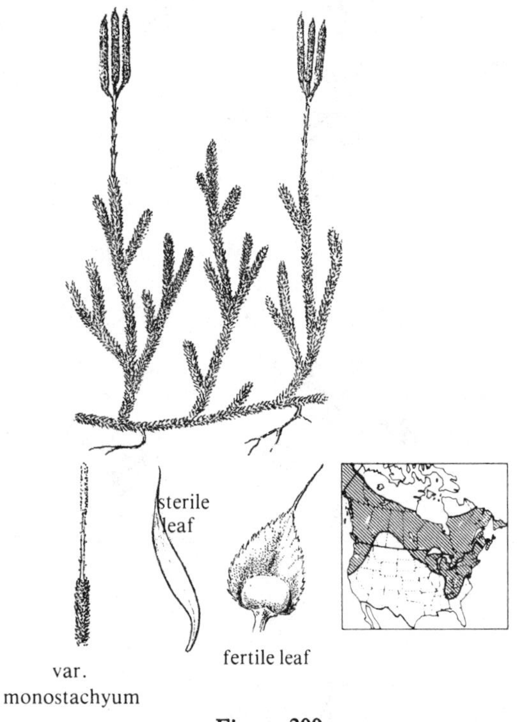

Figure 200 *Lycopodium clavatum*

Stems wide-creeping and highly branched, forming extensive low colonies. Leaves in many rows, ascending, one-eighth to one-fourth inch long, each with a hair-like tip. Cones on peduncles three to six inches long, usually two to three cones per peduncle. Cones one and one-half to three inches long. Evergreen. Open woods to bogs. Common. Northern North America; Europe, Asia.

Plants with only one cone per stalk are referred to var. *monostachyum* Desv. or even given species rank by some botanists.

12a Aerial stems much-branched, branches spreading, tree-like. 13

12b Aerial stems unbranched or only slightly branched, all branches erect. (Fig. 201) .... STIFF CLUBMOSS, *Lycopodium annotinum* L.

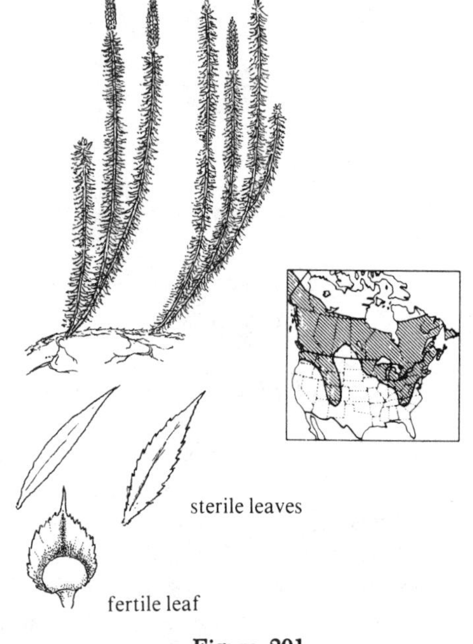

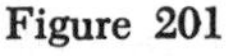

Figure 201

Figure 201 *Lycopodium annotinum*

Rhizome long-creeping on ground surface with erect, sometimes forked branches four to eight inches tall, the branches never spreading. Leaves spreading, one-fourth to one-half inch long, leaf margin toothed to nearly smooth. Cones about one inch long, borne without peduncles on tips of vegetative erect branches. Evergreen. Coniferous woods or deciduous woods. Frequent. Northern North America; Europe, Asia.

13a Leaves appressed; plant two to four inches tall; branches erect; leaves about one-sixteenth inch long. (Fig. 202) .... SITKA CLUBMOSS, *Lycopodium sitchense* Rupr.

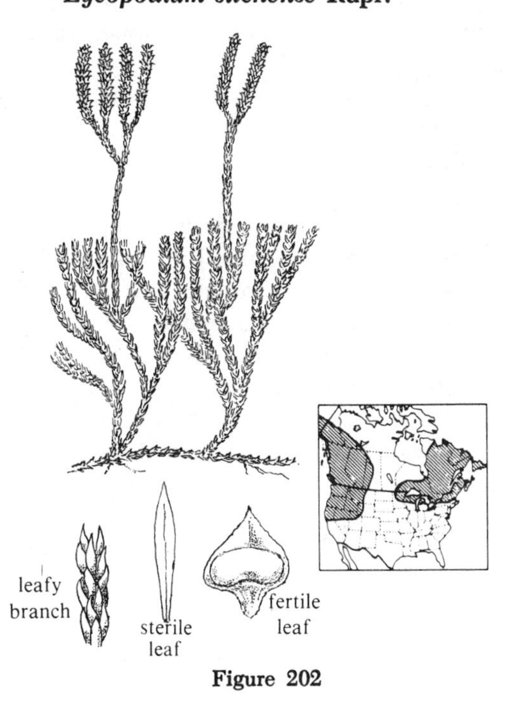

Figure 202

Figure 202 *Lycopodium sitchense*

Rhizome creeping on ground surface with upright, several-forked branches two to four inches tall. Leaves appressed to spreading, branches round in appearance. Cones one-fourth to five-eighths inch long, borne on extended leafy branches. Evergreen. Open rock, acid soil. Northern North America; eastern Asia.

13b Leaves spreading; plant six to twelve inches tall; branches spreading; leaves about one-eighth inch long. (Fig. 203). .... TREE CLUBMOSS, *Lycopodium obscurum* L.

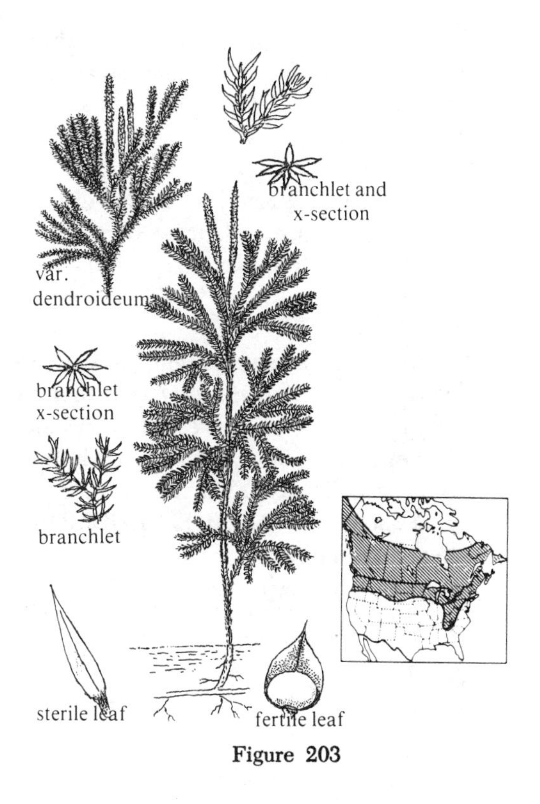

Figure 203

Figure 203 *Lycopodium obscurum*

Rhizome creeping, one to two inches below the ground, with erect, tree-like branches; leaves dark green, needle-like, one-eighth to one-fourth inch long, appressed to ascending, in six rows; cones one to two inches long, without peduncles. Evergreen. Moist evergreen or deciduous woods to bogs. Frequent. Northern North America; northeastern Asia.

Lycopodium dendroideum Michx. (sometimes given varietal rank under *L. obscurum*) is easily distinguished from *L. obscurum* by having the leaves of the main erect stem broadly spreading rather than tightly appressed. *Lycopodium dendroideum* also has two rows of leaves along the top of the branch instead of one, and is usually of a lighter green color. Although *L. dendroideum* may have a tendency to inhabit wetter sites, the two are often found growing together.

14a (9b) Cones lacking a peduncle or short-peduncled with sporangia straggling down the peduncle; branches not strongly flattened. 15

14b Cones long-peduncled with sharply defined cone base; plant strongly flat-branched and tree-like. 16

15a Lower leaves trowel-shaped; plant with dwarf alpine habit, usually two to four inches tall; leaves appressed; cone sessile; cone leaves spirally arranged in four rows. (Fig. 204). .... ALPINE CLUBMOSS, *Lycopodium alpinum* L.

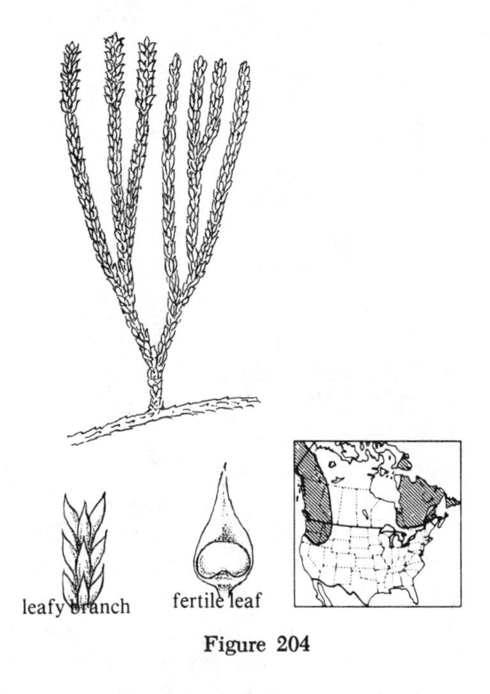

Figure 204

Figure 204 *Lycopodium alpinum*

Rhizome creeping on ground surface, with upright, several-forked branches two to four inches tall; branches appear somewhat flattened, the leaves in four rows, appressed, fused to the stem half their length; leaves of the lower surface trowel-shaped, not as strongly fused to the stem; cones one-fourth to one-half inch long, without peduncle. Evergreen. Northern acid slopes and meadows. Frequent. Northern North America; Europe, eastern Asia.

Lycopodium × *issleri* (Rouy) Lawalrée is the sterile hybrid between *L. alpinum* and *L. complanatum.*

15b Lower leaves linear; plants not compacted, usually four to six inches tall; cone leaves whorled; leaves usually spreading and somewhat curved; cone short-peduncled (about one inch) with one to two cones per peduncle. (Fig. 205). .... JUNIPER CLUBMOSS, *Lycopodium sabinifolium* Willd.

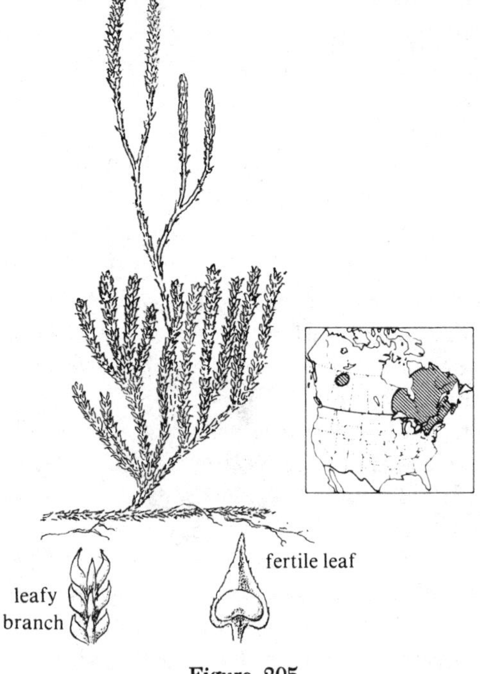

Figure 205

Figure 205 *Lycopodium sabinifolium*

Rhizome creeping just below or on ground surface with upright several-forked branches four to six inches tall; branches appear flattened; leaves in four rows, half fused to the stem, upper and lower rows appressed, lateral rows spreading to appressed, curved toward the branch tip; cone one-half to one inch long; peduncle one to two inches long, its leaves sparse and whorled; the lowest sporangia of the cone loosely running down the peduncle. Evergreen. Uncommon. Open woods and meadows. Northeastern North America.

This species is thought to have originated as a hybrid between *L. sitchense* and *L. tristachyum*, but it is now fertile and acts as a distinct species.

16a Leaves of underside of branch much reduced to a slight emergent tip; rhizome on ground surface or slightly underground; branchlets wide (one-sixteenth to one-eighth inch), green; annual constrictions present or absent. 17

16b Leaves of underside of branch scarcely reduced, about same size as lateral leaves; rhizome deeply underground; branchlets narrow (one thirty-second to one-sixteenth inch wide), bluish green; with conspicuous annual constrictions. (Fig. 206). .... GROUND CEDAR, *Lycopodium tristachyum* Pursh

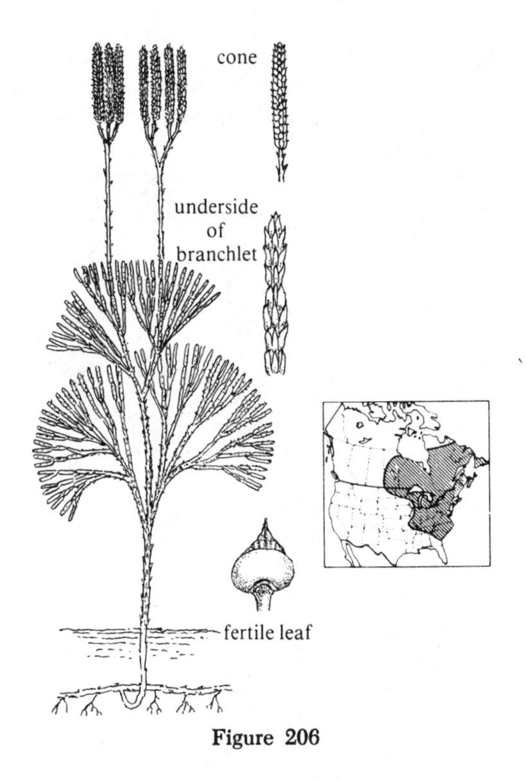

Figure 206

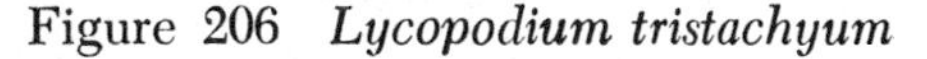

Figure 206 *Lycopodium tristachyum*

Rhizome creeping underground; erect branches repeatedly branching, five to ten inches tall; branches strongly flattened, bluish green and strongly ascending to make a somewhat funnel-shape when in the sun, more spreading and bright green in the shade; branches less than one-sixteenth inch wide, with annual constrictions evident; leaves in four rows with all leaves about the same size, lower row not especially smaller than the lateral leaves; cones about one inch long; peduncle two to four inches long, forking twice to bear four cones, the forks separated to give an H-shape as viewed from above. Evergreen. Open woods, sandy meadows, rocky barrens. Frequent. Northwestern North America; Europe.

Lycopodium × *habereri* House is the frequent hybrid between *L. tristachyum* and *L. flabelliforme*.

17a **Annual constrictions evident; branches irregularly divided, "spidery"; rhizome slightly underground; cones lacking sterile tips. (Fig. 207) .... NORTHERN RUNNING PINE, *Lycopodium complanatum* L.**

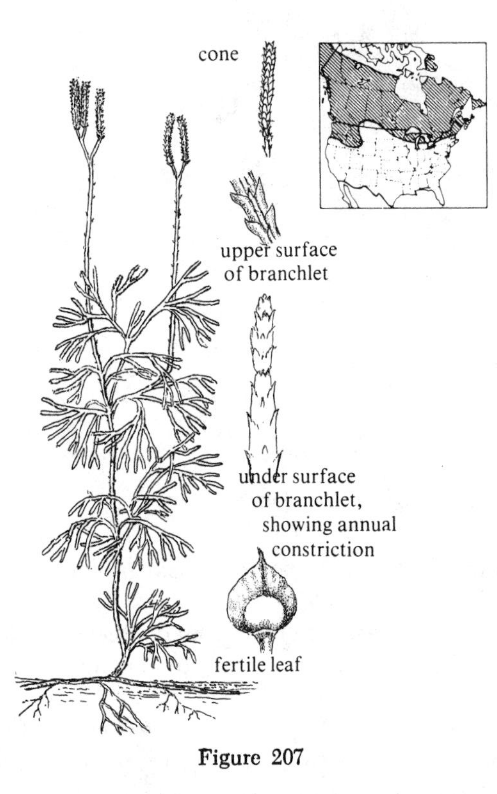

Figure 207

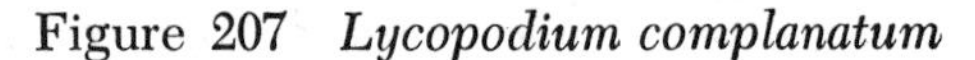

Figure 207 *Lycopodium complanatum*

Rhizome creeping slightly underground; erect branches much-branched, strongly flattened, five to ten inches tall; annual constrictions evident; branchlets one-sixteenth to one-eighth inch wide; leaves in four rows, reduced to small emergences; cones one-half to one inch long, typically four per peduncle, rarely one or two; peduncles one to two and one-half inches long, forking twice to give an H-shape as viewed from above (as in *L. tristachyum*). Evergreen. Open woods and barrens. Frequent. Northern North America; Europe, Asia.

17b **Annual constrictions absent; branches regularly divided, fan-shaped; rhizome on soil surface or just under leaf litter; cones usually having sterile tips. (Fig. 208). .... RUNNING CEDAR, *Lycopodium flabelliforme* (Fernald) Blanch.**

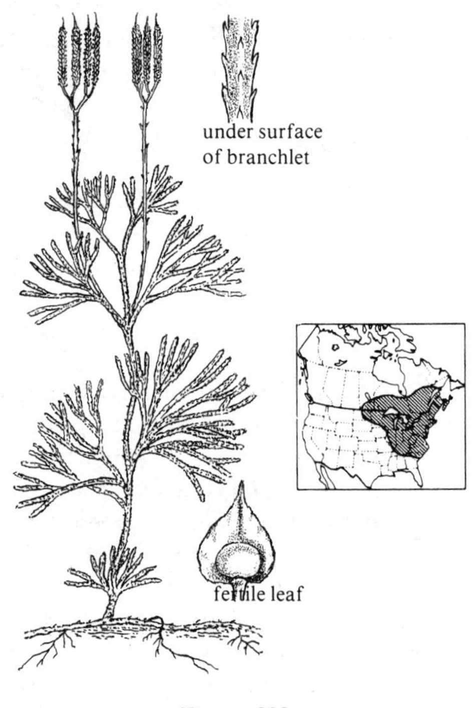

Figure 208

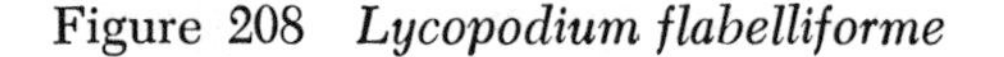

Figure 208 *Lycopodium flabelliforme*

Rhizome long-creeping on soil surface or just under leaf litter; erect stems five to ten inches tall, repeatedly branched, strongly flattened and spreading fan-like, without annual constrictions, one-sixteenth to one-eighth inch wide; leaves in four rows, bottom row much smaller than the lateral leaves; cones one to two inches long, often with a slight tail-like sterile tip; peduncles two to four inches long, usually forking twice, bearing four cones, the forkings close to one another to give an X-shape as viewed from above. Evergreen. Dry open woods or meadows. Common. Northeastern North America.

This has often been lumped with *L. tristachyum* and *L. complanatum* under the name of the latter, but studies have shown it to be a distinct species.

LYGODIUM

Climbing Fern

Rhizome slender, creeping, subterranean, hairy; fronds medium-sized to very long, the rachis twining, climbing on shrubs and small trees; pinnae forked in half, each half looking like one pinna; the sporangia borne on reduced pinnules or finger-like projections at the segment margins. Forty species, largely of the tropics.

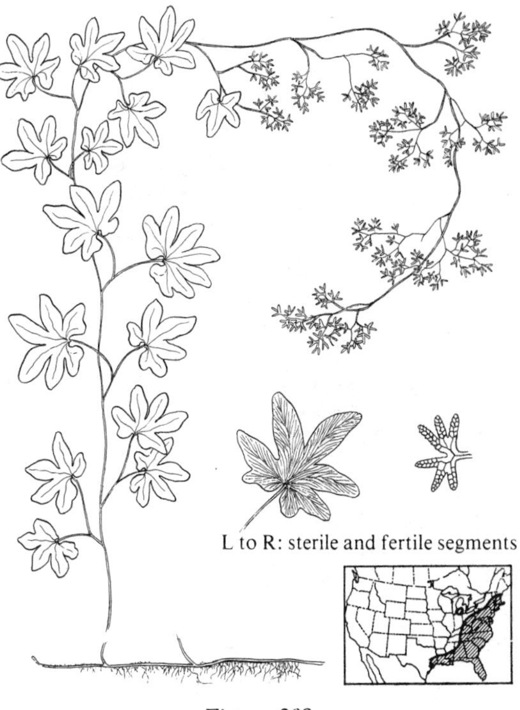

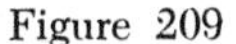
Figure 209

1a Sterile segments hand-shaped, arising in pairs on short stalks from the main rachis; strongly dimorphic, the fertile pinnae much smaller and more dissected than the sterile. (Fig. 209) .... HARTFORD FERN, CLIMBING FERN, *Lygodium palmatum* (Bernh.) Sw.

Figure 209 *Lygodium palmatum*

Fronds three to four feet long, three to six inches wide, sterile pinna halves hand-shaped with six narrow lobes; fertile pinna halves divided into several small discrete segments. Sandy bogs and swamps. Rare and local. Eastern and southern United States.

1b Sterile pinnules pinnately dissected into discrete segments; sporangia borne on finger-like projections on the segment margins; otherwise the fertile and sterile segments alike. (Fig. 210) .... JAPANESE CLIMBING-FERN, *Lygodium japonicum* (Thunb.) Sw.

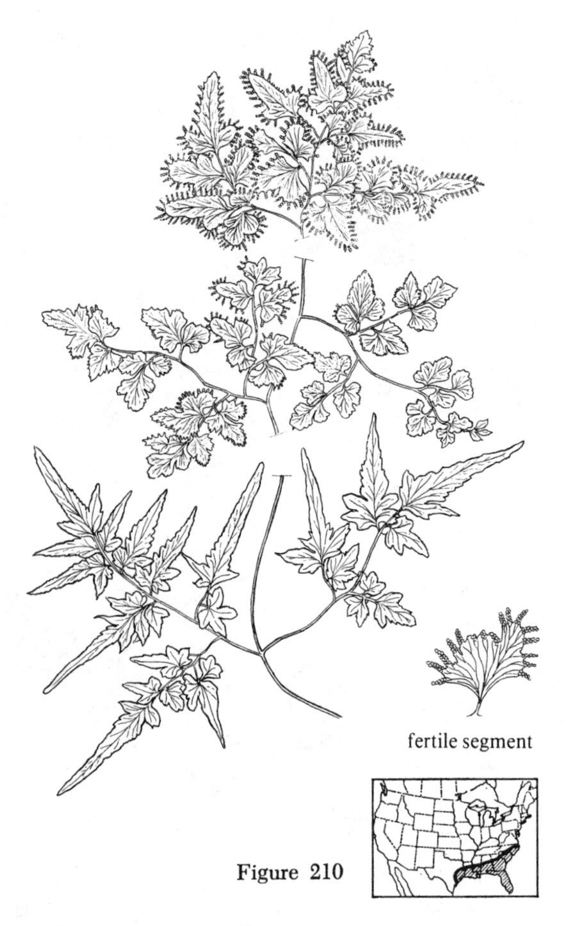

Figure 210

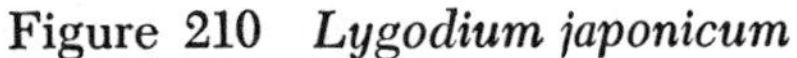

Figure 210 *Lygodium japonicum*

Fronds several feet long, eight to twelve inches wide; pinna halves long-triangular, pinnately divided into several triangular segments; fertile pinnae similarly dissected with several finger-like projections bearing the sporangia on each segment. Open woods and more exposed sites. Abundant, sometimes a pest. Southeast coastal states (though marginally hardy north to Connecticut); native to eastern Asia.

Another species, *L. microphyllum* (Cav.) R. Brown, has also become sparingly naturalized in southern Florida. Its segments are narrowly oval, the segments not dimorphic.

MARSILEA

Water Clover

Rhizome slender, long-creeping, naked or hairy, rooted in mud, either under water or at water's edge, some species appearing only seasonally during wet periods; leaves long-stiped with four wedge-shaped pinnae close together at the tip of the stipe, appearing like a four-leaf clover; sporangia borne in special nut-like sporocarps at or near the base of the stipe; heterosporous. Sixty species of temperate and tropical regions.

1a Leaflets one-half to two inches wide, the tips not hooked; apex smooth-margined or nearly so. 2

1b Leaflets one-eighth to three-eighths inch wide, often hooked at the tip; apices toothed or wavy-margined. (Fig. 211). .... NARROW-LEAVED WATER-CLOVER, *Marsilea tenuifolia* Kunze

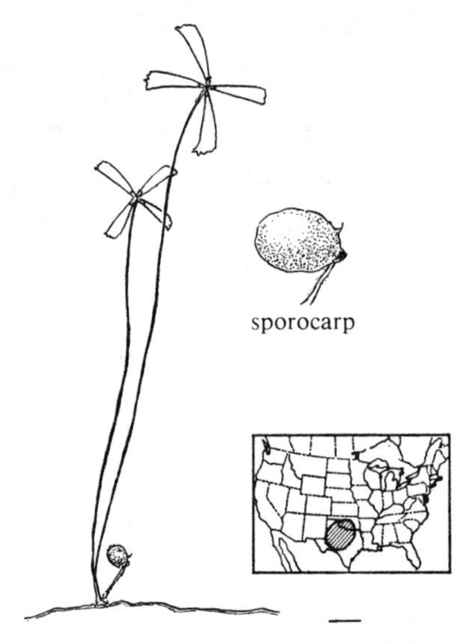

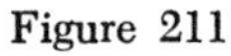

Figure 211

Figure 211 *Marsilea tenuifolia*

Pinnae one-half to one inch long, one-eighth to three-eighths inch wide, naked; sporocarps solitary, one-fourth to three-eighths inch long, divergent teeth more or less equal. Lake margins, wet depressions. Uncommon. Texas, Oklahoma.

2a **Leaves naked or sparsely hairy. 3**

2b **Leaves usually covered with long, golden hairs, especially at the top of the stipe. (Fig. 212). .... GOLDEN WATER-CLOVER, *Marsilea macropoda* Engelm. ex A. Braun**

Figure 212 *Marsilea macropoda*

Pinnae three-eighths to three-fourths inch long and wide; sporocarps two to six per peduncle, three-sixteenths to five-sixteenths inch long, teeth short and obscure. Frequent. Texas.

3a **Sporocarps one per peduncle, arising from juncture of stipe and rhizome, or several on a very short peduncle arising from the rhizome; southern or western. 4**

3b **Sporocarps two to three per peduncle, arising from the stipe base; northeastern. (Fig. 213) .... EUROPEAN WATER-CLOVER, *Marsilea quadrifolia* L.**

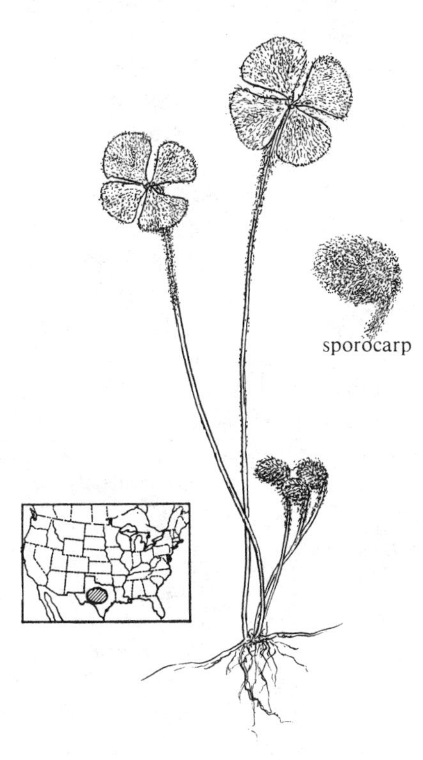

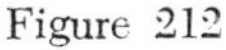
Figure 212

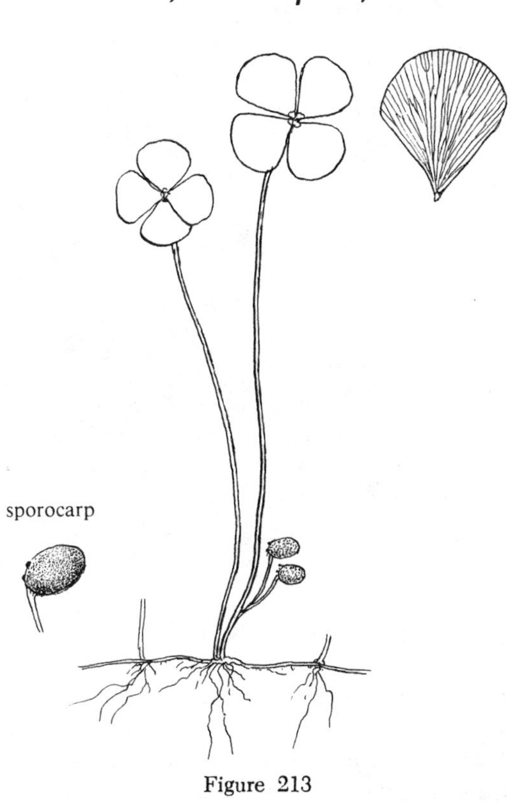

Figure 213

Figure 213 *Marsilea quadrifolia*

Pinnae one-half to one inch long and wide, naked; sporocarps one-eighth to three-sixteenths inch long, teeth short, inconspicuous. Ponds and slow-moving streams. Frequent. Escaped and naturalized in southern New England; native of Europe.

4a Peduncle two times as long as the sporocarp; upper sporocarp tooth often hooked; leaves usually naked. (Fig. 214) .... HOOKED WATER-CLOVER, *Marsilea uncinata* A. Braun

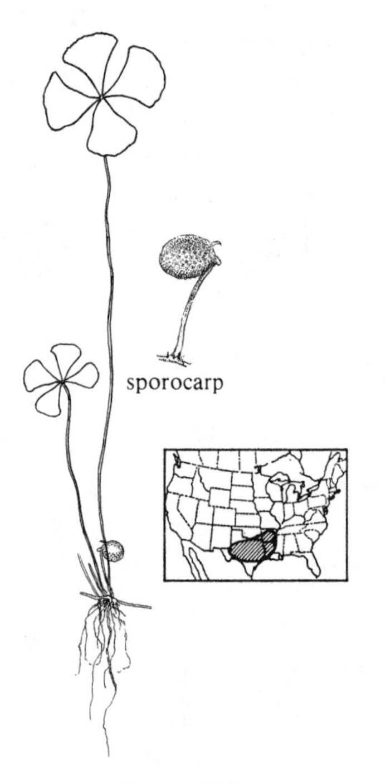

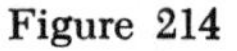
Figure 214

Figure 214 *Marsilea uncinata*

Pinnae three-eighths to three-fourths inch long, three-eighths to five-eighths inch wide, naked or with sparse hairs; sporocarps three-sixteenths to one-fourth inch long, teeth relatively long, the upper one hooked. Wet ditches and water margins. Frequent. Southern United States.

This species closely resembles *M. vestita,* and there are some specimens that are difficult to place with certainty in one species or the other.

Marsilea mexicana A. Braun occurs rarely in southern Texas and south to Honduras. It is easily identified by its naked leaves possessing short, reddish streaks between the veins on the lower surface of the pinnae, and the sporocarps being massed on a very short peduncle arising from the rhizome.

4b Peduncle about as long as the sporocarp; both teeth short and blunt; leaves sparsely hairy. (Fig. 215). .... HAIR WATER-CLOVER, *Marsilea vestita* Hook. & Grev. [*M. oligospora* Goodd., *M. mucronata* A. Braun]

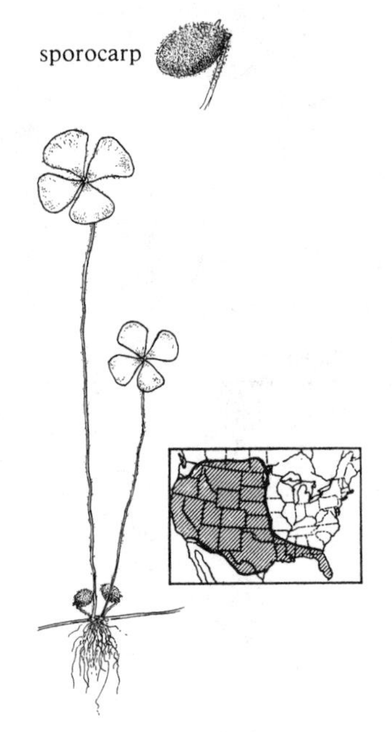

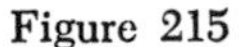
Figure 215

Figure 215 *Marsilea vestita*

Pinnae one-fourth to one-half inch long and wide, lightly hairy; sporocarps one-eighth to three-sixteenths inch long, teeth short and blunt, upper more conspicuous. Edges of ponds and rivers, ditches and wet meadows. Frequent. Southern and western North America.

Marsilea mucronata and *M. oligospora* are doubtfully distinct and are included here. The former is supposed to have broad, appressed hairs shorter and fewer than in *M. vestita,* and occurs in the southern and midwestern states north to the Canadian prairies. *Marsilea vestita* in the more restricted sense has abundant, long, slender spreading hairs, and is of the west coast. *Marsilea oligospora* is a synonym of *M. vestita.*

MATTEUCCIA

Ostrich Fern

Rhizome stout, ascending, almost trunk-like; fronds large, clustered to form a vase-like plant; dimorphic; sterile fronds pinnate-pinnatifid, short-stiped, tapering gradually at the base to form a plume-shaped blade; veins free; fertile fronds woody, shorter; sori hidden and protected by curled margin of the bead-like segments. Three species of north temperate regions. (Fig. 216) .... OSTRICH FERN, *Matteuccia struthiopteris* (L.) Todaro [*Matteuccia pensylvanica* (Willd.) Raymond, *Pteretis nodulosa* (Michx.) Nieuwl.]

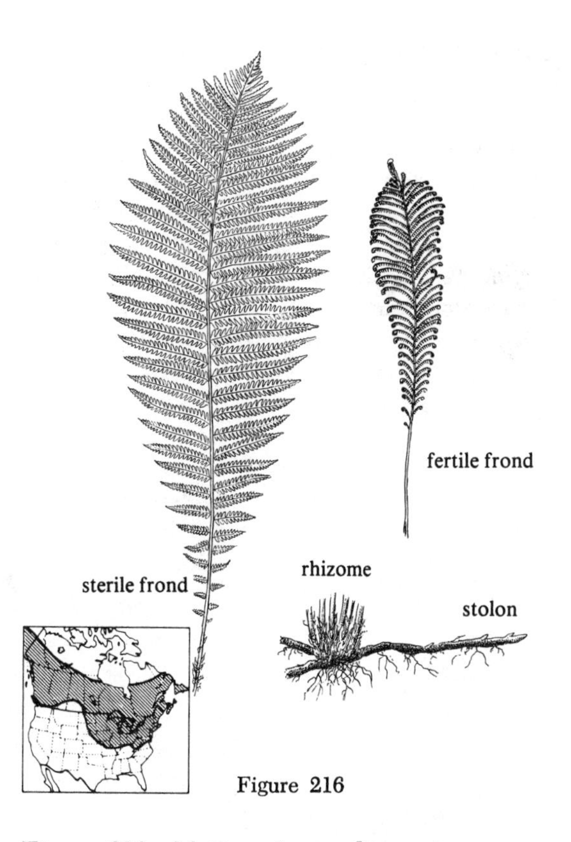

Figure 216

Figure 216 *Matteuccia struthiopteris*

Rhizome stout, erect, fronds clumped, vase-like; slender rhizome branches (stolons) go laterally underground to form new plants nearby; fronds dimorphic, sterile fronds large, two to five feet tall, to fifteen inches wide, plume-like, broadest above the middle, pinnate-pinnatifid, tapering gradually to the base of the frond; stipe very short, nearly lacking, naked; fertile fronds arising in mid-summer, six to twelve inches tall, tough, becoming woody and brown in fall, segments tightly rolled to protect the sori within, with the spores to be released early the following spring. Wooded river bottomlands and swamps in neutral to alkaline muck. Abundant. Northern North America; Europe, Asia.

MAXONIA

Climbing Wood-Fern

Rhizome rooted in ground and climbing trees, stout, scaly; fronds bipinnate to tripinnate, dimorphic; fertile fronds with much narrower segments than the sterile; sori round; indusium nearly round, appearing umbrella-shaped. One species, of the Caribbean region. (Fig. 217) .... CLIMBING WOOD-FERN, *Maxonia apiifolia* (Sw.) C.Chr.

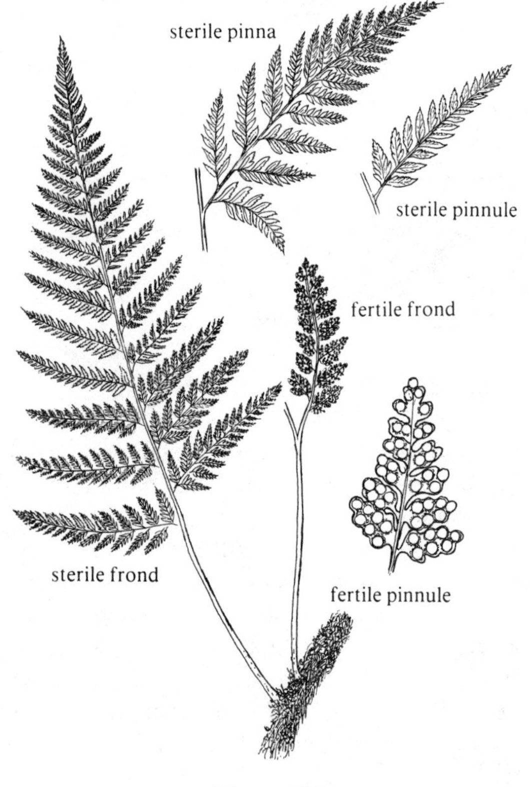

Figure 217

Figure 217 *Maxonia apiifolia*

Rhizome stout, densely brown-scaly, climbing trees; fronds large, spaced on the rhizome; one and one-half to three feet long, ten to twenty inches broad, bipinnate to tripinnate; sterile segments oblong, toothed; fertile segments just broad enough to carry the sori; sori round; indusium appearing round but is kidney-shaped with overlapping bases. Moist woods. Very rare, possibly escaped from cultivation. Florida (Dade County); West Indies, northern South America.

NEPHROLEPIS

Sword Fern

Rhizome ascending or erect, scaly; with long, thread-like stolons producing young plants along their length; fronds medium to large size; pinnate; veins free; sori medial, round to semicircular; indusium present. Thirty species of tropical regions.

The species of this genus are often difficult to distinguish, and a great deal of study is needed to resolve the problems. Mr. Clifton E. Nauman, who is working on this group in Florida, has kindly provided the following key.

1a Upper surface of midveins of central pinnae moderately to densely covered with short, erect hairs (often also with scales). 2

1b Upper surface of midveins of central pinnae not clothed with short, erect hairs (sometimes with a few scales). 3

2a Basal portions of mature stipes clothed with dark brown, appressed scales with pale margins. (Fig. 218). .... ASIAN SWORD-FERN, *Nephrolepis multiflora* (Roxb.) Jarrett ex Morton

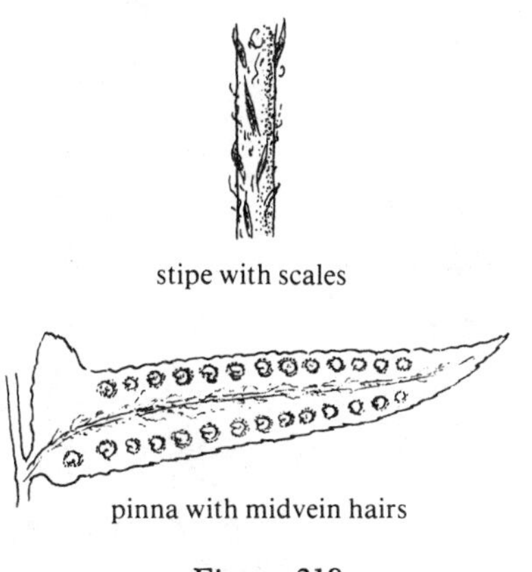

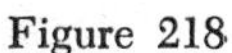
Figure 218

Figure 218 *Nephrolepis multiflora*

Fronds one to five feet tall, two and one-half to five inches wide; pinnae usually auricled, margin smooth; sori round to kidney-shaped. Terrestrial or epiphytic. Frequent. Naturalized in southern Florida, West Indies, and Mexico; native of Asia.

Due to its great variability, this species may be confused with *N. biserrata* and *N. exaltata,* but the short, erect hairs of the pinna midveins and the dark, appressed stipe scales readily distinguish it.

2b Basal portions of mature stipes not clothed with dark brown, appressed scales, but often with a few loose reddish-brown scales, or naked. (Fig. 219) .... GIANT SWORD-FERN, *Nephrolepis biserrata* (Sw.) Schott

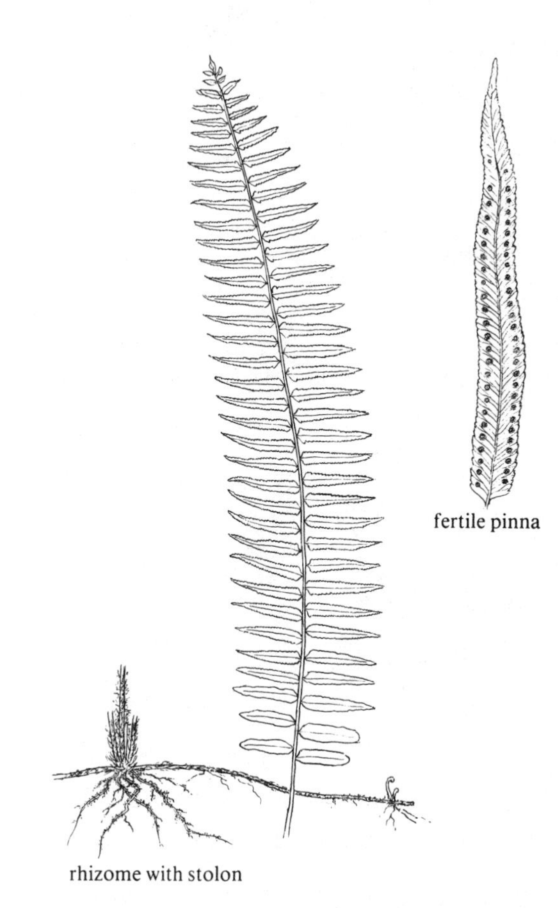

Figure 219

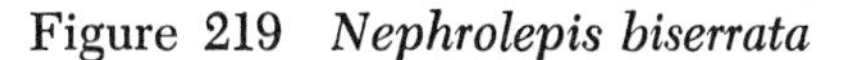
Figure 219 *Nephrolepis biserrata*

Fronds three to seven feet tall, six to twelve inches wide; pinnae not strongly auricled, margin wavy; sori round or nearly so. Terrestrial in moist woods. Frequent. South Florida; pantropical.

3a Pinnae often slightly curving upward, with slightly to strongly pointed tips; stolons never bearing tubers. (Fig. 220) .... SWORD FERN, *Nephrolepis exaltata* (L.) Schott

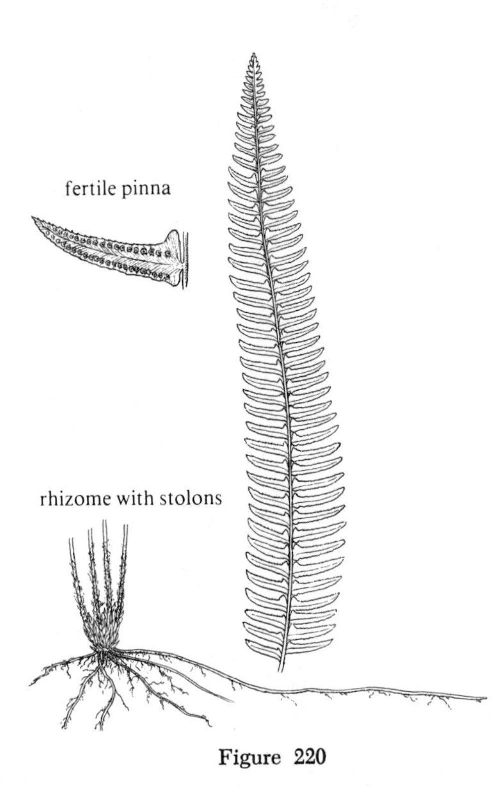

Figure 220

Figure 220 *Nephrolepis exaltata*

Fronds one to seven feet tall, two and one-half to five inches wide; pinnae strongly auricled, margin smooth to slightly toothed; sori usually kidney-shaped or semicircular. Terrestrial or epiphytic, especially on palmettos. Common. Naturalized in southern Florida; native to Old World tropics.

3b Pinnae never curving upward; their tips blunt; stolons often bearing tubers. (Fig. 221). ...... TUBEROUS SWORD-FERN, *Nephrolepis cordifolia* (L.) Presl

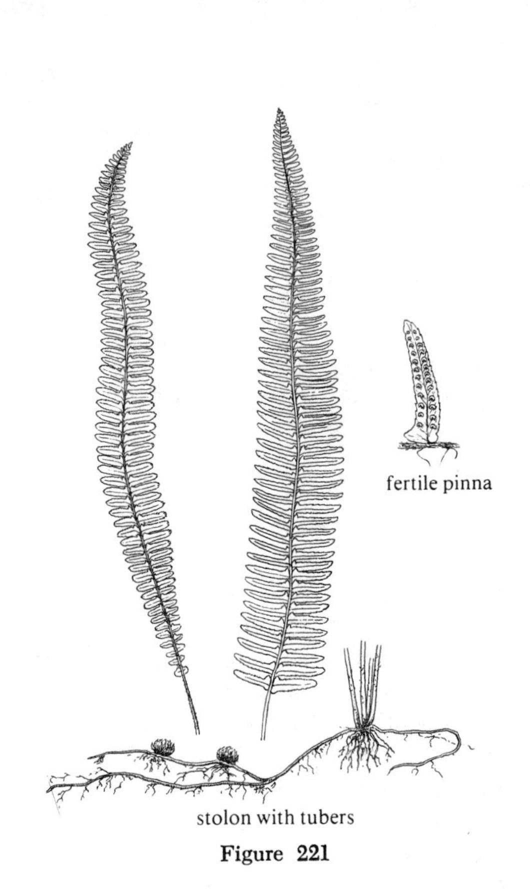

Figure 221

Figure 221 *Nephrolepis cordifolia*

Fronds ten to thirty inches tall, one to two and one-half inches wide; pinnae strongly auricled, margin smooth to slightly toothed; sori semicircular. Terrestrial or epiphytic, especially on palmettos. Frequent. Escaped from cultivation in southern Florida; native of Africa and Asia.

Nephrolepis pectinata Schott resembles *N. cordifolia* in being narrow, smallish, and epiphytic but lacks tubers and its pinnae lack auricles and usually bear white lime dots at the vein endings on the upper pinna surface; the sori are semicircular; very rare in southern Florida; but widespread in tropical America. There are no recent records of this species in Florida.

NEURODIUM

Ribbon Fern

Rhizome long-creeping on branches of trees, covered with broad scales; fronds small, undivided; veins netted; sori in a line along the margin toward the tip of the frond; indusium lacking; the leaf margin not recurved. One species. (Fig. 222) .... RIBBON FERN, *Neurodium lanceolatum* (L.) Fée [*Paltonium lanceolatum* (L.) Presl]

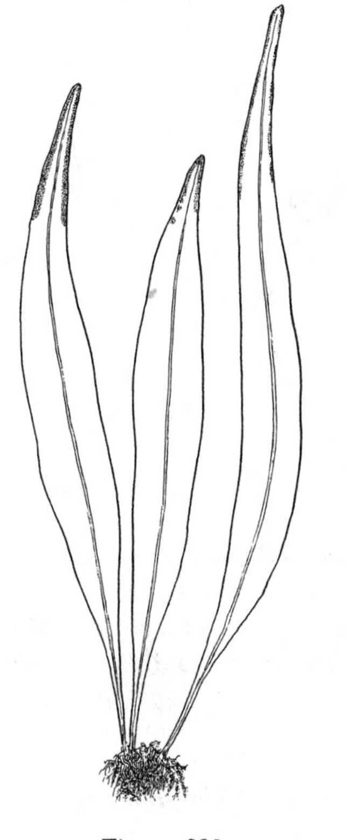

Figure 222

Figure 222 *Neurodium lanceolatum*

Fronds eight to sixteen inches long, one-half to one inch wide; stipe one to one and one-half inches long, brown; blade narrowly oval, texture firm; veins netted but not evident. Epiphyte in moist woods. Very rare. Southernmost Florida; tropical America.

ONOCLEA

Sensitive Fern

Rhizome slender, long-creeping at soil surface, scaly; fronds distant, long-stiped, net-veined; dimorphic; sterile fronds pinnatifid; fertile fronds woody, with bead-like segments, brown and persistent through the winter. One species, of wet open places. (Fig. 223) .... SENSITIVE FERN, BEAD FERN, *Onoclea sensibilis* L.

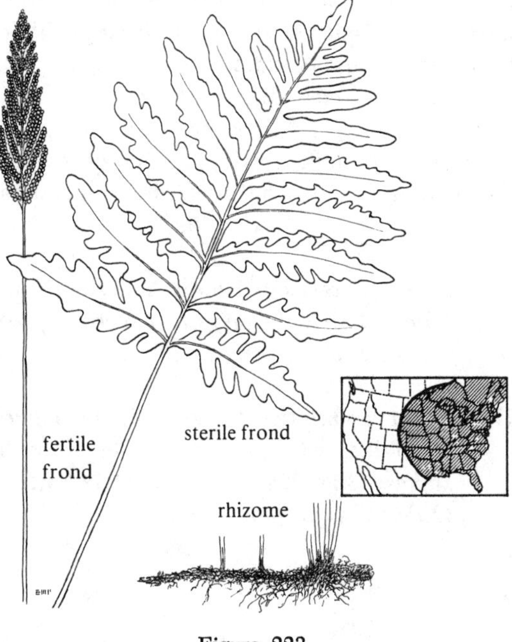

Figure 223

Figure 223 *Onoclea sensibilis*

Rhizome long-creeping, green, at soil surface; fronds spaced along the rhizome, dimorphic; sterile fronds fourteen to thirty inches tall, stipe about one-half the frond length, blade pinnatifid, the pinnae margins wavy to shallowly lobed, naked; fertile fronds arising in mid-summer, ten to sixteen inches tall, tough, becoming woody and brown in fall, segments tightly rolled, bead-like, to protect the sori within; the spores released early the following spring. Marshes, ditches, swamps, wooded or exposed. Abundant. Eastern North America.

OPHIOGLOSSUM

Adder's-Tongue

Stem subterranean, fleshy, naked, erect; fronds small, undivided to forking, naked, veins netted; sporangia large, borne on one (rarely more) fertile spike arising from the stipe. Thirty species of largely temperate regions.

Many species appear above ground only during rainy periods, especially in the spring (February through April), and the best habitat for finding terrestrial species is cemeteries.

1a **Plants terrestrial; fronds undivided, unlobed, with only one erect fertile spike. 2**

1b **Plant epiphytic; fronds deeply lobed, hand-like, with several pendant fertile spikes. (Fig. 224). .... HAND FERN, *Ophioglossum palmatum* L.**

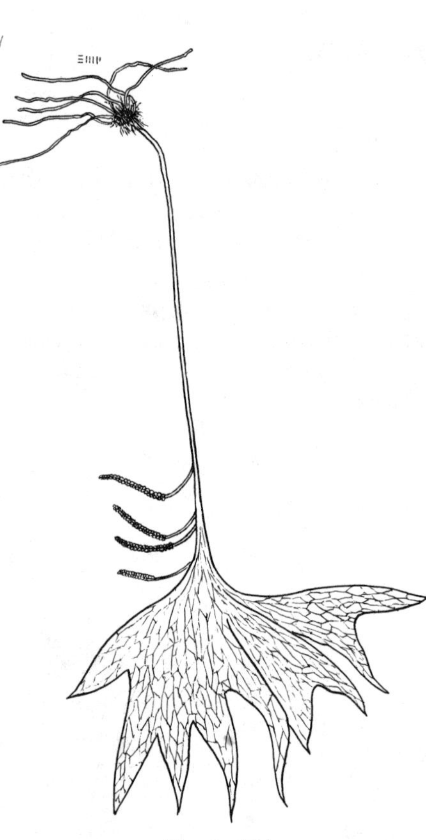

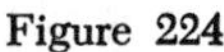
Figure 224

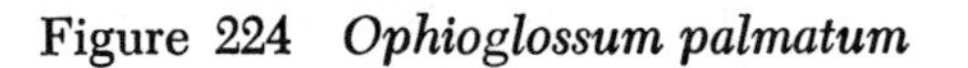
Figure 224 *Ophioglossum palmatum*

Fronds five to thirty inches long, deeply lobed, with several fertile spikes arising from the stipe near the base of the blade. Epiphytic on trunks of palmettos. Rare. Southern Florida; tropical America.

2a **Stem cylindrical, not bulbous. 3**

2b **Stem globose, bulbous. (Fig. 225) .... BULBOUS ADDER'S-TONGUE, *Ophioglossum crotalophoroides* Walt.**

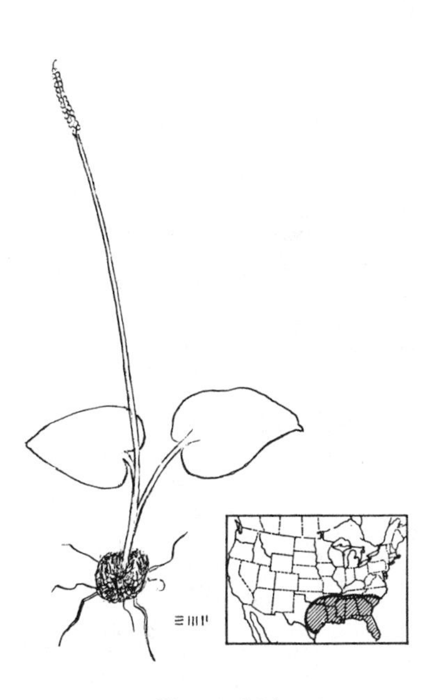

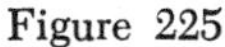

Figure 225

Figure 225 *Ophioglossum crotalophoroides*

Stem globose; fronds one to four inches tall; blade heart-shaped or broadly oval, held horizontally, about one-half inch long; fertile stalk one-half to two and one-half inches tall. Moist fields and other grassy places. Common. Southeastern United States; Middle and South America.

3a Fronds small, mostly under two inches long. 4

3b Fronds larger, more than two inches long. 5

4a Blade fleshy; California. (Fig. 226). .... CALIFORNIA ADDER'S-TONGUE, *Ophioglossum lusitanicum* L.

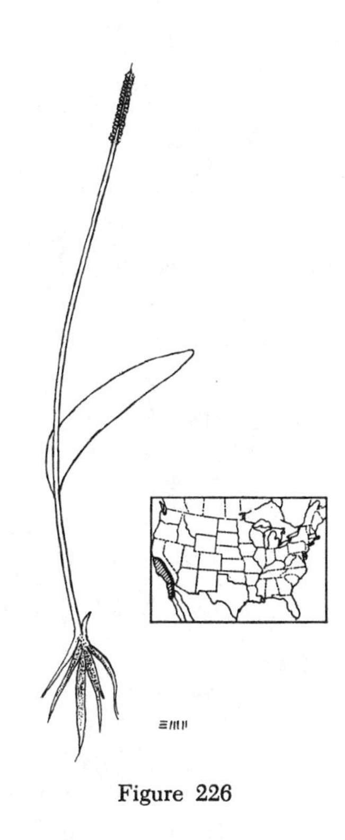

Figure 226

Figure 226 *Ophioglossum lusitanicum*

Stem cylindrical; fronds one to five inches long; blade oblong or lance-shaped, one-half to two inches long, fleshy; fertile stalk one-half to two and one-half inches long, not far exceeding the blade. Moist sandy or grassy places. Rare. California; found sparingly on all continents.

Our material is subspecies *californicum* (Prantl) Clausen [*O. californicum* Prantl].

4b Blade thin; southeastern (west to Arizona). (Fig. 227) .... LEAST ADDER'S-TONGUE, *Ophioglossum nudicaule* L. f.

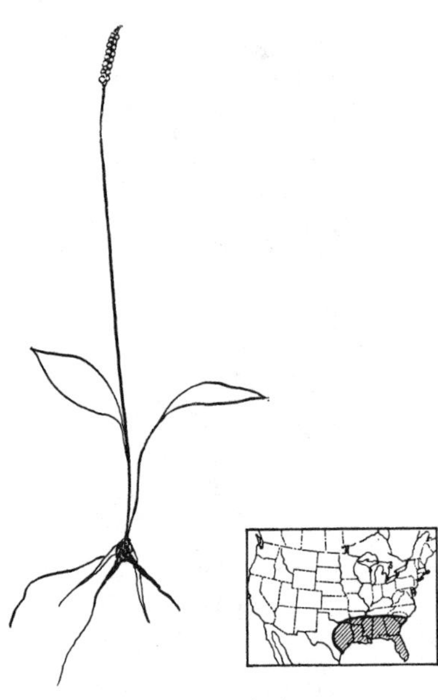

Figure 227

Figure 227 *Ophioglossum nudicaule*

Stem subglobose; fronds one and one-half to five inches tall; blade oval or oblong, one-fourth to one and one-half inch long; fertile stalk four to eight times longer than the sterile blade. Grassy areas. Frequent during spring rains. Southeastern United States; tropical America, Africa, Asia, Australia.

Sometimes a larger form, *Ophioglossum dendroneuron* E. St. John (of Florida), is distinguished by having more complex venation and a pronounced midvein at the base of the blade whereas in *O. nudicaule* the midvein is not evident.

5a **Veins forming only one set of areoles, with free veins inside. 6**

5b **Veins forming two sets of areoles—fine areoles bounded by larger areoles. (Fig. 228). .... ENGELMANN'S ADDER'S-TONGUE, *Ophioglossum engelmannii* Prantl**

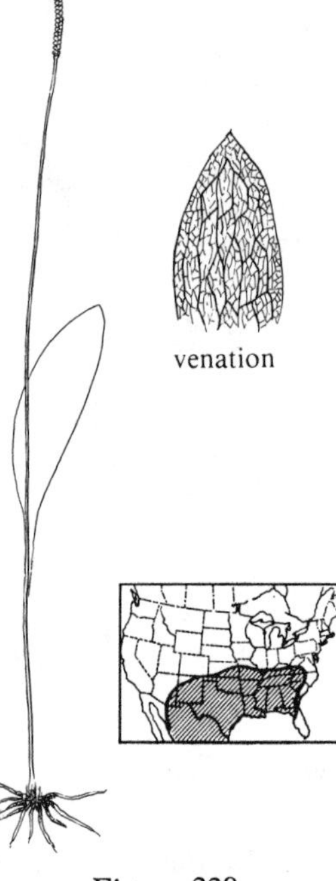

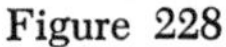

Figure 228

Figure 228 *Ophioglossum engelmannii*

Stem cylindrical; fronds two to eight inches tall; blade oblong, slightly pointed at both ends, one-half to four and one-half inches long; veins forming large areoles which contain smaller networks within; fertile stalk one to five inches tall. Wet depressions in limestone regions. Locally common to abundant. Southern United States; Mexico.

6a **Blades lance-shaped; veins forming few areoles, without included veinlets. (Fig. 229). .... STALKED ADDER'S-TONGUE, *Ophioglossum petiolatum* Hooker**

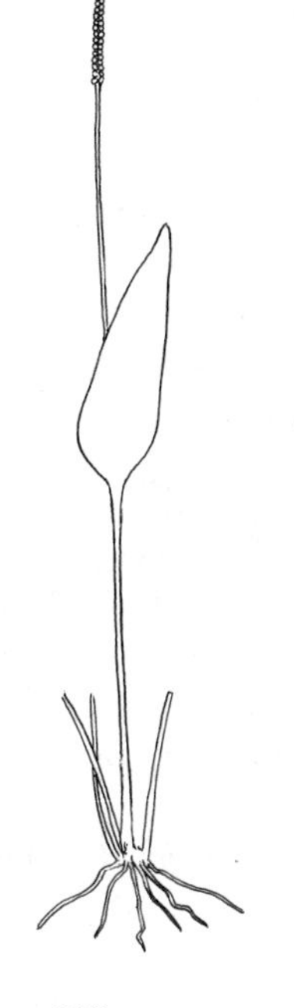

Figure 229

Figure 229 *Ophioglossum petiolatum*

Stem cylindrical; fronds two to eight inches tall; blade lance-shaped, one to two and one-half inches long; fertile stalk one to three and one-half inches long. Common. Moist meadows. Florida and Arkansas; pantropical.

This is one species of adder's-tongue that is easily cultivated and can become a greenhouse weed.

6b Blades oval or oblong; veins forming many areoles with included veinlets. (Fig. 230). .... ADDER'S-TONGUE, *Ophioglossum vulgatum* L.

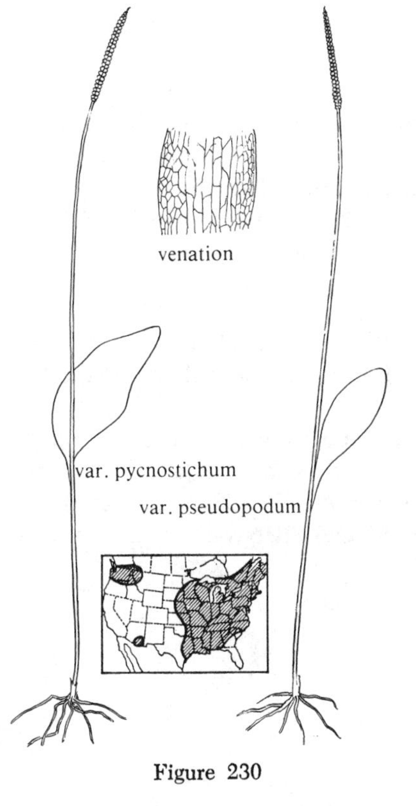

Figure 230

Figure 230 *Ophioglossum vulgatum*

Stem cylindrical; fronds three to twelve inches tall; blade one to four inches long; fertile stalk one to seven inches long. Moist grassy areas and open woods. Frequent but commonly overlooked. Much of North America; Mexico, Europe, Asia.

The SOUTHERN ADDER'S-TONGUE, *O. vulgatum* var. *pycnostichum* Fern., has an oval sterile blade and occurs largely in the South and West, but occurs as far north as Michigan. The NORTHERN ADDER'S-TONGUE, var. *pseudopodum* (Blake) Farw., on the other hand, has an oblong sterile blade and occurs in the North, south to Virginia in the East and Washington in the West. Present evidence suggests that they may actually be distinct species.

OSMUNDA

Rhizome stout, short-creeping, hairy; fronds large, pinnate-pinnatifid to bipinnate, hairy or naked; veins free; sporangia borne on highly reduced pinnae that lack leafy tissue. Fourteen species of temperate and subtropical regions, ours occurring mostly in marshes and swamps.

1a **Blade pinnate-pinnatifid; sporangia borne elsewhere than just the tip. 2**

1b **Blade fully bipinnate; sporangia borne on reduced pinnae at top of vegetative fronds. (Fig. 231). .... ROYAL FERN, FLOWERING FERN, *Osmunda regalis* L.**

Figure 231

Figure 231 *Osmunda regalis*

Fronds two to five feet tall, ten to twenty inches wide; stipe half the frond length; blade bipinnate, the segments oblong, spaced apart on pinna rachises, naked except for a few scattered hairs on the rachises; pinnae dimorphic; fertile pinnae lacking leaf tissue, located at tip of frond. Swamps and other wet sites. Common. Eastern North America; tropical America, Europe.

Osmunda × *ruggii* R. Tryon is a spectacular hybrid of *O. claytoniana* and *O. regalis.* Exceedingly rare.

2a **Blades completely fertile or completely sterile; sterile blade with a tuft of rusty hairs at the base of each pinna; stipe and rachis with abundant cinnamon hairs, especially in spring. (Fig. 232). .... CINNAMON FERN, *Osmunda cinnamomea* L.**

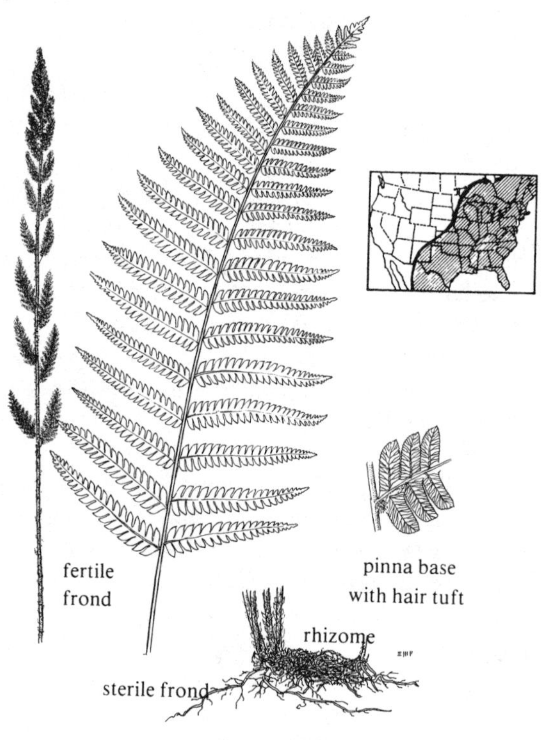

Figure 232

Figure 232 *Osmunda cinnamomea*

Fronds dimorphic, sterile leaves two and one-half to four feet tall, five to ten inches wide; stipe one-fourth to one-third the frond length; blade pinnate-pinnatifid, with dense tuft of rusty hairs beneath the base of each pinna; stipe and rachis also hairy; fertile fronds totally lacking leafy tissue, arising in late spring and collapsing by mid-summer. Swamps and other wet areas. Abundant. Eastern North America; Mexico, West Indies.

2b Blades normally fertile only in the middle of the frond; sterile pinnae lacking dense tuft of hairs at base. (Fig. 233). .... INTERRUPTED FERN, *Osmunda claytoniana* L.

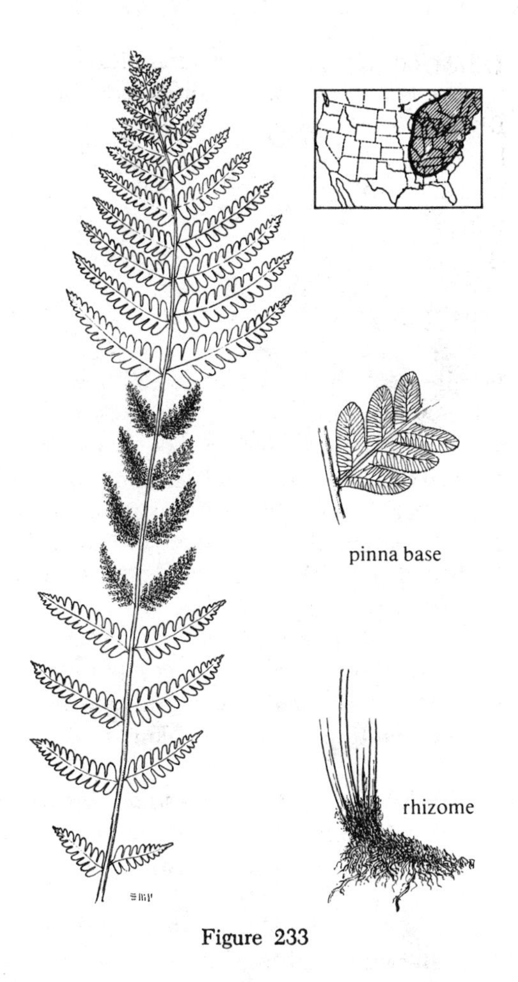

Figure 233

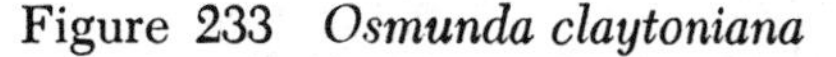

Figure 233 *Osmunda claytoniana*

Fronds two to three and one-half feet tall, six to twelve inches wide; stipe about one-third the frond length; pinnae dimorphic, sterile pinnae pinnatifid, naked or with a few scattered hairs at the base, no dense tuft of hairs; fertile pinnae in the middle of the frond, dark brown, lacking leafy tissue. Moist woods, more open areas in northern part of range. Common. Eastern North America; eastern Asia.

This may be confused with the cinnamon fern, especially the sterile fronds. The interrupted fern generally prefers somewhat drier sites, i.e., moist woods and not in or by the water; it also has broader segments and has a lighter yellow-green color, in addition to lacking the dense tufts of cinnamon hairs beneath.

PELLAEA

Cliff Brake

Rhizome short- to long-creeping; scales linear, usually toothed, often bicolorous—brown with a heavy black central streak; fronds medium-sized, pinnate to tripinnate, naked or with only a few scattered hairs; sori marginal or rarely elongate along the veins in one species, the leaf margin recurved to protect the sporangia. Eighty species of temperate and subtropical dry regions.

1a Stipe and rachis yellow or tan; pinnules round-tipped, oval to oblong. 2

1b Stipe and rachis chestnut brown to black; pinnules usually pointed, elongate. 4

2a Rachis straight. 3

2b Rachis zig-zag, the pinnae often pointing slightly downward. (Fig. 234) .... *Pellaea ovata* (Desv.) Weath.

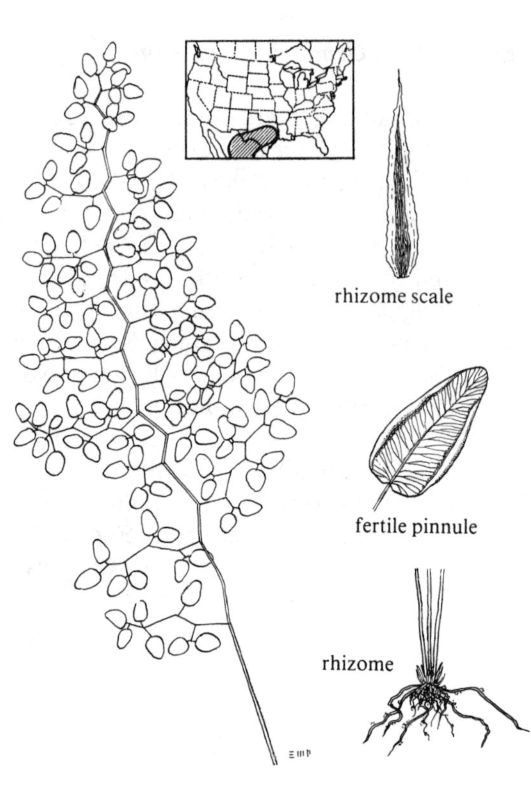

Figure 234

Figure 234 *Pellaea ovata*

Rhizome slender, short-creeping; fronds close; older rhizome scales bicolorous; apical scales mostly brown or tan; fronds eight to twenty-four inches tall; stipe and rachis hairy; stipe yellow, round; blade tripinnate, the pinnae reflexed downward; pinnules oval, their bases rounded or slightly heart-shaped. On noncalcareous soil or rocks. Frequent. Texas; widespread in tropical America.

3a Veins visible; stipe and rachis naked, with a slight whitish bloom. (Fig. 235). .... COFFEE FERN, *Pellaea andromedifolia* (Kaulf.) Fée

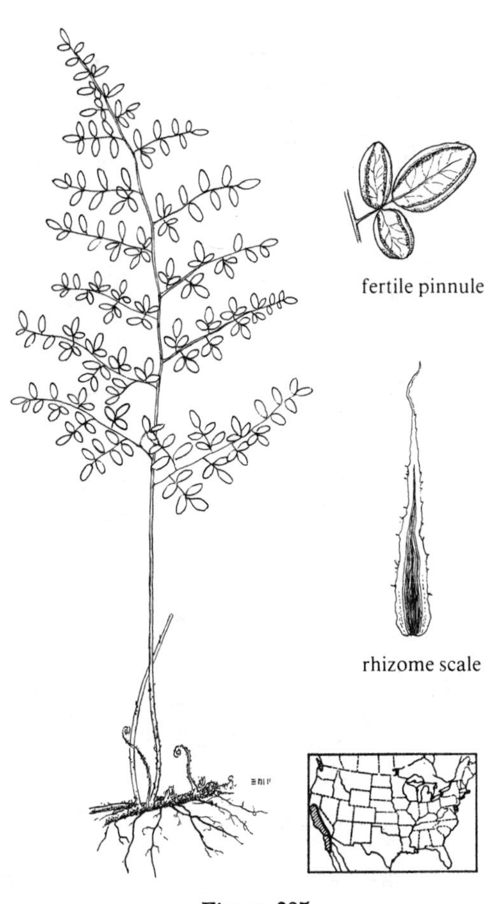

Figure 235

Figure 235 *Pellaea andromedifolia*

Rhizome slender, mostly long-creeping, the fronds distant; scales bicolorous; fronds four to twenty-four inches tall; stipe round, yellow, naked (or rarely short-hairy); blade mostly tripinnate (rarely quadripinnate at base), triangular; pinnules oval, their margins smooth or lobed, their bases rounded. On noncalcareous rocks. Common. California; Baja California.

Pellaea cardiomorpha Weath. [*P. sagittata* (Cav.) Link var. *cordata* (Cav.) A. Tryon] is common in Mexico and barely gets into the United States in the Big Bend region of Texas. It resembles *P. andromedaefolia* but has a stout, compact, short-creeping rhizome with the fronds clumped; the scales on the stipe base are broad and chaffy, and the pinnules are heart-shaped.

3b Veins obscure; stipe and rachis hairy. (Fig. 236). .... *Pellaea intermedia* Mett. ex Kuhn

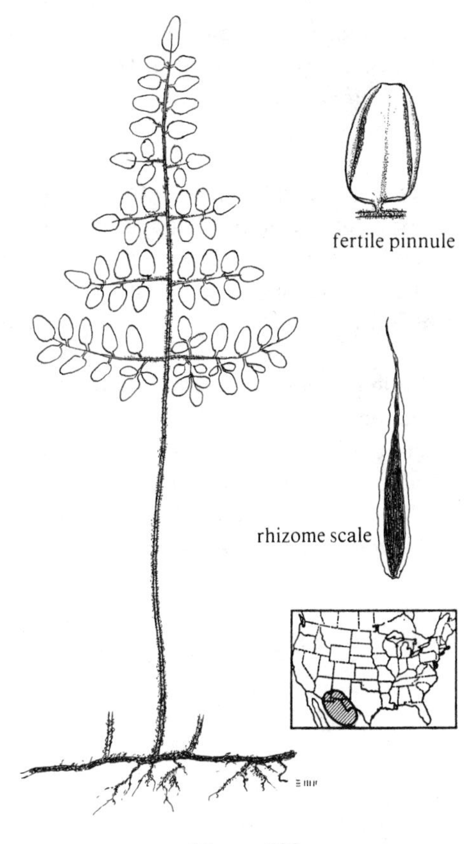

Figure 236

Figure 236 *Pellaea intermedia*

Rhizome long-creeping, cord-like, the fronds distant; scales bicolorous; fronds four to twenty-four inches tall; stipe round, yellow to brown, short-hairy; blade triangular, pinnate to tripinnate, mostly bipinnate; segments oval or oblong, stalked. Calcareous or noncalcareous rocks and slopes. Frequent. Southwestern United States; northern Mexico.

4a **Rhizome scales uniformly rust-brown or slightly bicolorous with age in *P. bridgesii;* stipe and rachis round (or flat on top side in some). 5**

4b **Rhizome scales bicolorous; stipe and rachis grooved on top side (flat in *P. ternifolia*). 8**

5a **Blade pinnate to bipinnate, at least the basal pinnae deeply lobed or pinnate; margin generally at least slightly modified and recurved; segments not folded in half. 6**

5b **Blade pinnate, pinnae not divided, heart-shaped; margins not modified and not recurved but segments often folded in half; sori elongate along veins. (Fig. 237) .... BRIDGES' CLIFF-BRAKE, *Pellaea bridgesii* Hooker**

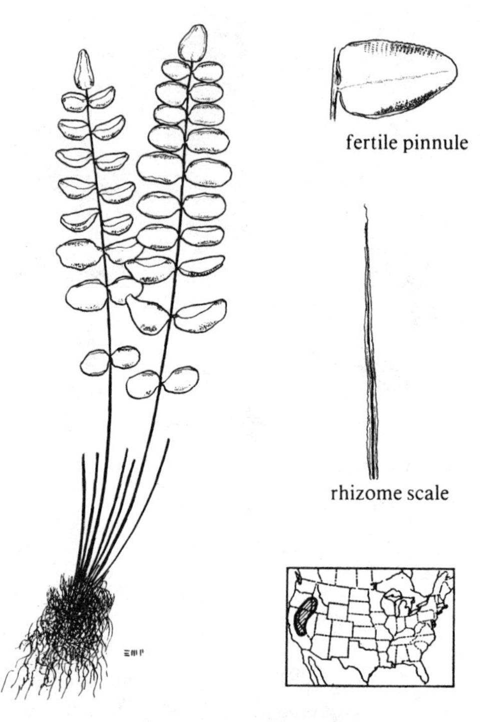

Figure 237

Figure 237 *Pellaea bridgesii*

Rhizome short-creeping; scales bicolorous with age; fronds clumped, to fourteen inches tall; stipe chestnut brown; blade very narrow-oblong, pinnate; pinnae round to oblong, not staled; fertile pinnae often folded in half; sori extending slightly along the veins. Rocky slopes at high elevations. Rare. Western United States.

6a **Stipes not articulated, lacking horizontal lines or grooves; stipe and rachis purplish to black; rachis dark to the tip. 7**

6b **Stipes articulated, with horizontal lines or grooves; stipe and rachis chestnut brown; rachis green toward the tip. (Fig.**

238). .... **BREWER'S CLIFF-BRAKE,**
***Pellaea breweri* D. C. Eaton**

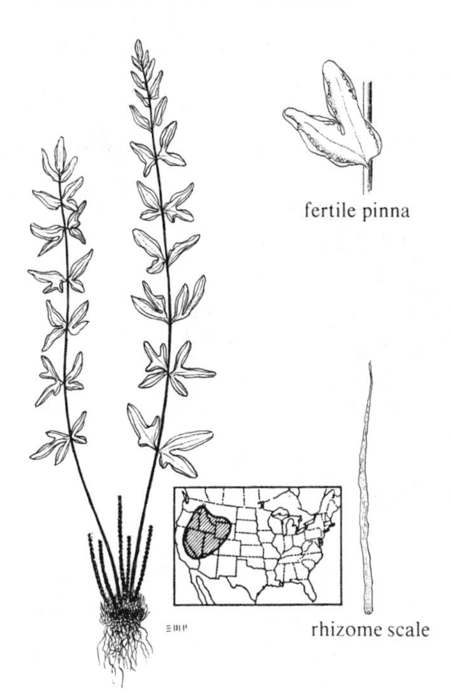

Figure 238

Figure 238 *Pellaea breweri*

Rhizome ascending; scales rust-brown, concolorous, their margins wavy; fronds clumped, two to six inches tall; stipes chestnut brown, articulated, breaking off along horizontal lines, round, with a few sparse hairs; blade pinnate to pinnate-pinnatifid; pinnae smooth-margined, deeply two-lobed, naked. Calcareous and noncalcareous rocks. Rare. Western United States.

7a **Stipe and rachis moderately hairy; rhizome scales tan at the tip of the rhizome; fronds somewhat dimorphic. (Fig. 239). .... PURPLE CLIFF-BRAKE, *Pellaea atropurpurea* (L.) Link**

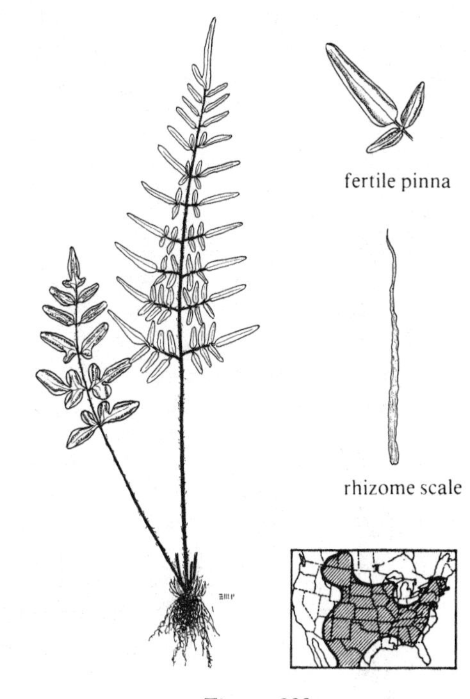

Figure 239

Figure 239 *Pellaea atropurpurea*

Rhizome short-creeping; scales rust-brown when mature, tan at tip of rhizome, concolorous; fronds to twenty inches tall; stipe round, hairy, dark purple to black; blade bipinnate to nearly tripinnate, triangular; pinnae with two to five pairs of pinnules. Calcareous rocks. Common. Much of North America; south to Guatemala.

7b **Stipe and rachis naked or only sparsely hairy; rhizome scales uniformly rust-brown; fronds not at all dimorphic. (Fig. 240). .... SMOOTH CLIFF-BRAKE, *Pellaea glabella* Mett. ex Kuhn**

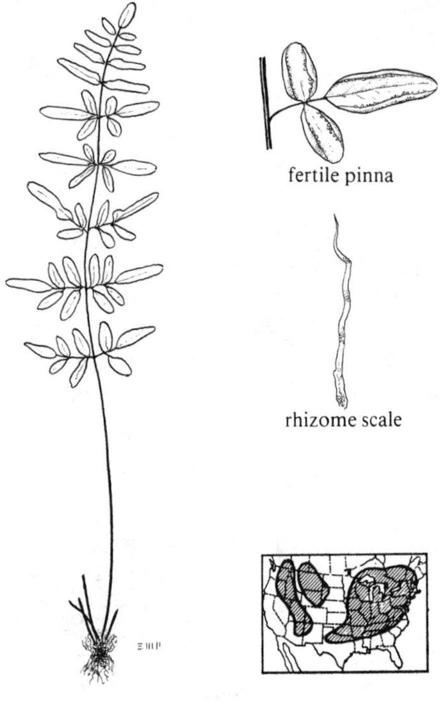

Figure 240

Figure 240 *Pellaea glabella*

Rhizome short-creeping or ascending; scales rust-brown, concolorous, their margins wavy to toothed; fronds to fourteen inches tall; stipes round, naked or with a few hairs, chestnut brown (occasionally with lines of articulation); blade pinnate to bipinnate, one to three pairs of lobes or pinnules, narrowly oblong. Calcareous, rarely noncalcareous rocks. Frequent. Much of North America.

There are three rather distinct varieties of this species with distinct geographical ranges. Their differences have not been tested experimentally, and they might be environmentally induced. Var. *glabella* reaches fourteen inches in height and has pinnae with one pair of pinnules; it occurs in the East. Var. *simplex* Butters is up to eight inches tall, has pinnae barely lobed, and has the westernmost range. Var. *occidentalis* (E. Nelson) Butters reaches only six inches in height, has unlobed pinnae, and occupies the central distribution.

8a Pinnae pinnate. 9

8b Pinnae with three unstalked pinnules, or undivided. (Fig. 241). .... TERNATE CLIFF-BRAKE, *Pellaea ternifolia* (Cav.) Link

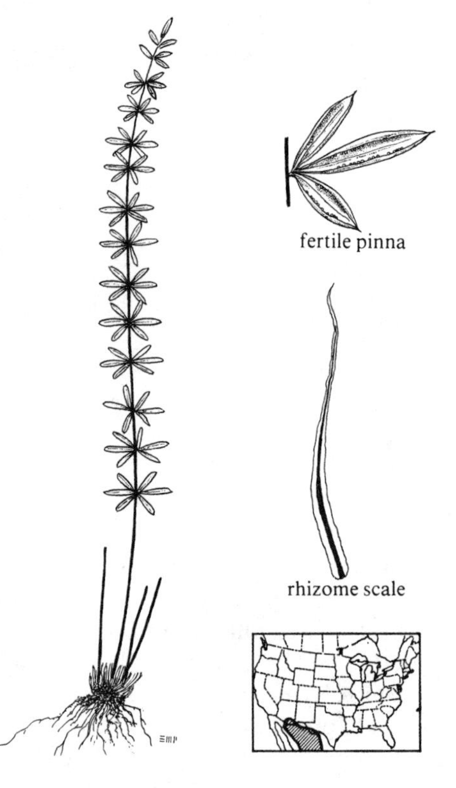

Figure 241

Figure 241 *Pellaea ternifolia*

Rhizome stout, short-creeping; scales bicolorous; fronds four to eighteen inches tall; stipe round to flat on top, brown to black, no articulation lines; blade narrowly oblong, bipinnate near base, pinnate above; pinna rachis very short, each pinna consisting of three unstalked pinnules. Noncalcareous rocks. Frequent. Southwestern United States; tropical America.

9a **Pinnules elongate, about three times as long as broad, distant from each other; pinna rachis longer than the segments, with four to many pairs of pinnules. 10**

9b **Pinnules linear, ten to fifteen times as long as broad, crowded; the pinna rachis shorter than the segments; two to six pairs of pinnules per pinna. (Fig. 242) .... SIERRA CLIFF-BRAKE, *Pellaea brachyptera* (Moore) Baker**

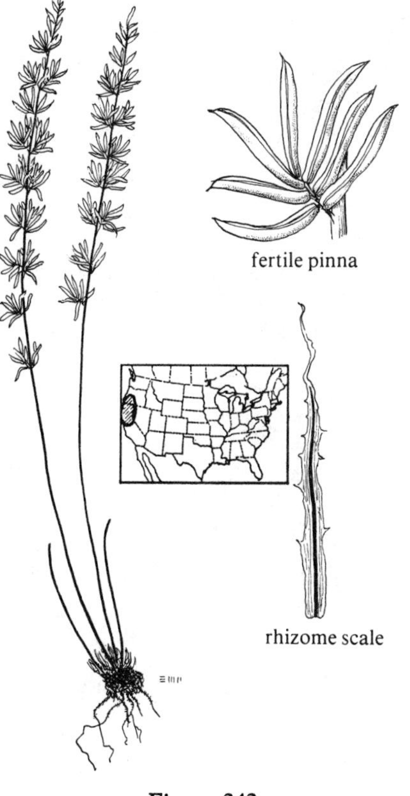

Figure 242

Figure 242 *Pellaea brachyptera*

Rhizome short-creeping; scales bicolorous; fronds four to fifteen inches tall; stipe grooved, chestnut brown, no articulation lines; blade very narrowly oblong, bipinnate; pinna rachis short, the two to six pairs of pinnules crowded, longer than the pinna rachis; segments linear, one-half to five-eighths inch long, their margins strongly revolute, the tip sharp-pointed. Basalt or serpentine rocks. Frequent. Western United States.

10a **Pinnules six to eight pairs per pinna, spaced well apart; pinna rachis usually well over one inch long; sori with pale yellow wax. 11**

10b **Pinnules one to five pairs per pinna; pinna rachis up to one inch long; sori lacking any wax. (Fig. 243). .... WRIGHT'S CLIFF-BRAKE, *Pellaea wrightiana* Hooker**

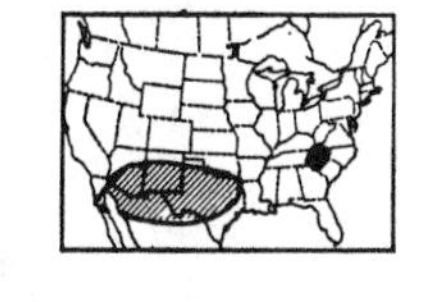

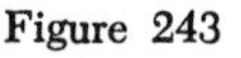

Figure 243

Figure 243 *Pellaea wrightiana*

Rhizome short-creeping; scales bicolorous; fronds four to fifteen inches tall; stipe grooved, chestnut brown; blade narrowly triangular, bipinnate; upper pinnae divided from their bases into three segments, lower pinnae with two to four pairs of pinnules on elongated pinna rachis. Among rocks. Frequent. Southwestern United States and North Carolina; northern Mexico.

This species probably originated as a hybrid between *P. ternifolia* and *P. truncata,* and subsequently became fertile.

11a Blade bipinnate; pinnules not folded in half; southwestern United States except California. (Fig. 244). .... SPINY CLIFF-BRAKE, *Pellaea truncata* Goodd. [*P. longimucronata* of many authors, not of Hooker, which = *P. mucronata*]

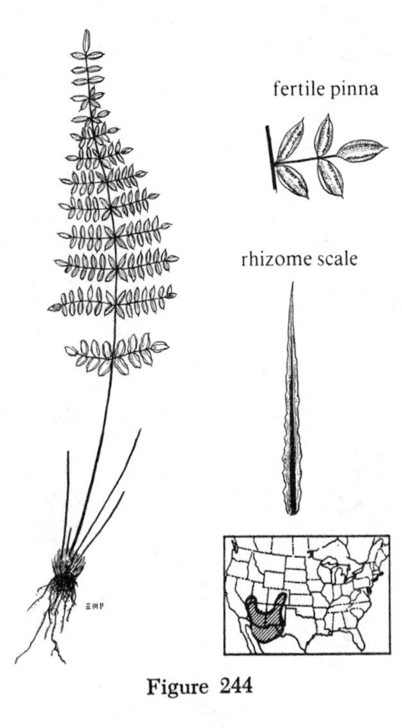

Figure 244

Figure 244 *Pellaea truncata*

Rhizome stout, short-creeping; scales bicolorous; fronds five to fifteen inches tall; stipe grooved, naked, chestnut brown, without articulation lines; blade triangular, bipinnate, rarely tripinnate; pinnae with up to ten pairs of pinnules; segments slender, sharp-pointed. Noncalcareous rocks. Common. Southwestern United States except California; northern Mexico.

11b Blade tripinnate, or bipinnate, the pinnules usually folded in half; California. (Fig. 245). .... BIRD'S-FOOT FERN, *Pellaea mucronata* (D. C. Eaton) D. C. Eaton

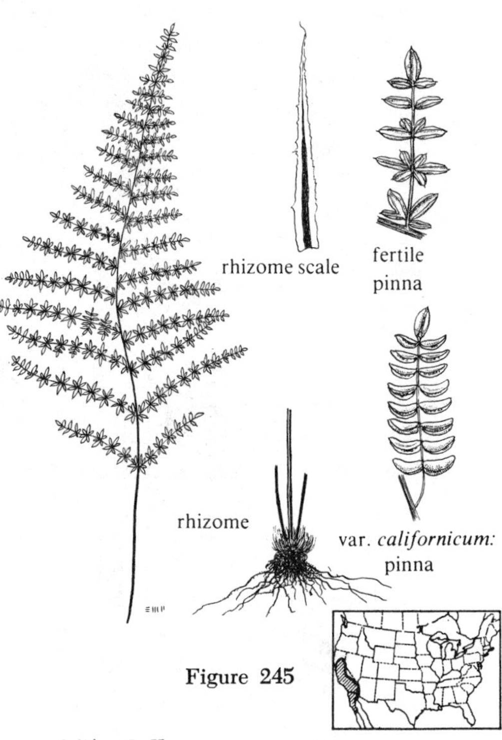

Figure 245

Figure 245 *Pellaea mucronata*

Rhizome short-creeping; scales bicolorous; fronds three to sixteen inches tall; stipe grooved, naked, chestnut brown, without articulation lines; blade bipinnate to tripinnate; segments often folded in half. Noncalcareous rocks. Frequent. California; Baja California.

Variety *mucronata* is mostly or partly tripinnate, the pinnules often being divided into threes, pinnae at more or less right angles to the rachis, whereas var. *californica* (Lemmon) Munz & Johnston [*P. compacta* (Davenp.) Maxon] is entirely bipinnate, the pinnules not divided; the pinnae often folded in half and curving slightly upward.

PHANEROPHLEBIA

Holly Fern

Rhizome ascending, scales tan; leaves clumped; blade once pinnate, leathery; sori scattered on under surface, round, with an umbrella-shaped indusium. Eight species of the New World, largely of Mexico. Closely related to *Polystichum* and the commonly cultivated Old World *Cyrtomium*, the Japanese holly fern.

1a Pinnae strongly auriculate, three to twelve pairs, often with an incised margin; rachis strongly scaly; indusium not raised in the center. (Fig. 246). .... EARED HOLLY FERN, *Phanerophlebia auriculata* Underw.

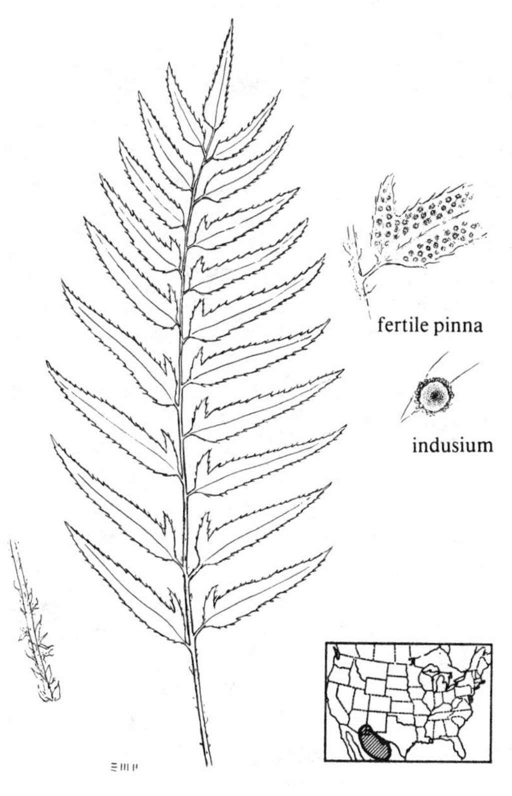

Figure 246

Figure 246 *Phanerophlebia auriculata*

Fronds six to twenty-seven inches tall; pinnae three to twelve pairs, each broadly wedge-shaped, usually with a large auricle on the upper side of the base, the margins finely toothed with bristle tips or sometimes incised; indusium delicate, often soon disintegrating. Damp canyon walls and cliffs. Frequnt. Southwestern United States; northern Mexico.

1b Pinnae not auriculate, twelve to eighteen pairs, margin finely bristle-toothed but not incised; rachis not strongly scaly; indusium with a raised central knob (umbo), persistent. (Fig. 247). .... UMBONATE HOLLY FERN, *Phanerophlebia umbonata* Underw.

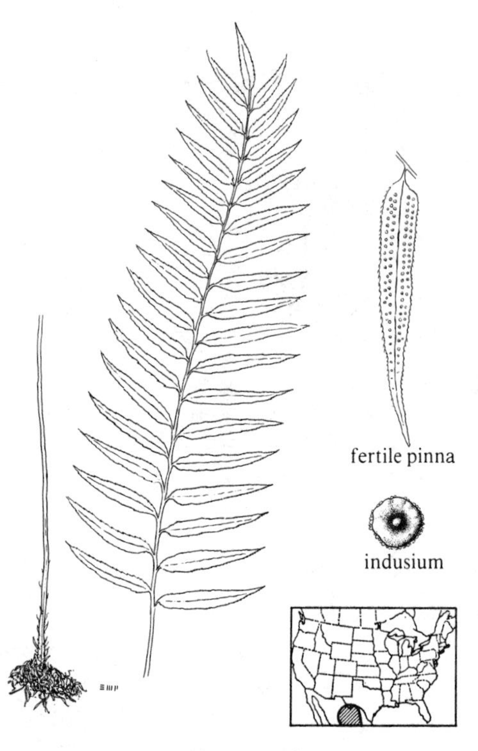

Figure 247

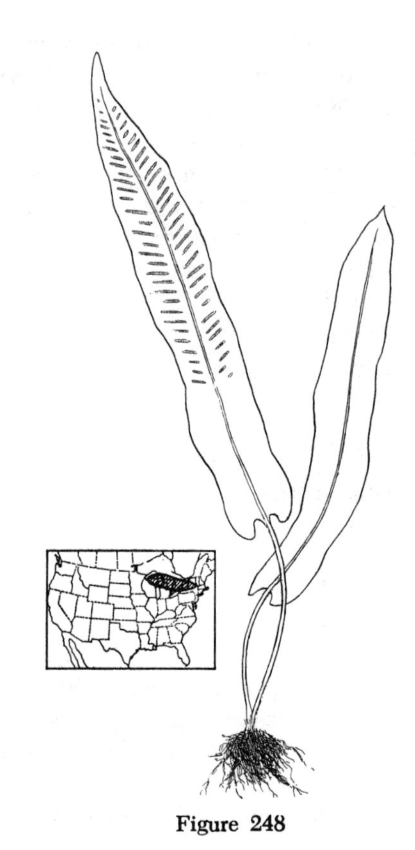

Figure 248

Figure 247 *Phanerophlebia umbonata*

Fronds ten to thirty inches tall; pinnae twelve to eighteen pairs, never auricled or incised, finely toothed margin with bristle teeth; indusium firm, persistent, and with a central depression and raised central projection or umbo. Damp canyon walls and cliffs. Frequent. Texas; northern Mexico.

PHYLLITIS

Hart's-Tongue

Rhizome ascending, scaly; fronds clumped, simple, strap-shaped, leathery, auricled at the base; sori linear along the veins, in pairs facing each other. Two species of temperate regions. (Fig. 248) .... HART'S-TONGUE, *Phyllitis scolopendrium* (L.) Newm.

Figure 248 *Phyllitis scolopendrium*

Fronds eight to sixteen inches long, one to two inches wide, leathery texture, base strongly auricled (eared). Cool, moist, shaded, dolomitic limestone. Rare. Northeastern North America and Tennessee; Europe, Asia.

The North American plants are considered a distinct variety, var. *americana* Fernald, from the European and Asian material, which is much more abundant than ours is. The Eurasian material is diploid and ours is a tetraploid.

PILULARIA

Pillwort

Rhizome naked, slender, widely creeping, forming colonies; fronds spaced apart, consisting only of the leaf stalk, lacking any leaf blade, i.e., no leaflets at all; sporangia borne in round, leathery or stony sporocarps formed at the base of the leaf stalks; heterosporous. Six species of warm temperate regions. (Fig. 249) .... AMERICAN PILLWORT, *Pilularia americana* A. Braun

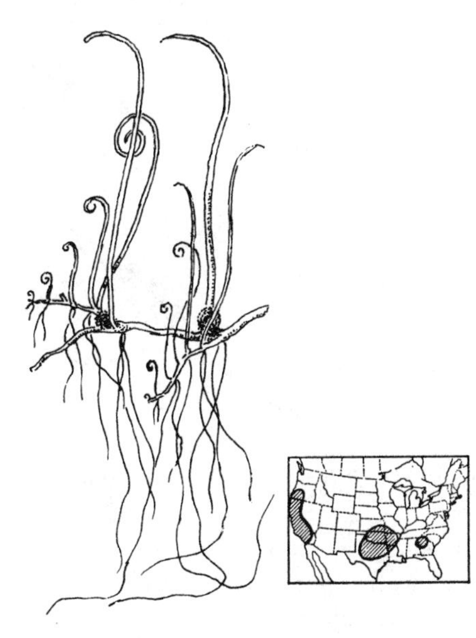

Figure 249

Figure 249 *Pilularia americana*

Leaves one and one-half to five inches tall; sporocarps one-sixteenth to one-eighth inch long. Wet soil by ponds or in seasonal pools. Uncommon. Georgia, Arkansas, Texas, Oklahoma, Kansas, California, Oregon. The peculiar distribution pattern is unexplained.

PITYROGRAMMA

Goldback Fern and Silverback Fern

Rhizome ascending, short, scaly; fronds clumped; stipe dark brown to black; blades pinnate to tripinnate, covered beneath with a white or yellow wax; sporangia running along the veins, appearing as black specks of pepper among the wax; indusium absent. Seventeen species, largely of tropical regions.

1a **Blade pinnately divided, triangular, bipinnate to tripinnate, one to three feet tall; central pinnae pinnate to pinnate-pinnatifid; white waxy beneath. Moist habitats. Florida. (Fig. 250). .... SILVERBACK FERN, *Pityrogramma calomelanos* (L.) Link**

.... **GOLDBACK FERN, SILVERBACK FERN,** ***Pityrogramma triangularis*** **(Kaulf.) Maxon**

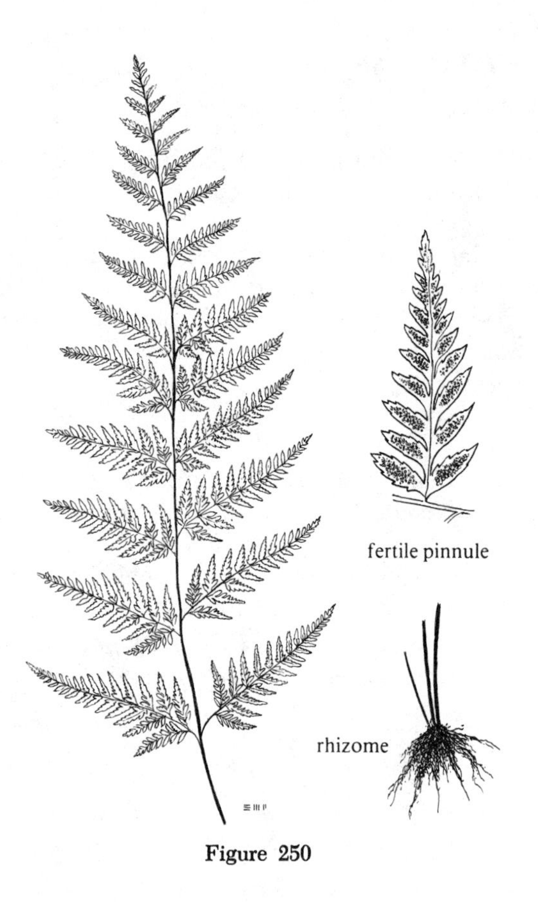

Figure 250

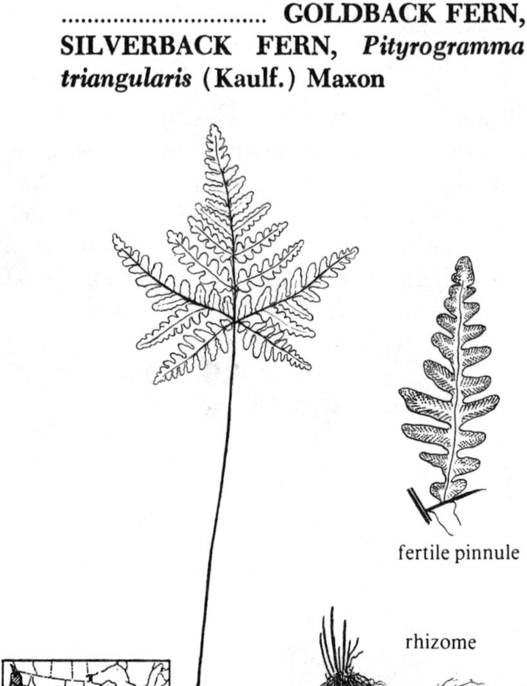

Figure 251

Figure 250 *Pityrogramma calomelanos*

Rhizome ascending; blade narrowly triangular, bipinnate to bipinnate-pinnatifid; one to three feet tall, five to ten inches wide; stipe nearly one-half the frond length; lower surface white waxy, upper surface naked. Open woods or disturbed soil. Escape from cultivation or rare native in southern Florida; common in tropical America.

1b Blade pentagonal, the lowest pinna pair large, nearly the size of the rest of the leaf; central pinnae pinnatifid; the downward pointing pinnules of the basal pinnae nearly as large as the upper pinnae; fronds four to eight inches tall; dry habitats; western. (Fig. 251). ....

Figure 251 *Pityrogramma triangularis*

Rhizome short-creeping to ascending; fronds tufted, mostly four to eight inches tall, rarely taller, two to four inches wide; stipe two-thirds the frond length; blade triangular or pentagonal, the lowest pair of pinnae more strongly developed on the lower side; lower surface covered with white or yellow wax, upper surface naked or with some wax glands; fronds often curl when dry. On or among rocks. Common. Western North America; northern Mexico.

There are four varieties recognized: var. *triangularis* has yellow wax below and is totally naked above (Canada to Baja California); the other three have glands and/or a sticky substance on the upper surface. Var. *viscosa* D. C. Eaton has the basal pinnules of the lowest pinnae undivided, not pinnatifid (southern California). Both var. *maxonii* Weath. and var. *pallida* Weath. have pinnatifid basal segments, but the former has a reddish-brown stipe like *triangularis* and *viscosa,* and occurs in southern California, Arizona, and Mexico. Var. *pallida* is the only variety with a black stipe and occurs in central California. These may actually prove to be distinct species.

POLYPODIUM

Polypody

Rhizome creeping, scaly; leaves small to medium-sized; blade pinnatifid, pinnate, or undivided; veins free or netted; sori medial, round, indusium lacking. Sori sometimes with sterile hairs (paraphyses). Nearly one thousand species, largely of tropical regions.

Botanists are divided on whether to split *Polypodium* into several small genera or maintain it as a single large genus. I have taken the latter view here but have placed in parentheses the splinter genus names.

1a Blade undivided. **2**

1b Blade divided (pinnatifid or pinnate). **4**

2a Fronds over ten inches long. **3**

2b Fronds less than six inches long. (Fig. 252). .... **VINE FERN, *Polypodium* (*Microgramma*) *heterophyllum* L.**

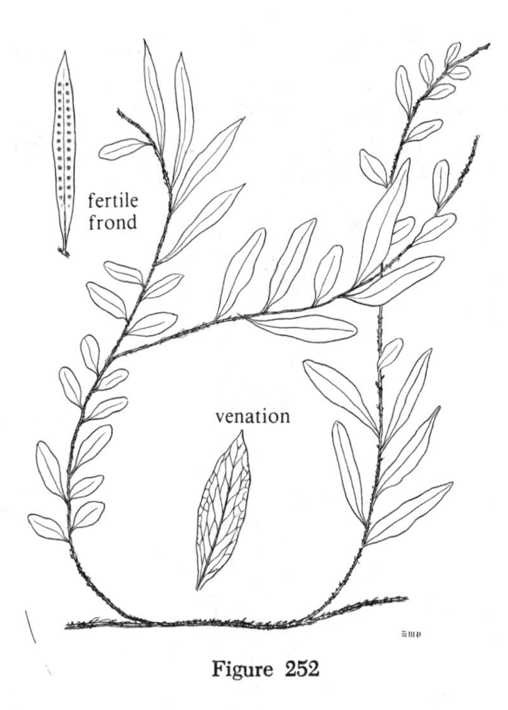

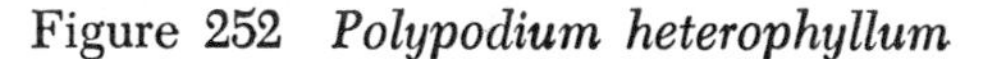

Figure 252

Figure 252 *Polypodium heterophyllum*

Rhizome long-creeping; the fronds distant, two to four inches tall; stipe short, nearly lacking; blade undivided, slender, narrow, naked, slightly wavy-margined; some fronds may be smaller and rounder on the same rhizome; veins netted; sori in a single row on each side of the midvein. Epiphyte. Rare. Southernmost Florida; West Indies.

Polypodium (*Pleopeltis*) *erythrolepis* Weath. has the same general aspect of leaf form and the long-creeping rhizome, but the leaves are covered with small, round scales, are of more of a leathery texture, and occur in southwest Texas and northern Mexico; rare.

Polypodium (*Pleopeltis*) *astrolepis* Liebm. has a naked blade but is distinguished by its elongate sori; very rare in southernmost Florida, but widespread and common in the West Indies and Middle America.

3a Blade linear, usually less than one-half inch wide; rhizome white-waxy among the scales. (Fig. 253). .... NARROW STRAP FERN, *Polypodium* (*Campyloneuron*) *angustifolium* Sw.

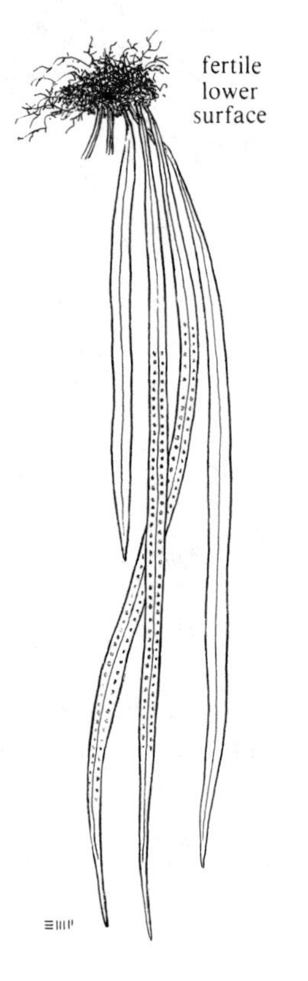

Figure 253

Figure 253 *Polypodium angustifolium*

Rhizome short-creeping; fronds twelve to twenty-four inches long, one-fourth to three-eighths inch wide, pendant; stipe nearly lacking; blade slender, ribbon-like; sori in a single line on each side of the rachis. Epiphyte. Rare. Southernmost Florida; tropical America.

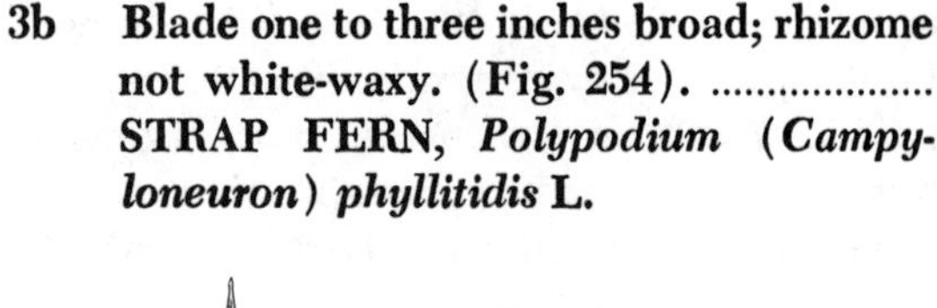

3b Blade one to three inches broad; rhizome not white-waxy. (Fig. 254). .... STRAP FERN, *Polypodium* (*Campyloneuron*) *phyllitidis* L.

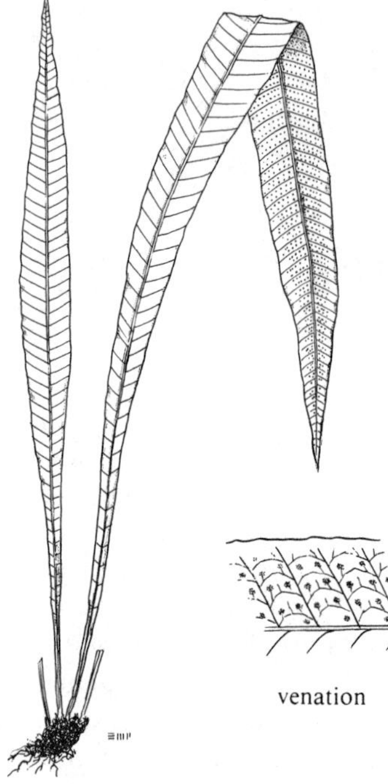

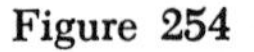

Figure 254

Figure 254 *Polypodium phyllitidis*

Rhizome short-creeping, scaly; fronds simple, leathery, erect; strap-like, fifteen to thirty inches long, one and one-half to two and one-half inches wide; stipe short; veins netted; sori in two rows between major veins, usually two sori per areole. Epiphyte, rarely terrestrial. Frequent. Southern Florida; tropical America.

Polypodium (*Campyloneuron*) *costatum* Kunze has fronds twelve to sixteen inches long with a slender, tail-like tip, and the secondary veins are obscure; rare; southern Florida and tropical America.

Polypodium (*Campyloneuron*) *latum* (Moore) Sod. with similar range has a stipe over four inches long, scales on the rachis, and each sorus is in its own areole; rare.

4a **Blade pinnatifid.** **5**

4b **Blade fully once pinnate. (Fig. 255).** .... ***Polypodium triseriale*** **Sw.**

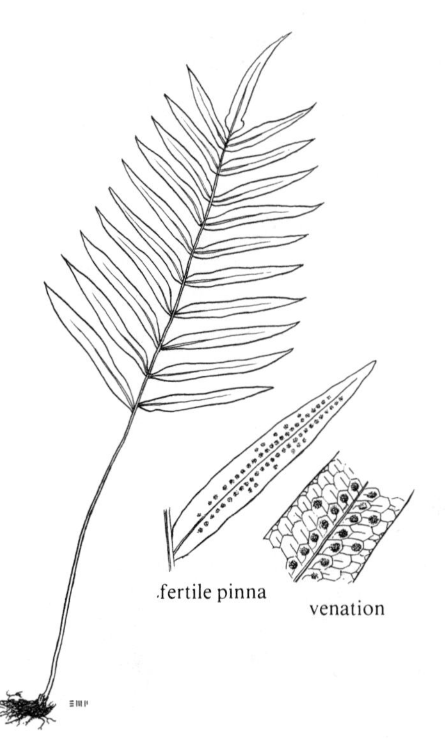

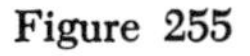

Figure 255

Figure 255 *Polypodium triseriale*

Rhizome short-creeping; fronds twelve to twenty-four inches tall; stipe brown, naked, about one-half the frond length; blade pinnate, the pinna bases not touching the ones directly above or below; pinnae with narrow attachment to the rachis in lower part of the frond, broader attachment near the frond tip; pinnae six to nine pairs; venation netted; sori in one to three rows parallel to the pinna midvein. Epiphyte. Very rare. Southernmost Florida; tropical America.

5a **Blade not densely scaly on lower surface.** **6**

5b **Blade densely scaly on lower surface. (Fig. 256).** .... **RESURRECTION FERN, GRAY POLYPODY, *Polypodium polypodioides* (L.) Watt**

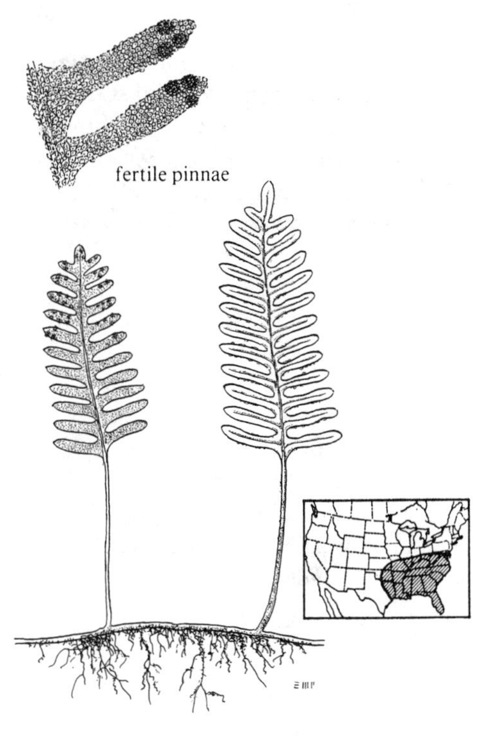

Figure 256

Figure 256 *Polypodium polypodioides*

Rhizome slender, long-creeping; fronds distant, four to eight inches tall; stipe one-third to one-half the frond length; blade deeply pinnatifid; pinnae seven to fourteen pairs, about one-eighth inch wide, densely scaly on lower surface, few or no scales on upper surface. On trees or rocks. Abundant. Southeastern United States; tropical America.

Our material is var. *michauxianum* Weath.

Polypodium thyssanolepis A. Braun, of western Texas, Arizona, and tropical America, has only three to seven pairs of pinnae that are three-sixteenths to three-eighths inch wide and tends to be larger (six to eighteen inches tall). Its upper surface has many finely dissected scales. Rare.

6a Pinnae two to thirty pairs, one-eighth to five-eighths inch wide, naked or scaly, but never with a whitish bloom; rhizome slender (up to three-eighths inch diameter); veins free, if netted, with only one vein serving the sorus; veins often totally obscure. 7

6b Pinnae four to nine pairs, three-fourths to one and one-fourth inch wide, naked, often with a whitish bloom on the blade; rhizome stout (about one-half inch thick) with long, reddish-orange scales; veins netted, with two veins serving each sorus, never obscure. (Fig. 257). .... GOLDEN POLYPODY, *Polypodium (Phlebodium) aureum* L.

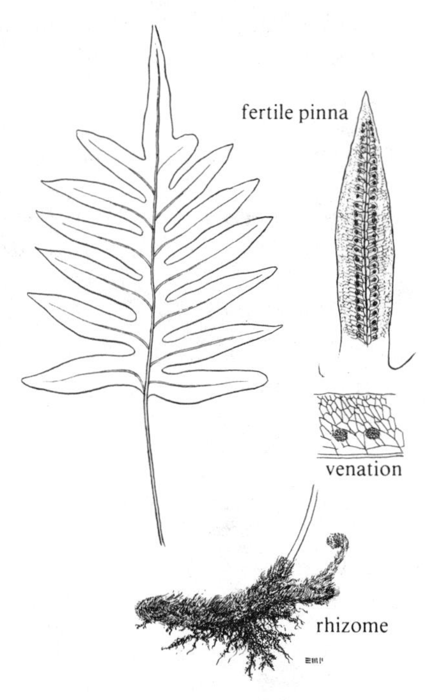

Figure 257

Figure 257 *Polypodium aureum*

Rhizome stout, creeping; fronds fifteen to thirty inches tall; stipe one-half the frond length; pinnae few, four to nine pairs, three-fourths to one and one-fourth inch wide; veins netted, with two veins feeding each sorus; sori in one or two rows on each side of the pinna midvein. Epiphyte. Common. Peninsular Florida; tropical America.

Polypodium (Microsorium) scolopendria Burm. f., the HOBNAIL FERN or WART FERN, is a rare introduction in southern Florida, widespread in the Old World tropics. It can be distinguished by the sori being irregularly placed and each served by a single vein, and also by having each sorus in a pocket that sticks out conspicuously on the upper surface of the frond.

7a **Stipe and rachis dark brown or black; pinnae quite narrow (up to three-sixteenths inch wide), twenty-five to sixty pairs, readily curling on drying; blade often tapering at the base; Florida. 8**

7b **Stipe and rachis green or straw-colored; pinnae broader (three-sixteenths to five-eighths inch wide), three to twenty pairs, not readily curling; blade not tapering at the base. 9**

8a **Rachis black; pinnae less than one-fourth inch wide; blade tapering abruptly at the base, rarely with one or two pairs of minute pinnae. (Fig. 258) .... COMB FERN, *Polyodium plumula* Humb. & Bonpl. ex Willd.**

fertile pinnae

Figure 258

Figure 258 *Polypodium plumula*

Rhizome short-creeping; fronds clumped, seven to twenty-four inches tall; stipe one-eighth to one-fourth the frond length; rachis black; blade pinnatifid, stopping abruptly at the base or rarely with one to two pairs of very short pinnae; pinnae less than one-eighth inch wide. On limestone rocks and tree trunks. Frequent. Peninsular Florida; tropical America.

Polypodium dispersum A. M. Evans with similar distribution has pinnae one-eighth to one-fourth inch wide, has short hairs on both surfaces of the blade, and occurs on limestone rocks and tree trunks.

8b Rachis brown; pinnae usually one-fourth to one-third inch wide; blade tapers gradually to the base, usually with several pairs of very small pinnae at the base. (Fig. 259). ...... GREATER COMB FERN, *Polypodium ptilodon* Kunze [*P. pectinatum* of authors]

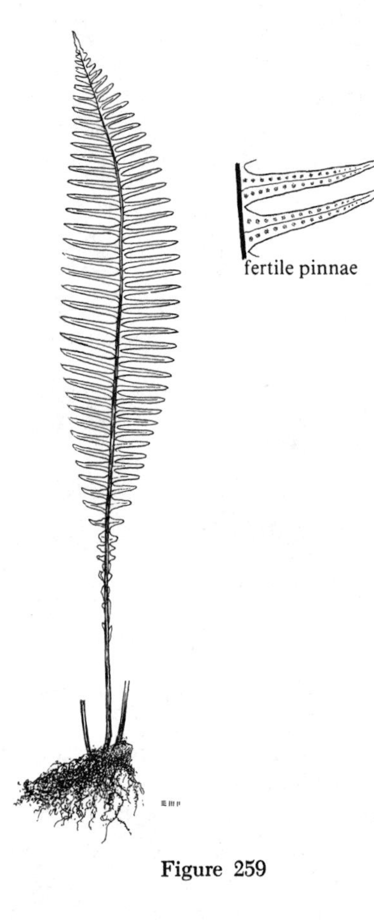

Figure 259

Figure 259 *Polypodium ptilodon*

Rhizome short-creeping; fronds clumped, ten to thirty-six inches tall; stipe less than one-sixth the frond length; rachis brown; blade pinnatifid, tapering gradually at the base to several pairs of minute pinnae; pinnae one-fourth to one-third inch wide. On soil, limestone rocks, or tree bases. Frequent. Peninsular Florida; tropical America.

Our material is var. *caespitosum* (Jenm.) A. M. Evans.

9a Blade not leathery; rhizome not whitish; blade scales lacking; sori less than one-eighth inch broad, away from the midveins, between the midveins and the margins; pinnae eight to twenty-five pairs, rarely as few as four. 10

9b Blade rigidly leathery; rhizome whitish; scales on rachis and pinna midveins; mature sori generally one-eighth inch or more broad, crowded against the pinna midveins; pinnae only four to five pairs, rarely to nine. (Fig. 260). .... LEATHERY POLYPODY, *Polypodium scouleri* Hook. & Grev.

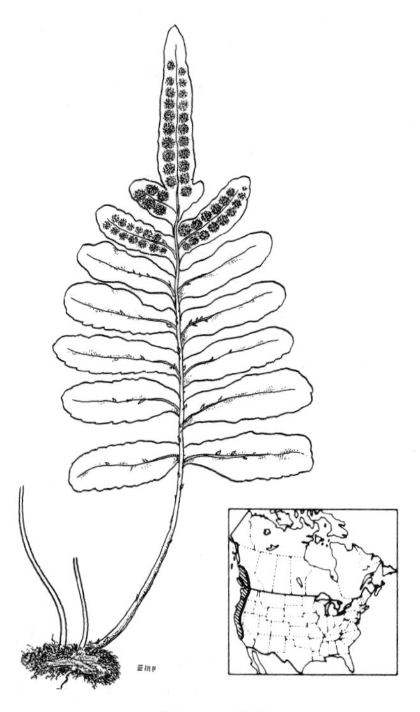

Figure 260

Figure 260 *Polypodium scouleri*

Rhizome creeping, whitish under the scales; fronds mostly four to eighteen inches tall; stipe one-third to one-half the frond length; blade pinnatifid, very leathery, often with scales scattered on rachis and pinna midveins; pinnae usually two to six pairs, rarely to nine, with rounded tips. On rocks and trees, usually within a mile of the ocean. Frequent. Far western North America.

10a Sori midway between the midvein and margin; sori lacking paraphyses; western. 11

10b Sori often nearer the margin than the midvein; sori with paraphyses mixed with the sporangia; mostly eastern and central. (Fig. 261). .... COMMON POLYPODY, ROCKCAP FERN, *Polypodium virginianum* L.

Figure 261 *Polypodium virginianum*

Rhizome creeping; fronds four to fourteen inches tall; stipe one-third to one-half the frond length; blade pinnatifid; pinnae eleven to eighteen pairs, tips narrowly rounded to pointed, about three-sixteenths inch wide. On or among moist, shaded rocks. Common. Eastern and central North America.

11a Pinnae five-sixteenths to five-eighths inch wide, rounded at the tip; rhizome not smelling like licorice when fresh. .. 12

11b Pinnae one-fourth to three-eighths inch wide, attenuate at the tip; the fresh rhizome smelling like licorice. (Fig. 262) .... LICORICE FERN, *Polypodium glycyrrhiza* D. C. Eaton

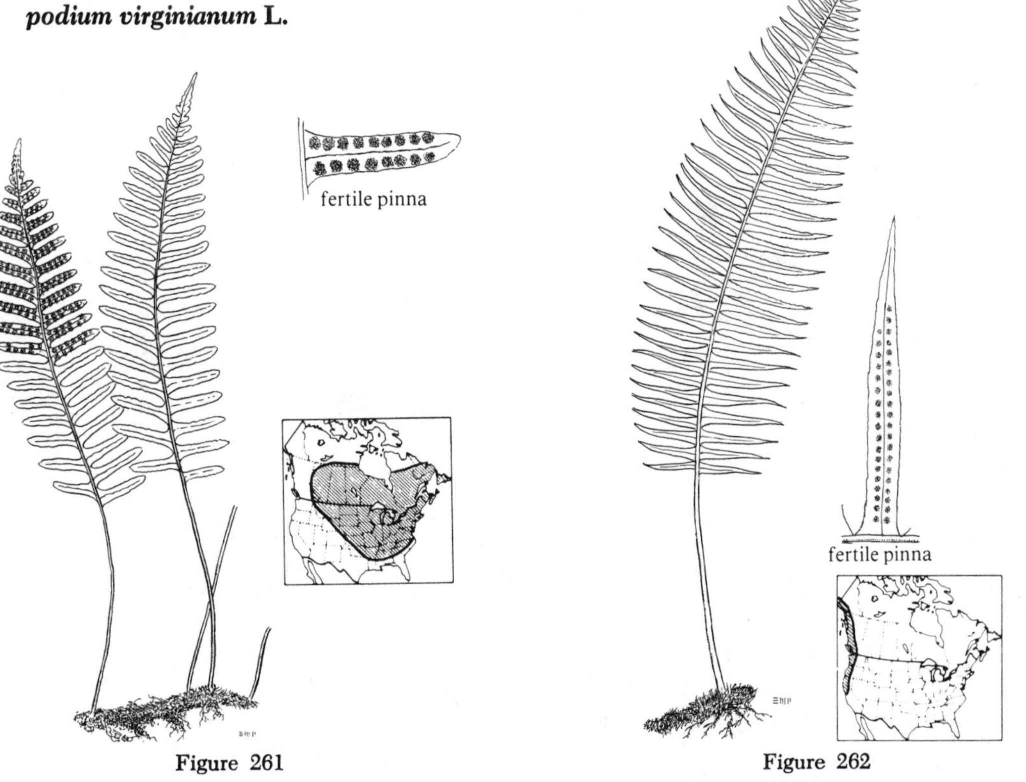

Figure 261

Figure 262

Figure 262 *Polypodium glycyrrhiza*

Rhizome creeping; fronds five to twenty-four inches tall, two to four inches wide; stipe about one-third the frond length; blade pinnatifid; pinnae ten to twenty-five pairs, sharp-pointed, finely toothed. On trees and rocks. Common. Western North America.

12a Blade one to one and three-fourths inch wide; margin smooth or slightly wavy-toothed. (Fig. 263). .... WESTERN POLYPODY, *Polypodium hesperium* Maxon

Figure 263

Figure 263 *Polypodium hesperium*

Rhizome creeping; fronds four to fifteen inches tall, one to one and three-fourths inches wide; stipe about one-third the frond length; blade pinnatifid; pinnae four to fourteen pairs, with rounded tips. Rock crevices. Frequent. Western North America.

This wide-ranging species takes slightly different forms in different habitats. *Polypodium amorphum* Suksd., for example, is a diminutive form.

12b Blade two to three and one-half inches wide; margin sharply toothed. (Fig. (264). ...... CALIFORNIA POLYPODY, *Polypodium californicum* Kaulf.

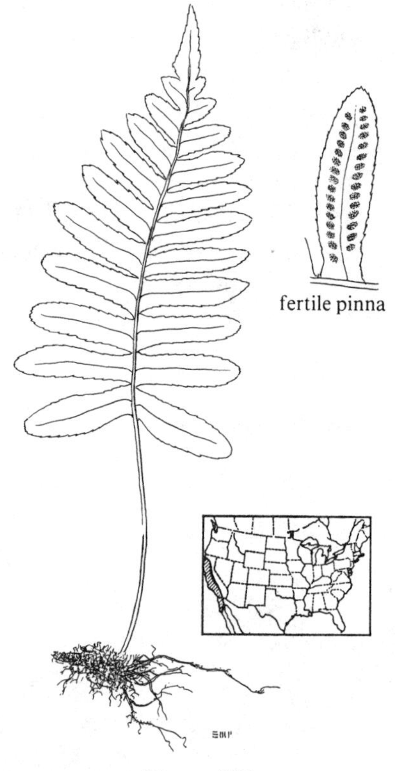

Figure 264

Figure 264 *Polypodium californicum*

Rhizome creeping; fronds seven to fifteen feet tall, two to three and one-half inches wide; stipe one-third to one-half the frond length; blade pinnatifid; pinnae nine to thirteen pairs with rounded to slightly pointed tips and sharply toothed margins. On rocks. Common. California; Baja California.

Polypodium australe Fée, which is distinguished from *P. californicum* by having inflexed lower pinnae and paraphyses in the sori, has been reported from San Clemente Island; it is widespread in Europe.

POLYSTICHUM

Holly Fern

Rhizome ascending, stout, scaly; fronds medium-sized to large, pinnate to bipinnate, leathery; veins free; sori medial, round; indusium umbrella-shaped. One hundred seventy-five species of temperate regions.

Figure 265 Relationships of some western holly ferns. Hybrids and backcrosses are indicated by formulae. Each letter represents one set of chromosomes. Unknowns are in parentheses. (From W. H. Wagner, Jr., Reticulation of holly ferns (*Polystichum*) in the western United States and adjacent Canada, American Fern Journal 63: 99-115 (1973), with permission of the author.)

1a **Fronds once pinnate; pinnae not at all dissected. 2**

1b **Fronds pinnate-pinnatifid to bipinnate. 4**

2a **Fertile pinnae not noticeably different from the sterile pinnae. 3**

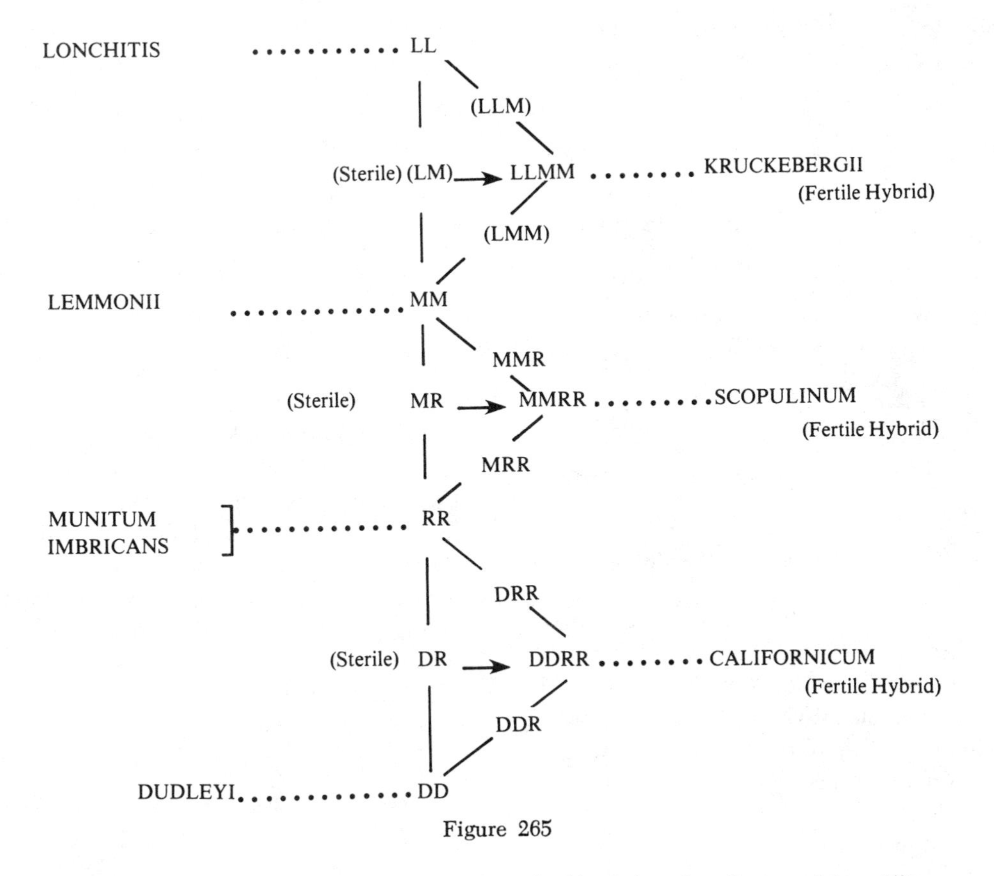

Figure 265

2b **Fertile pinnae markedly smaller and narrower than the sterile pinnae, fertile portion limited to the tip of the frond. (Fig. 266). .... CHRISTMAS FERN, *Polystichum acrostichoides* (Michx.) Schott**

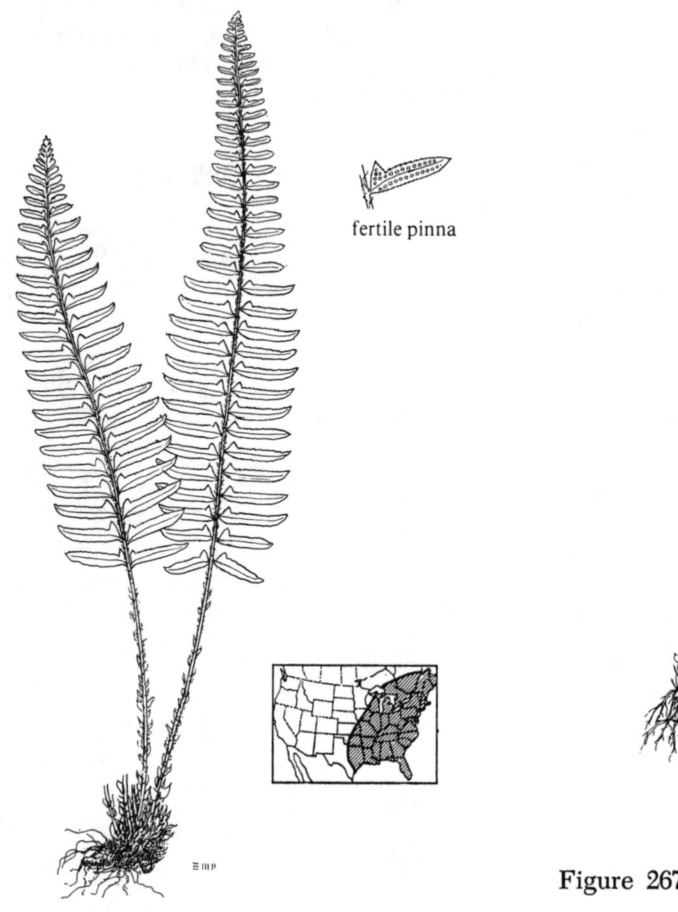

Figure 266

Figure 266 *Polystichum acrostichoides*

Fronds twelve to twenty-four inches tall; blade linear, broadest at base, pinnate; stipe and rachis scaly; pinnae auricled, with marginal bristle teeth, occasionally deeply toothed; fertile portion only the tip of the frond, the fertile pinnae contracted. Moist woods. Abundant. Eastern North America.

3a **Pinnae one to one and one-half inches long; blade reduced at the base. (Fig. 267). ...... NORTHERN HOLLY-FERN, *Polystichum lonchitis* (L.) Roth**

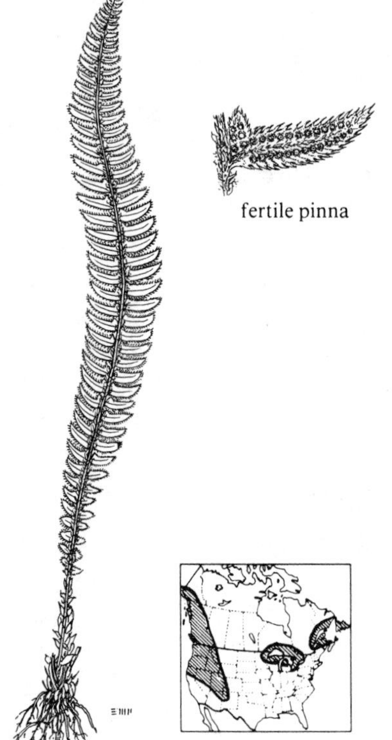

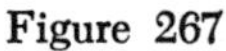

Figure 267

Figure 267 *Polystichum lonchitis*

Fronds six to twenty-four inches tall; blade pinnate, linear, reduced at the base; stipe essentially lacking, less than one-tenth the frond length; stipe and rachis sparsely scaly; pinnae short, with bristle teeth. Moist, shaded rocks. Frequent. Northern and western North America; Europe, Asia.

3b **Pinnae one and one-half to five inches long; blade not reduced at base. (Fig. 268). .... WESTERN SWORD-FERN, *Polystichum munitum* (Kaulf.) Presl**

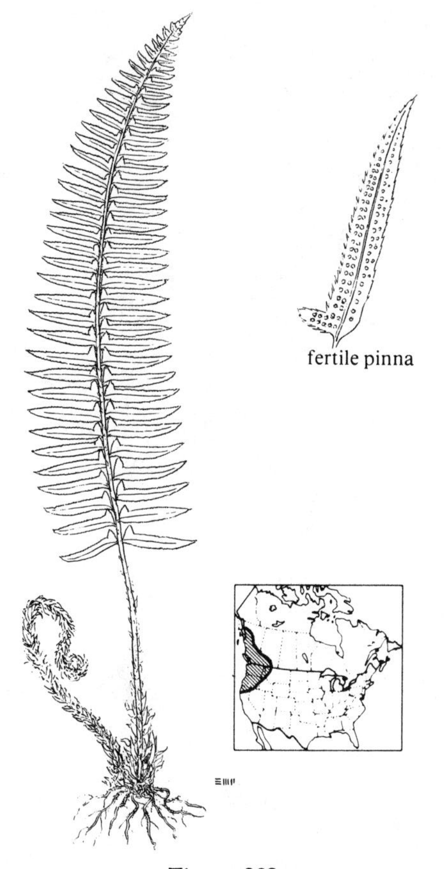

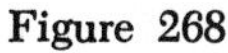

Figure 268

Figure 268 *Polystichum munitum*

Fronds eighteen to seventy-two inches tall (rarely only twelve); stipe scaly with large and small scales, less than one-fourth the frond length; blade pinnate, only slightly reduced at base; pinnae slender, with bristle teeth. Moist woods. Abundant. Northwestern North America.

Polystichum imbricans (D. C. Eaton) D. H. Wagner, also of the Northwest, has the pinnae often overlapping, the rhizome and stipe base scales long and red-orange, the stipe is naked, one-third to one-half the frond length, and the rachis is only sparsely scaly.

4a **Fronds less than twenty inches tall, mostly about twelve inches tall. 5**

4b **Fronds sixteen to forty-two inches tall. 6**

5a **Pinnae generally with only one free pinnule, with sharp teeth. (Fig. 269). .... WESTERN HOLLY-FERN, *Polystichum scopulinum* (D. C. Eaton) Maxon**

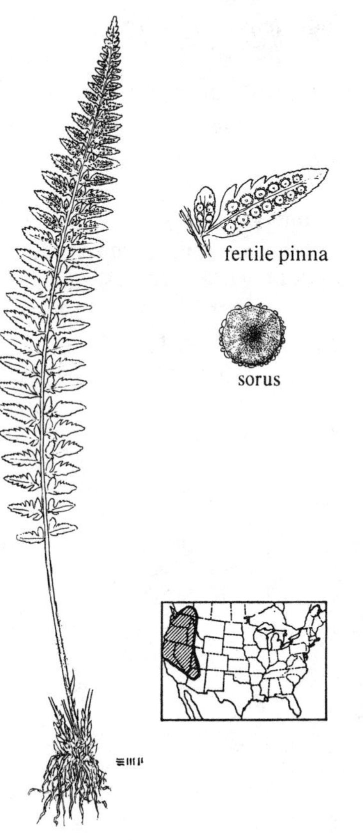

Figure 269

Figure 269 *Polystichum scopulinum*

Fronds six to fourteen inches tall; stipe one-fourth to one-third the frond length, sparsely scaly; blade pinnate-pinnatifid, linear, usually only one basal lobe free on each pinna, with about four pairs of lobe-like teeth per pinna, the teeth mosty not bristle-tipped. Noncalcareous cliffs and rock crevices. Common. Western North America and rare in Gaspé Peninsula of eastern Canada.

This is probably a fertile hybrid of *P. lemmonii* and *P. imbricans.*

Polystichum kruckebergii W. H. Wagner of the western states has short, somewhat triangular pinnae, each with only one to two pairs of lobe-teeth. This is presumably a fertile hybrid between *P. lemmonii* and *P. lonchitis.*

Polystichum aleuticum C. Chr., known from only one collection from Atka Island in the Aleutian Archepelago, is six inches tall and its pinnae are one-half inch long, lobed, and lack teeth.

5b Pinnae generally with three to five pairs of free pinnules, rounded, not sharp-toothed. (Fig. 270). .... SHASTA HOLLY-FERN, *Polystichum lemmonii* Underw. [*P. mohrioides* of authors]

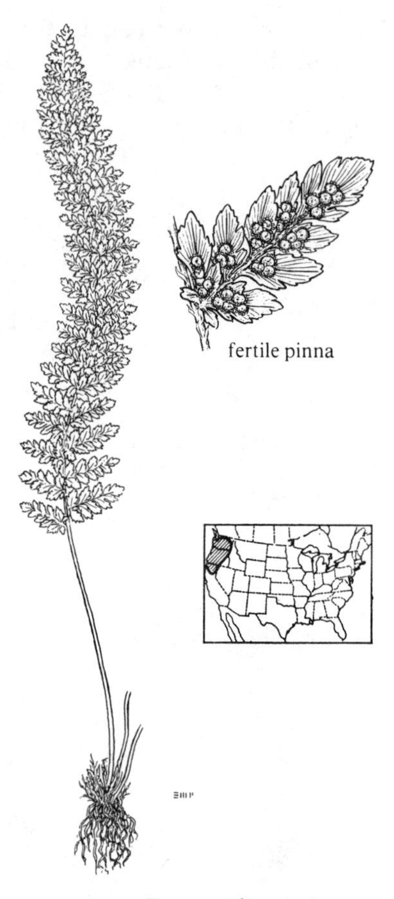

Figure 270

Figure 270 *Polystichum lemmonii*

Fronds four to sixteen inches tall, mostly less than twelve inches; blade bipinnate to bipinnate-pinnatifid, sparsely scaly; pinnae often overlapping; pinnules without spine-tipped teeth; pinnae not reduced at blade base. Among serpentine rocks. Frequent. Western United States.

6a All pinnules narrowly attached. 7

6b Most pinnules attached to the pinna rachis by most of their width. 8

7a **Blade not strongly narrowed at the base, with scattered scales; pinnules lobed, often pinnatifid; California. (Fig. 271). .... DUDLEY'S HOLLY-FERN, *Polystichum dudleyi* Maxon**

7b **Blade narrowed at the base, densely clothed with scales; pinnules not lobed except for one auricle per pinnule; northern. (Fig. 272) .... BRAUN'S HOLLY-FERN, *Polystichum braunii* (Spenner) Fée**

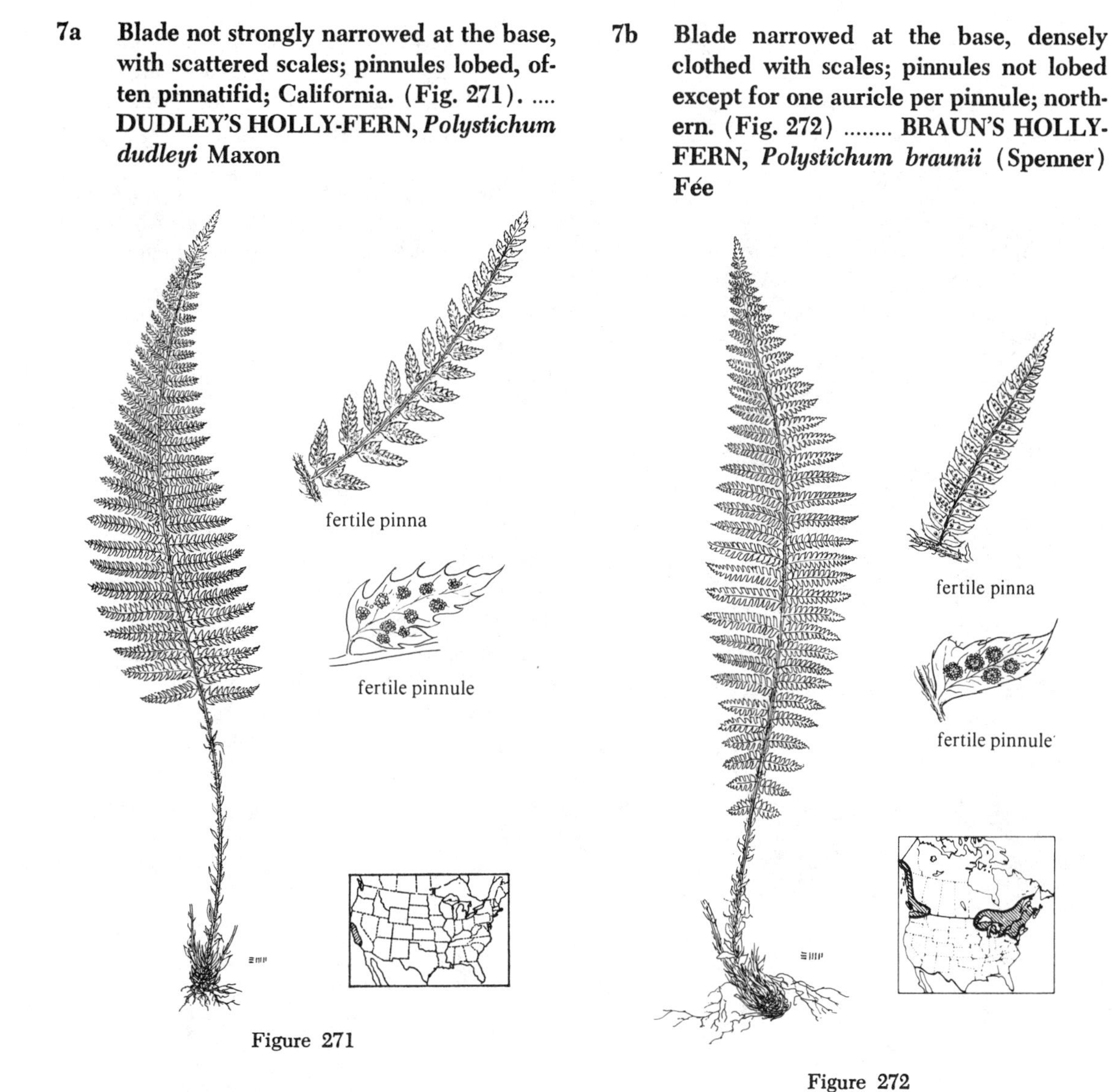

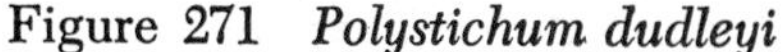
Figure 271 *Polystichum dudleyi*

Fronds sixteen to forty inches tall; stipe one-third the frond length; blade bipinnate, not strongly tapering at base; stipe and rachis scaly; pinnules auricled or pinnatifid. Moist woods. Frequent. California.

Figure 272 *Polystichum braunii*

Fronds ten to twenty-eight inches tall; stipe heavily scaly, about one-fourth the frond length; blade bipinnate, tapering to the base; rachis and pinna rachises scaly; the pinnules with bristle teeth; fertile and sterile parts all alike. Moist woods. Frequent. Northeastern and northwestern North America; Europe, Asia.

Polystichum muricatum Fée, of southernmost Florida and tropical America, does not taper at the base of the blade.

8a Rachis with a bud near the tip on the under side. (Fig. 273) .... ANDERSON'S HOLLY-FERN, *Polystichum andersonii* Hopkins

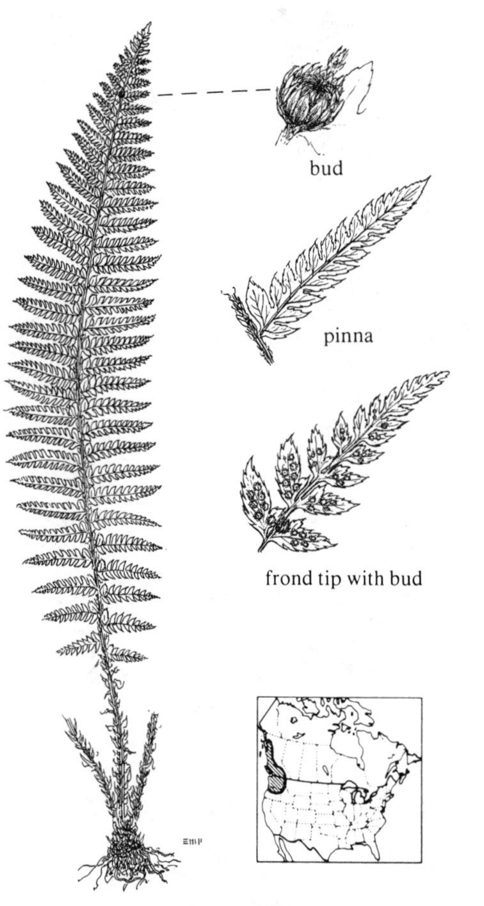

Figure 273

Figure 273 *Polystichum andersonii*

Fronds twenty to forty-two inches tall; stipe about one-sixth the frond length; blade pinnate-pinnatifid to bipinnate, somewhat reduced at the base; pinnules with bristle teeth. Cool, moist, rocky slopes. Frequent. Northwestern North America.

8b Rachis lacking buds. .... ALASKA HOLLY-FERN, *Polystichum setigerum* (Presl) Presl [*P. alaskense* Maxon]

Fronds twenty to forty inches tall; stipe about one-fourth the frond length; blade pinnate-pinnatifid to bipinnate, only slightly reduced at the base; pinnules with bristle teeth. Cool, moist woods. Rare. British Columbia and southeast Alaska.

This is probably a hybrid between *P. andersonii* and *P. braunii;* some plants are sterile, some fertile.

The CALIFORNIA HOLLY-FERN, *Polystichum californicum* (D. C. Eaton) Diels, of western coastal states, is pinnate-pinnatifid to barely bipinnate. It originated as a hybrid between *P. dudleyi* and *P. munitum.* Some forms are sterile, others fertile. Uncommon.

This is a mixture of two hybrids—*P. dudleyi* × *munitum* and *P. dudleyi* × *imbricans*—which cannot be distinguished from each other. Some specimens are sterile, others fertile. Uncommon.

Polystichum microchlamys (Christ) Matsum. of eastern Asia has been found on Attu Island of the Aleutian Archepelago. It has deeply incised pinnae.

PSILOTUM

Whisk Fern

Plant with neither roots nor leaves; slender stems creeping underground and also erect, green stems branching several times in half, bearing tiny, widely separated, scale-like bracts; sporangia solitary and three-chambered. (Fig. 274) .... WHISK FERN, *Psilotum nudum* (L.) Pal.

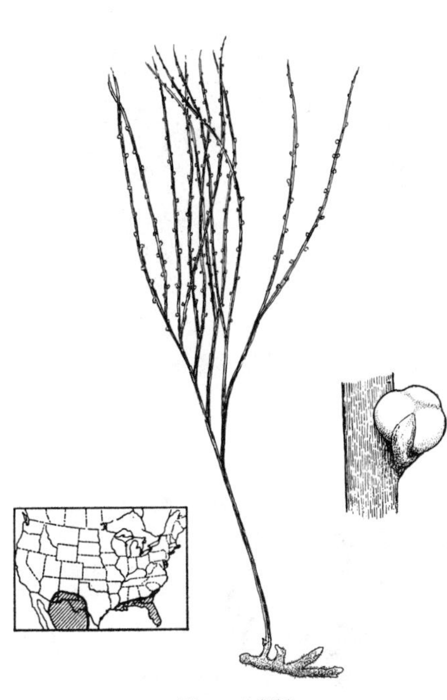

Figure 274

Figure 274 *Psilotum nudum*

Plants six to sixteen inches tall, arising from a slender, creeping, branched, dark, hairy rhizome. At tree bases and on logs and hummocks in low wet woods. Uncommon. Southeastern and southwestern United States; tropical America, Asia, Africa.

PTERIDIUM

Bracken

Rhizome long-creeping, subterranean, deep in the soil, hairy; fronds medium-sized to large; blade bipinnate to tripinnate, broadly triangular; veins free; sori marginal, protected by recurved margin. Often regarded as one species of world-wide distribution with a number of regional varieties. (Fig. 275- .... BRACKEN, BRAKE, *Pteridium aquilinum* (L.) Kuhn

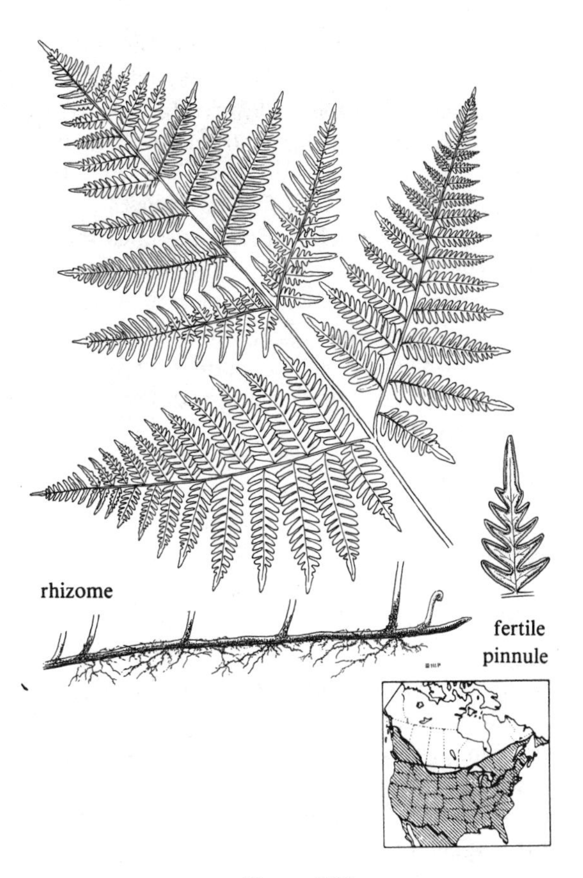

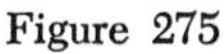

Figure 275

Figure 275 *Pteridium aquilinum*

Fronds one and one-half to fifteen feet tall, one to two feet wide; blade broadly triangular, bipinnate to tripinnate. Open woods and meadows; weedy. Abundant. Most of North America; tropical America, Europe, Asia, Africa.

There are four varieties recognized in North America. Var. *latiusculum* (Desv.) Underw. ex Heller of the Northeast is sparsely hairy beneath and has pinnule tips less than four times as long as wide (mostly about one-fourth inch broad). Var. *pseudocaudatum* (Clute) Heller of the Southeast is naked beneath and has pinnule tips more than six times as long as wide (mostly about one-eighth inch broad). Var. *caudatum* (L.) Sadebeck is least common, in southern Florida, and has the terminal segments distant from one another. Var. *pubescens* Underw. is widespread in the West, is densely hairy beneath and has the pinnules perpendicular to the midveins. The first two varieties are one to four feet tall, but the other two tend to be much larger, sometimes reaching fifteen feet in height. Not all plants are easily placed in a particular variety; var. *latiusculum* intergrades frequently with *pubescens* and *pseudocaudatum*. Var. *caudatum* may be a distinct species; it never intergrades with other varieties.

PTERIS

Brake

Rhizome short-creeping or ascending, scaly; fronds medium-sized to large, mostly pinnate to pinnate-pinnatifid, rarely to five times pinnate; veins free or netted; sori marginal, protected by the recurved margin. Two hundred fifty species of tropical and subtropical regions.

1a **Veins all free; fronds one to three feet tall.** **2**

1b **Veins netted; fronds three to nine feet tall. (Fig. 276)** .... ***Pteris tripartita*** **Sw.**

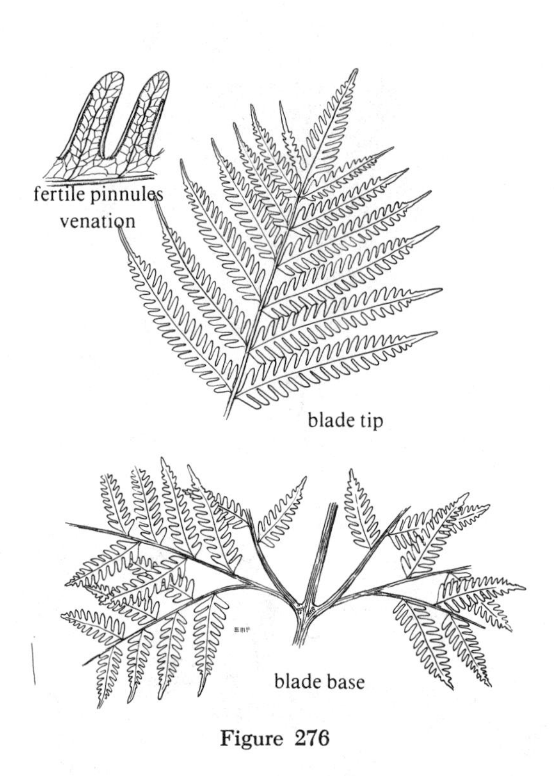

Figure 276

Figure 276 *Pteris tripartita*

Rhizome stout, ascending; fronds three to nine feet tall; stipe one-third the frond length, naked; blade pinnate-pinnatifid in upper part, the lowest pair of pinnae much divided. Moist woods. Naturalized occasionally in southern Florida; native of Africa.

Pteris grandifolia L. is once pinnate, and the venation is free near the midveins, netted near the margin. It is rare in southern Florida; tropical America.

2a **Blade pinnate throughout, the lowest pinnae undivided.** **3**

2b **Blade pinnate above, the lowest pinna pair with large pinnules.** **4**

3a **Stipe and rachis densely clothed with hair-like scales; sterile pinna margin toothed; pinnae strongly ascending; terminal pinna very long, usually longer than any of the lateral pinnae; rhizome scales light brown. (Fig. 277). .... LADDER BRAKE, *Pteris vittata* L.**

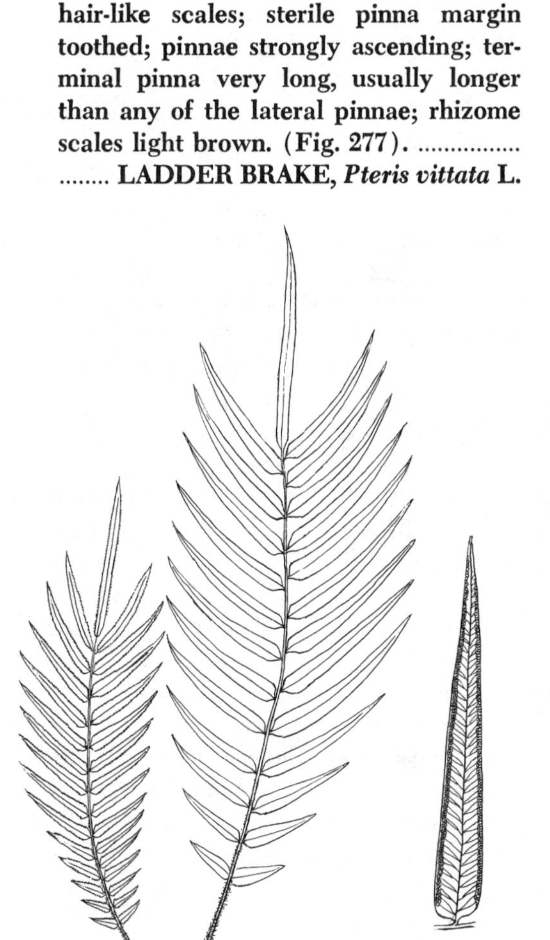

Figure 277

Figure 277 *Pteris vittata*

Rhizome stout, short-creeping; fronds one to three feet tall; stipe one-sixth to one-third the frond length, densely scaly; blade pinnate; sterile pinnae sharply toothed. Frequently escaped and naturalized, especially on limestone and mortar. South Carolina to Louisiana, southern California; native of Asia.

3b **Stipe and rachis slightly scaly or naked; sterile pinna margin obscurely toothed to smooth; pinnae at nearly right angles to the rachis; terminal pinna generally not the longest pinna; rhizome scales dark brown. (Fig. 278). .... BAHAMA BRAKE, *Pteris bahamensis* (Agardh) Fée [*P. longifolia* var. *bahamensis* (Agardh) Hieron.]**

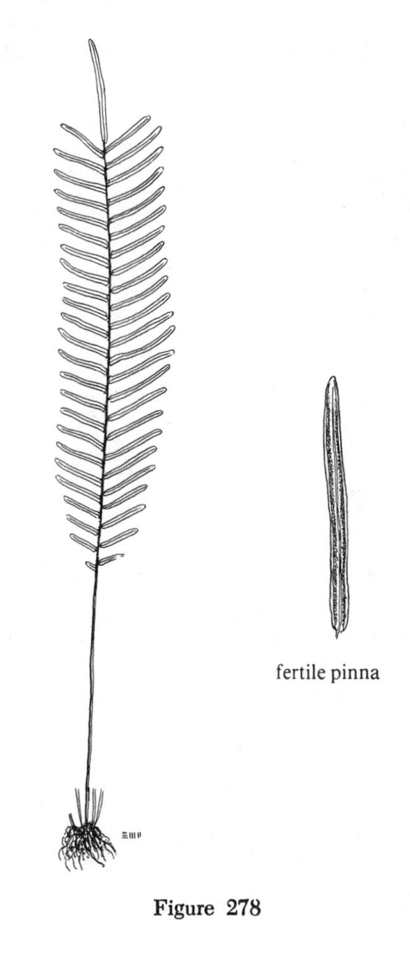

Figure 278

Figure 278 *Pteris bahamensis*

Rhizome short-creeping, fronds one to three feet tall; stipe one-sixth to one-third the frond length, lightly scaly but scales falling off; blade pinnate; pinnae undivided, linear; sterile pinnae ony finely toothed or smooth-margined. Shaded limestone outcrops. Peninsular Florida; tropical America.

Pteris longifolia L. has a scaly rachis, longer pinnae and is rare.

4a Pinnae one to three pairs, only the uppermost broadly attached and running down the rachis; lowest pair of pinnae each with a single, large, downward pointing pinnule; upper blade surface lacking tiny streaks. (Fig. 279) .... CRETAN BRAKE, *Pteris cretica* L.

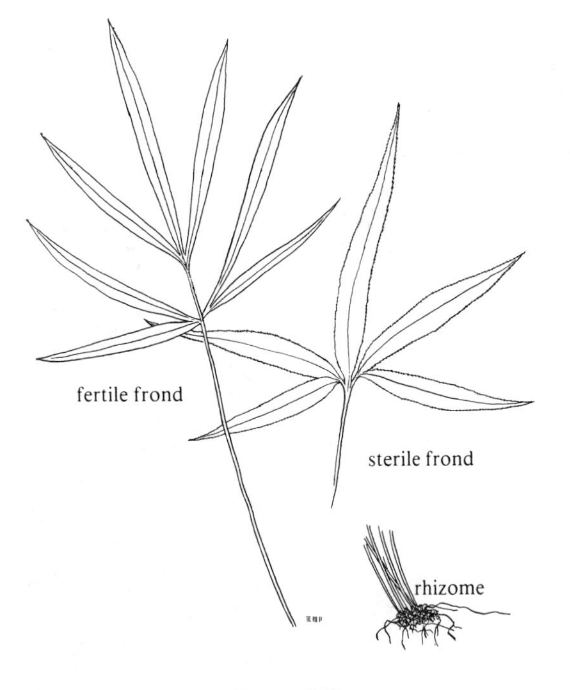

Figure 279

Figure 279 *Pteris cretica*

Rhizome short-creeping; fronds eight to twenty-four inches tall; stipe more than one-half the frond length, naked; blade pinnate, broadly triangular; pinnae one to three pairs, the lowest pinnae each with a large, downward pointing pinnule; sterile pinnae finely toothed. Escaped and naturalized on limestone ledges, shaded slopes, and rocky meadows. Frequent. Florida to Louisiana, southern California; Mexico, West Indies, Old World tropics.

Var. *albolineata* Hooker has a white streak in the middle of each pinna.

Pteris ensiformis Burman var. *victoriae* Baker is naturalized in peninsular Florida and resembles *P. cretica* but has more and narrower pinnae, with white streaks on the pinnae; native of Asia.

4b Pinnae three to seven pairs, mostly broadly attached and running down the rachis; lowest one to three pairs of pinnae with one to two pairs of large pinnules; upper blade surface with tiny longitudinal streaks (seen through a hand lens). (Fig. 280). ...... SPIDER BRAKE, *Pteris multifida* Poiret

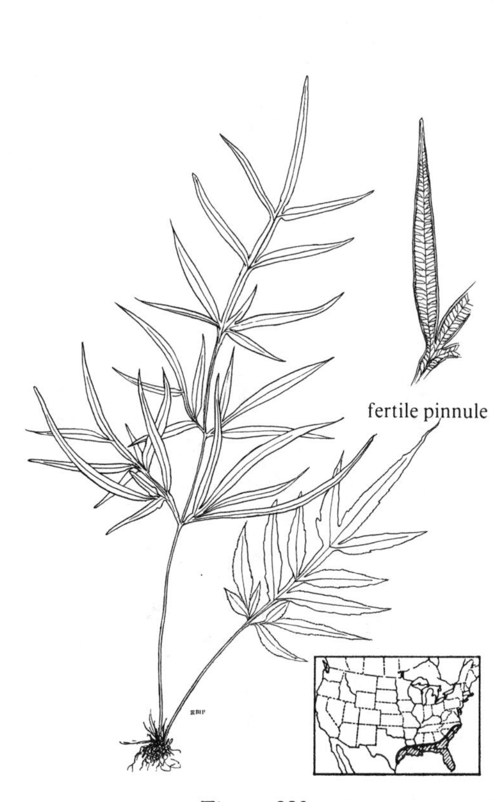

Figure 280

Figure 280 *Pteris multifida*

Rhizome short-creeping; fronds ten to twenty-four inches tall; stipe about one-third the frond length; blade pinnate, oblong, pinnae three to seven pairs, most undivided but the lowest one to three pairs with one to two pairs of large pinnules; sterile margins finely toothed. Escaped and naturalized on limestone and mortar of walls. Frequent. North Carolina to Texas; native of Asia.

SALVINIA

Water Spangles

Rhizome floating on water surface or creeping on mud; fronds opposite, round or oval, undivided, upper leaf surface clothed with branched hairs; sporangia produced in round sporocarps borne on trailing, dissected, root-like organs (actually leaves). Ten species of tropical regions. (Fig. 281) .... WATER SPANGLES, FLOATING FERN, *Salvinia minima* Baker [*S. rotundifolia* Willd.]

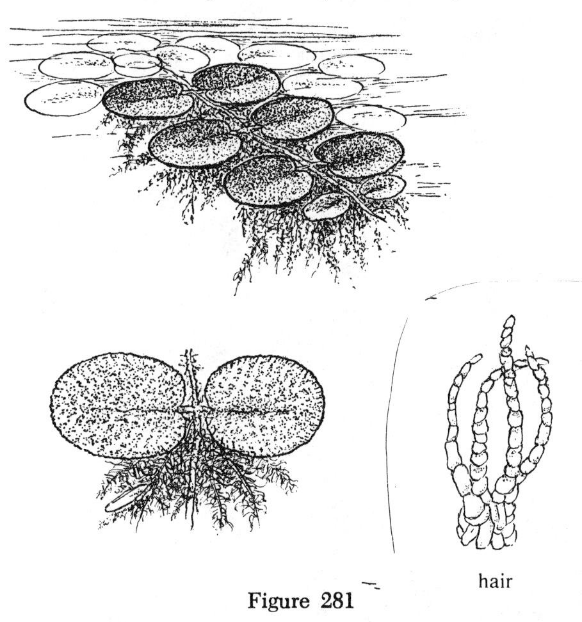

Figure 281

Figure 281 *Salvinia minima*

Leaves round or oval, about one-fourth to one-half inch long; leaf hairs with four spreading branches. Floating on quiet water of ditches, ponds, slow-moving streams. Frequent. Southeastern United States, also reported from Minnesota; tropical America.

SCHIZAEA

Rhizome short-creeping, hairy; fronds small, simple, grass-like; sporangia borne on small, finger-like projections from the end of the frond. Thirty species of tropical or rarely of temperate regions. (Fig. 282) .... CURLY-GRASS FERN, *Schizaea pusilla* Pursh

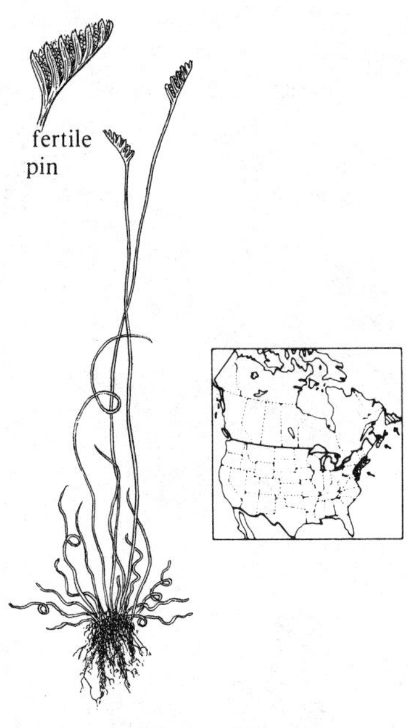

Figure 282

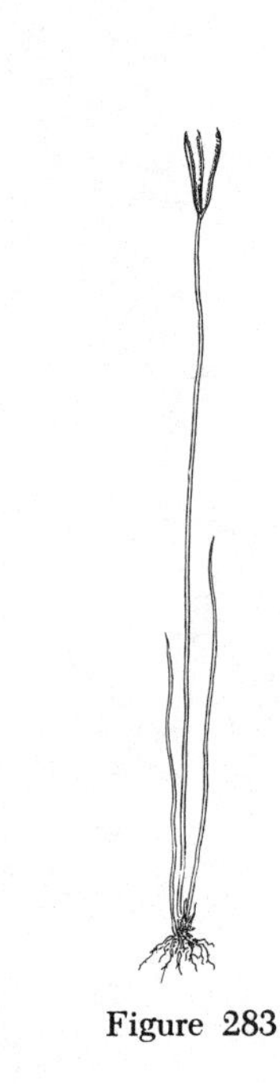

Figure 283

Figure 282 *Schizaea pusilla*

Fronds dimorphic; sterile fronds one to two and one-half inches long, one sixty-fourth inch wide, curling, often in a low spiral like a pig's tail; fertile fronds erect, two to four inches tall; blade pinnatifid, the several tiny yellow to brown pinnae close to the very tip, one to six pairs, less than one-eighth inch long. Acid bogs. Northeastern North America.

The small slender plants, often growing among grasses, are extremely difficult to see, even when your eye is close to the ground.

Schizaea germanii (Fée) Prantl (Fig. 283) is a very rare species occurring in southern Florida and tropical America. It is four to six inches tall, its sterile fronds are upright, and the fertile fronds are like the sterile but tipped with a few erect, narrow, sporangia-bearing segments one-half inch long arising from one point.

Figure 283 *Schizaea germanii*

SELAGINELLA

Spikemoss

Plants mostly small, creeping, and moss-like; leaves very small, round to linear, with a single vein and an inconspicuous ligule, leaves often ending in a hair-like tip (seta), borne spirally or in four rows; sporangia borne in the axils of special leaves, forming a four-sided terminal cone (except in one species with cylindrical cones); heterosporous, usually with four megaspores per megasporangium and very many microspores in each microsporangium. Seven hundred species, largely of tropical regions.

1a **Leaves oval or oblong, spreading, in four rows of two types—two rows of spreading lateral leaves and two rows of appressed median leaves on top of the stem. 2**

1b **Leaves all linear, in several rows spirally arranged. 6**

2a **Plants creeping, not rosette-forming; leaves often distant, spreading, thin, delicate; of moist areas. 3**

2b **Plants forming rosettes; leaves overlapping; growing in dry areas. (Fig. 284) .... RESURRECTION PLANT, *Selaginella lepidophylla* (Hook. & Grev.) Spring**

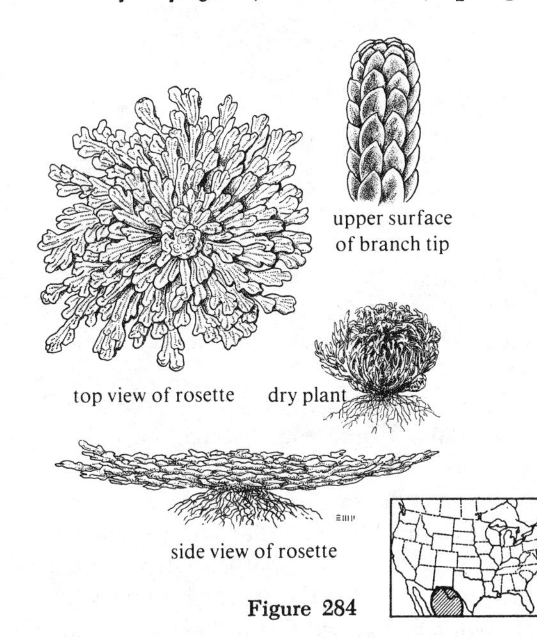

Figure 284

Figure 284 *Selaginella lepidophylla*

Plants rosette-forming, flat when moist, inrolled in a brown ball when dry; each stem three to four inches long; dark green on top, white to reddish brown beneath; leaves in four overlapping rows, two lateral, two median; leaves with rounded tips, white margined, and lacking hairs. Limestone ledges and slopes. Frequent. Southwestern United States; Mexico.

Selaginella pilifera A. Braun, of similar distribution, has leaves with bristle tips and hairy, green margins; rare.

3a **Plants of the East. 4**

3b **Plants of the Northwest. (Fig. 285) .... *Selaginella douglasii* (Hook. & Grev.) Spring**

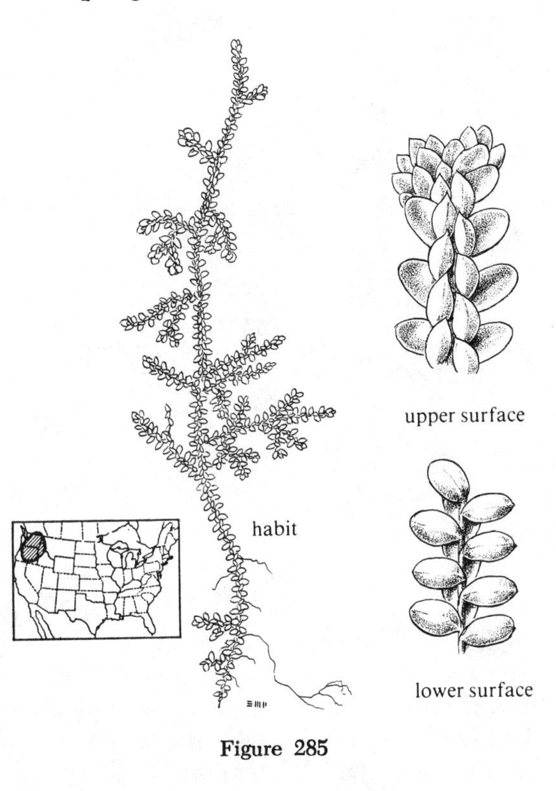

Figure 285

Figure 285 *Selaginella douglasii*

Stems creeping, prostrate; lateral leaves smooth-margined with rounded tip; median leaves smooth-margined with slender pointed tip and a few hairs at the base. Moist shaded rocks. Frequent. Northwestern United States.

4a **Plant green. 5**

4b **Plant bluish-green. (Fig. 286) .... *Selaginella uncinata* Spring**

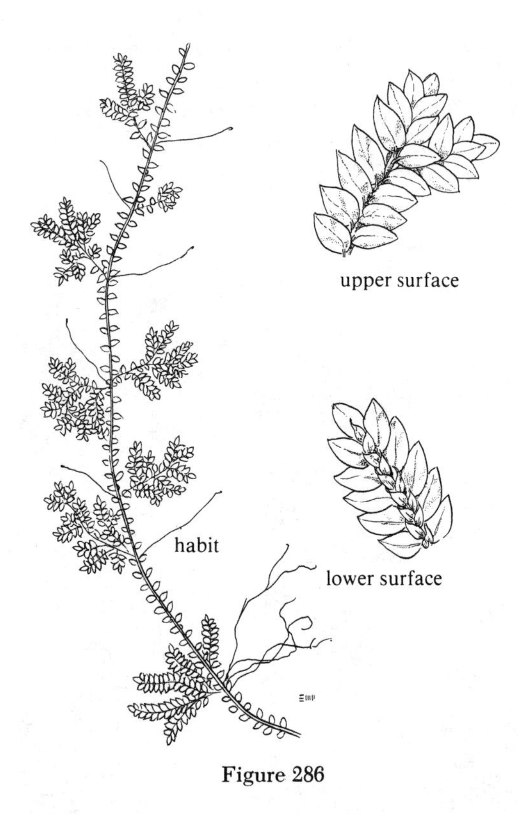

Figure 286

Figure 286 *Selaginella uncinata*

Plants creeping with flat, frond-like branches; leaves with a pale, shiny margin without teeth. Moist woods. Rare. Naturalized in Florida to Louisiana; native of Asia.

Selaginella willdenovii (Desv.) Baker is a rare escape in moist woods of southern Florida; it is native to Asia but escaped in many tropical regions. The plants have stout, vine-like stems that scramble on other plants and are several feet long. The leaves are all smooth-margined, the lateral leaves with an auricle overlapping the stem.

5a Median leaves linear. (Fig. 287) .... ***Selaginella eatonii*** **Hieron.**

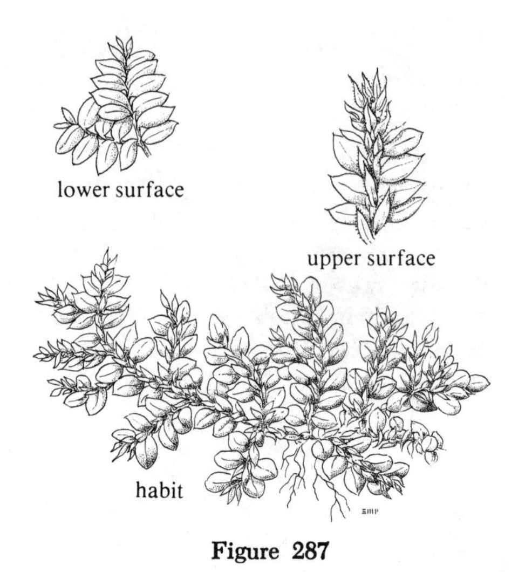

Figure 287

Figure 287 *Selaginella eatonii*

Plants creeping, prostrate, very small, generally only one inch long; leaves finely toothed. Moist woods. Uncommon. Southernmost Florida.

This species has sometimes been called S. *armata* Baker, but that species has ciliate leaves and occurs in the West Indies.

Selaginella kraussiana A. Braun, one of the most commonly cultivated species, is sometimes escaped in the Southeast. It too has linear median leaves but is much more wide-creeping than S. *eatonii,* with stems two to four inches long.

5b Median leaves narrowly oval or lance-shaped. (Fig. 288). .... **.... MEADOW SPIKEMOSS,** ***Selaginella apoda*** **(L.) Spring**

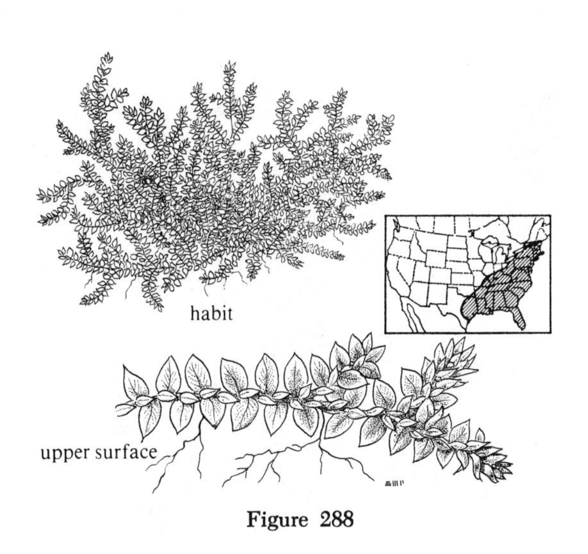

Figure 288

Figure 288 *Selaginella apoda*

Plants creeping, prostrate, forming large mats; leaves finely toothed; lateral leaves oval, median leaves narrower with a pointed tip; the tips, if attenuate, consisting of more than just the midvein; leaf margins not distinctly pale. Wet woods, swamps, or meadows; in lawns. Frequent. Eastern United States.

Selaginella eclipes Buck is closely allied to *S. apoda* but can be distinguished by its long-attenuate median leaf apices consisting of the midveins. It occurs frequently in the St. Lawrence River Valley and Great Lakes region southwest to Indiana, Illinois, to Oklahoma and Arkansas.

Selaginella ludoviciana A. Braun also closely resembles *S. apoda* but has a more erect habit and the leaves have distinct pale or clear margins. It occurs from northern Florida to Louisiana; frequent.

Selaginella braunii Baker is a frond-like erect plant, and the stem has many short, pointed hairs. It is rarely escaped in the Southeast and native of eastern Asia.

6a (1b) Cones four-sided, slender, less than one inch long; mostly of dry, exposed, rocky or sandy habitats. (The species of this group are extremely difficult to identify, especially in the Southwest.) 7

6b Cones cylindrical, often well over one inch long; leaves spreading, very coarsely and distinctly toothed; of northern bogs. (Fig. 289) .... *Selaginella selaginoides* (L.) Link

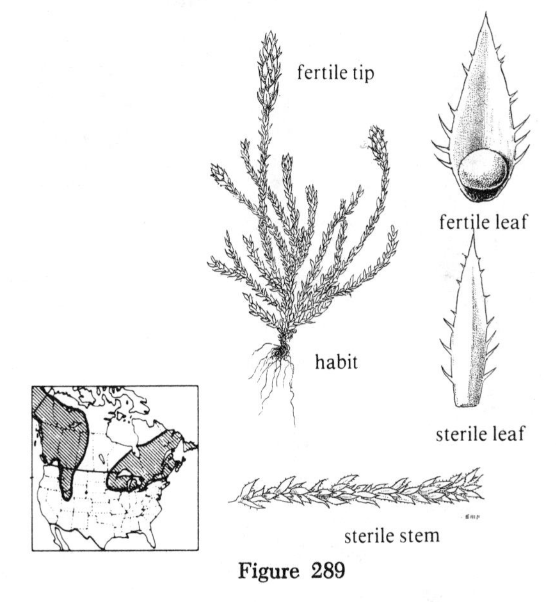

Figure 289

Figure 289 *Selaginella selaginoides*

Plants not forming large mats; leaves very coarsely and distinctly toothed. Wet rocks, banks, and bogs. Extremely local. Northern North America; Europe, Asia.

7a Main stems erect or ascending, rooting only near the base. 8

7b Main stems creeping and spreading, rooting along its length, or hanging from trees; branches may be erect. 11

8a **Plants of the Southeast (west to eastern Texas). 9**

8b **Plants of the Southwest (east to western Texas). 10**

9a **Leaf setae usually curled or twisted, often falling off with age; leaf margins smooth or with short teeth; Appalachian Mountains. (Fig. 290) .... *Selaginella tortipila* A. Braun**

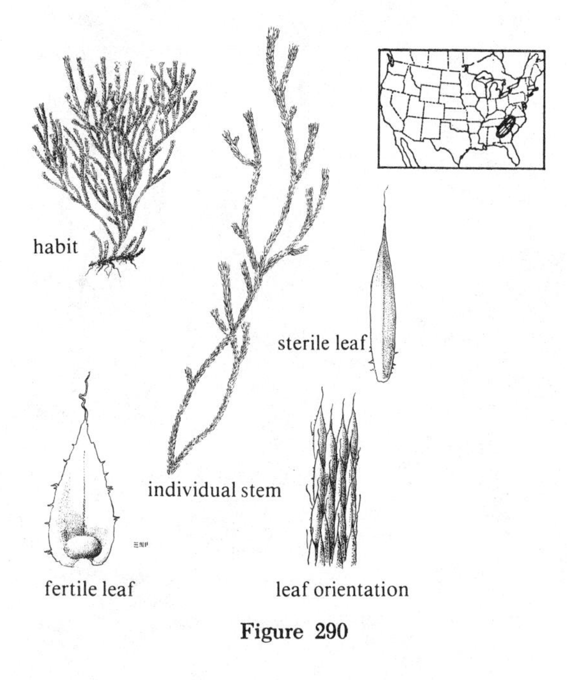

Figure 290

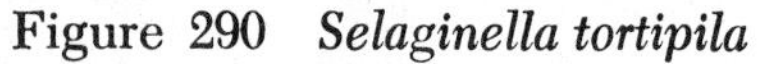

Figure 290 *Selaginella tortipila*

Forming mats of erect plants. On open gneiss and granite surfaces. Frequent. Southeastern United States.

9b **Leaf setae toothed, not curled or twisted; leaves with hair-like teeth; low elevation. (Fig. 291). .... *Selaginella arenicola* Underw.**

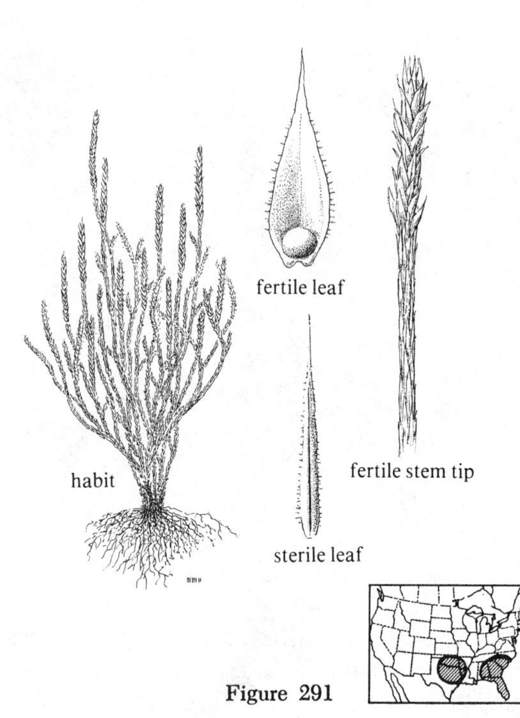

Figure 291

Figure 291 *Selaginella arenicola*

Plants forming dense mats of erect plants two to four inches tall. Open pine woods and dunes. Common. Southeastern United States.

Specimens of Texas to Alabama are often placed in subspecies *riddellii* (Van Eselt.) R. Tryon.

10a **Leaves with short teeth on the margin, none on the leaf surface; California. (Fig. 292). .... *Selaginella bigelovii* Underw.**

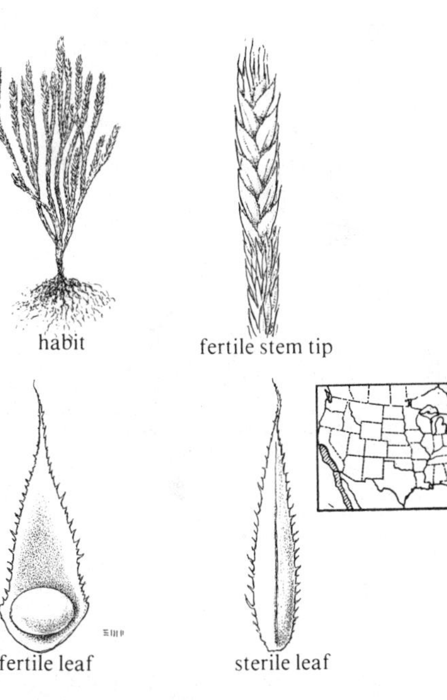

Figure 292

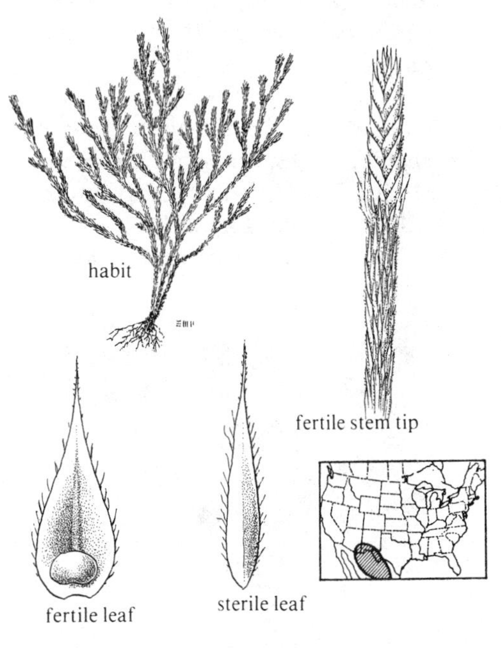

Figure 293

Figure 292 *Selaginella bigelovii*

Frequent. California; Baja California.

Selaginella viridissima Weath., of western Texas and Mexico, lacks a seta; leaf teeth are short; rare.

10b Leaves with long, hair-like teeth, some short ones on surface near the tip; east of California (Arizona, Texas, Colorado). (Fig. 293) ....
***Selaginella rupincola* Underw.**

Figure 293 *Selaginella rupincola*

Uncommon. Western Texas and Colorado to Arizona; northern Mexico.

Selaginella × *neomexicana* Maxon, of western Texas to Arizona, has a seta; cilia much shorter than in *S. rupincola.* It is probably a hybrid between *S. mutica* and *S. rupincola;* rare.

Selaginella weatherbiana R. Tryon, of Colorado and New Mexico, has a seta; leaf teeth very short; rare.

11a Branches with upper and lower sides noticeably different, leaves brown on lower side and strongly curving upward. 12

11b Branches with leaves essentially the same all around the stem, the leaves not strongly curving upward. 14

12a Plants spreading; branches often one-fourth inch apart at the base; cones inconspicuous. 13

12b Plants densely tufted with many short, erect branches; branches close to each other; cones abundant, erect, often one-half inch long. (Fig. 294) .... *Selaginella densa* Rydb.

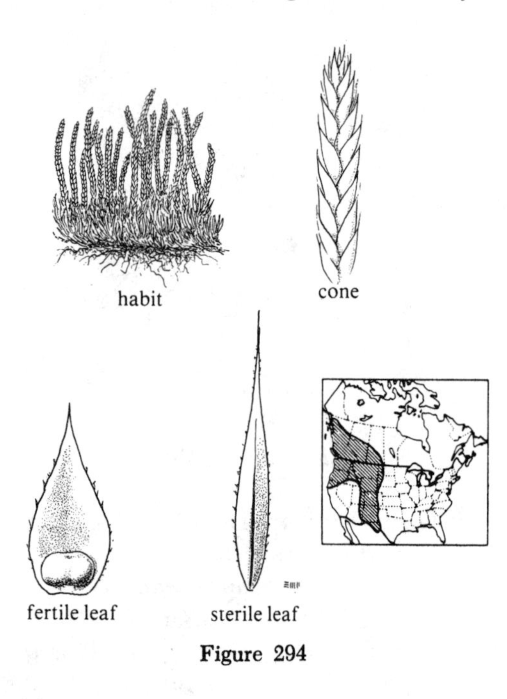

Figure 294

Figure 294 *Selaginella densa*

The most common *Selaginella* species in dry areas of the West. Meadows, dry rocks and slopes. Common. Western North America.

13a Plants of California. (Fig. 295) .... *Selaginella hansenii* Hieron.

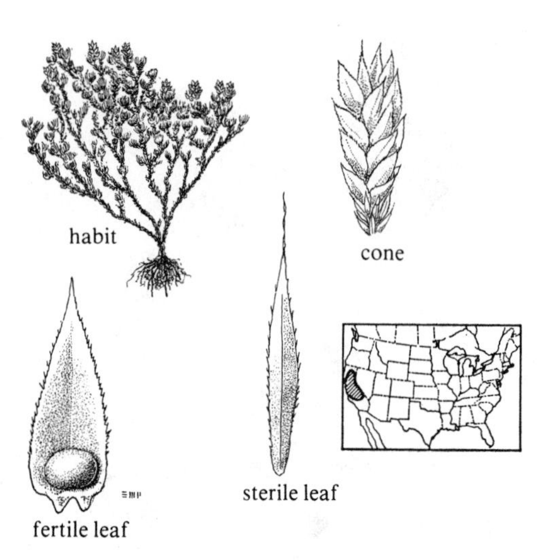

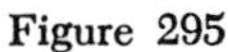

Figure 295

Figure 295 *Selaginella hansenii*

Leaves with hair-like teeth; setae straight, not falling off easily. Frequent. Central California.

Selaginella eremophila Maxon, of southwestern Arizona, southern California, and Baja California, has setae that are twisted and fall off early; rare.

13b Plants of Texas to eastern Arizona. (Fig. 296) .... *Selaginella wrightii* Hieron.

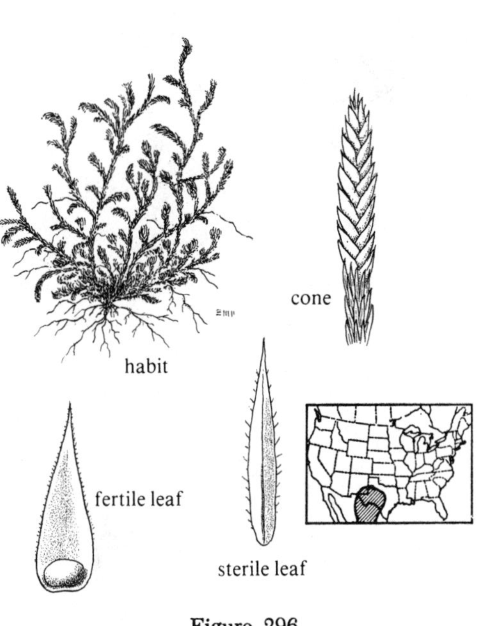

Figure 296

Figure 296 *Selaginella wrightii*

Setae short and stiff. Uncommon. Southwestern United States; Mexico.

Selaginella arizonica Maxon, of Texas to central Arizona and northern Mexico, has slender setae, often breaking off with age; uncommon.

Selaginella peruviana (Milde) Hieron., of Texas, Oklahoma, New Mexico, Mexico and South America, also has a long, slender seta, and also the spreading creeping habit of S. *wrightii;* rare.

14a Plants terrestrial. 15

14b Plants hanging from trees; long and trailing, often twelve inches or more long. (Fig. 297) ...... *Selaginella oregana* D. C. Eaton

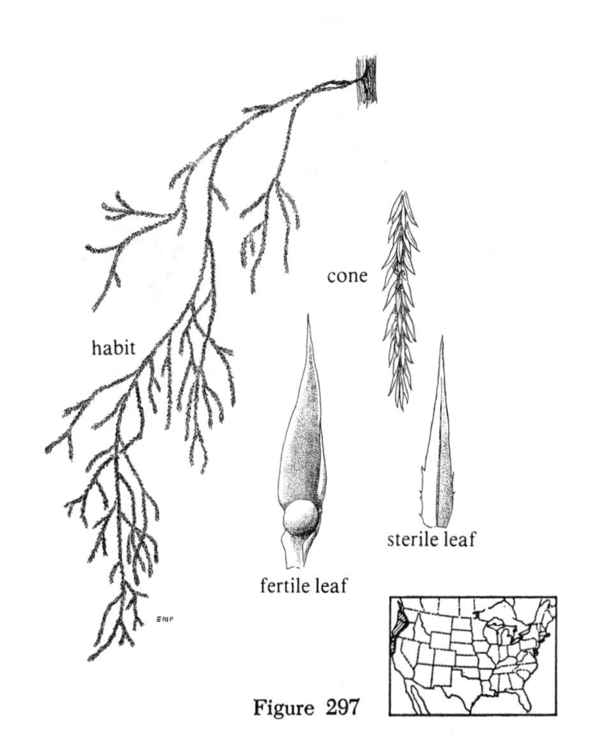

Figure 297

Figure 297 *Selaginella oregana*

Branches long and distant from each other; leaves with short hair-like teeth near leaf bases; setae short, green to white. Mossy tree trunks and branches, rarely on moist, shaded soil. Frequent. Northwestern United States.

15a Plants of the Rocky Mountains or farther west, or in the Southwest, east to Texas. 16

15b Plants of the Northeast, west to the prairie states and south to northern Georgia to Oklahoma. (Fig. 298) .... ROCK SPIKEMOSS, *Selaginella rupestris* (L.) Spring

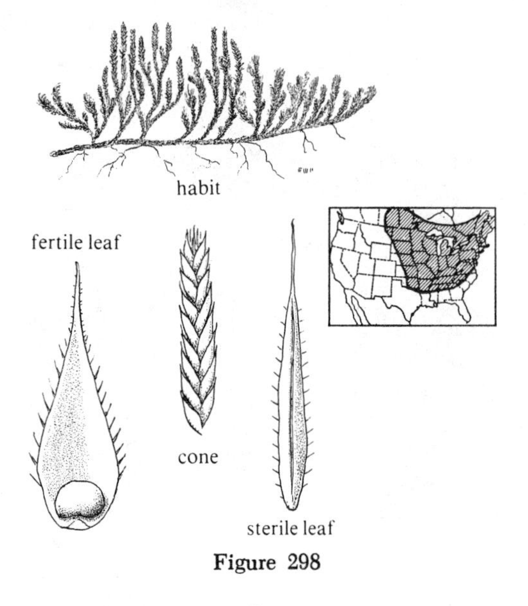

Figure 298

Figure 298 *Selaginella rupestris*

Leaves with hair-teeth; setae long. Frequent. Northeastern North America.

16a Leaves with teeth. **17**

16b Leaves lacking teeth. (Fig. 299). .... ***Selaginella watsonii* Underw.**

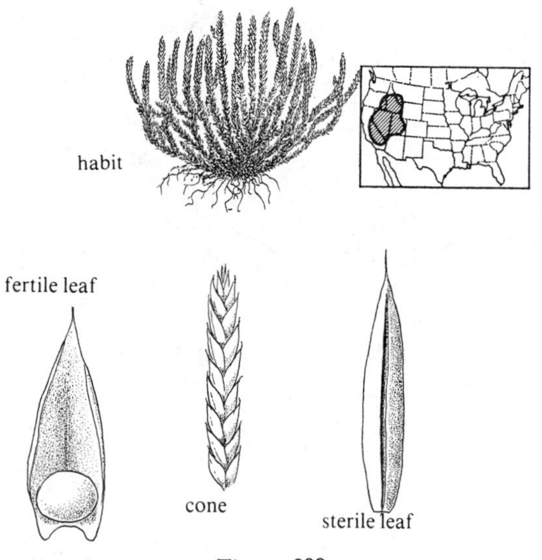

Figure 299

Figure 299 *Selaginella watsonii*

Frequent. Western United States.

17a Plants of the Southwest (Texas to Colorado and Arizona). **18**

17b Plants of the Far West (coastal states or Northwest). (Fig. 300) .... ***Selaginella wallacei* Hieron.**

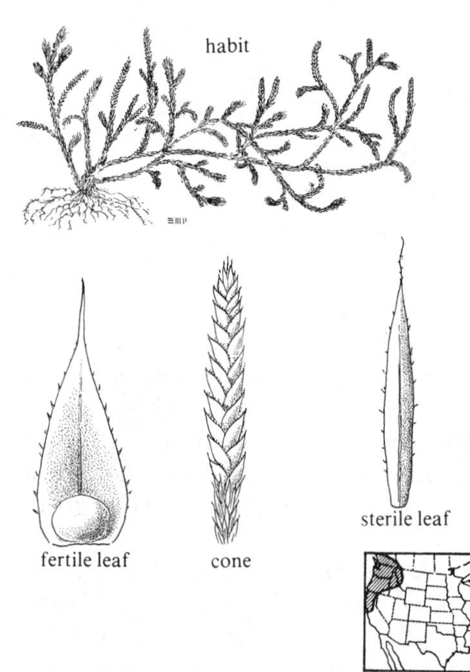

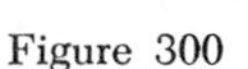
Figure 300

Figure 300 *Selaginella wallacei*

Leaves with or without hair-like teeth and long white setae. Frequent. Northwestern North America.

Selaginella sibirica (Milde) Hieron., of Alaska, Yukon, and eastern Asia, is much like *S. densa* but has white rather than yellowish setae and very long, hair-like teeth; uncommon.

18a Leaves with seta present. (Fig. 301) .... *Selaginella underwoodii* Hieron.

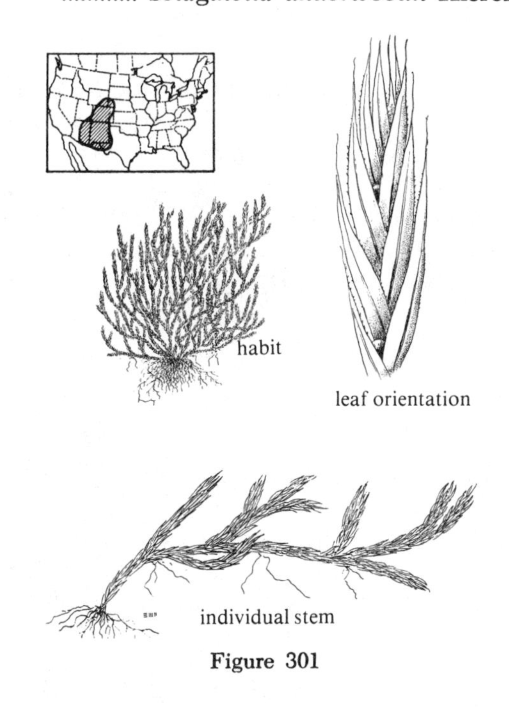

Figure 301

Figure 301 *Selaginella underwoodii*

Leaves small, narrow, overlapping, with long seta. Frequent. Southwestern United States.

Selaginella asprella Maxon, of southern California, is a distinctive gray-green color, has short, hair-like teeth and very long, white, toothed setae; rare.

18b Leaves without setae; leaves curved, boat-like. (Fig. 302). .... *Selaginella mutica* D. C. Eaton

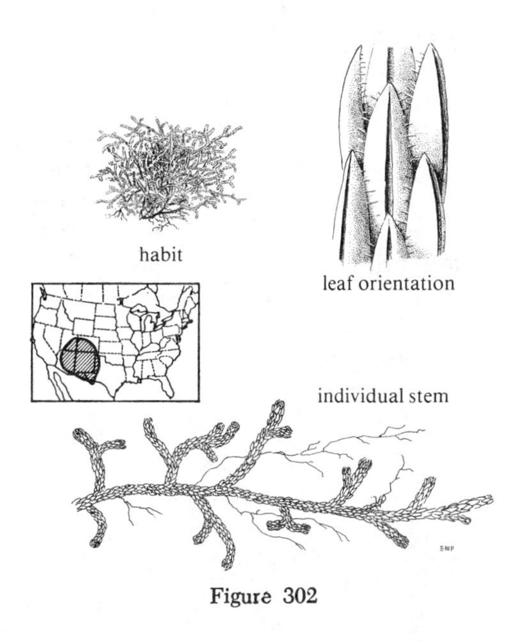

Figure 302

Figure 302 *Selaginella mutica*

Frequent. Southwestern United States.

The widespread var. *mutica* has leaves with long hair-like teeth and no seta whereas var. *limitanea* Weath. of New Mexico and western Texas has leaves with short, stiff, hair-teeth and a seta.

Selaginella utahensis Flowers, of southern Utah and Nevada, has setae very short or lacking; the plants are very short and compact; rare.

Selaginella cinerascens A. A. Eaton, of southern California and Baja California, is a minute plant with branches only about one thirty-second inch wide; leaves lacking setae; uncommon.

Selaginella leucobryoides Maxon, of southern California, Arizona, and Nevada, is very short and compact; the teeth are short and the seta is short or sometimes lacking; rare.

SPHENOMERIS

Rhizome short-creeping, hairy; fronds medium-sized, tripinnate to quadripinnate; segments very narrowly wedge-shaped; sporangia borne in small pockets at the tips of segments. Fifteen species of tropical regions, largely of the Old World. (Fig. 303) .. PARSLEY FERN, *Sphenomeris clavata* (L.) Maxon

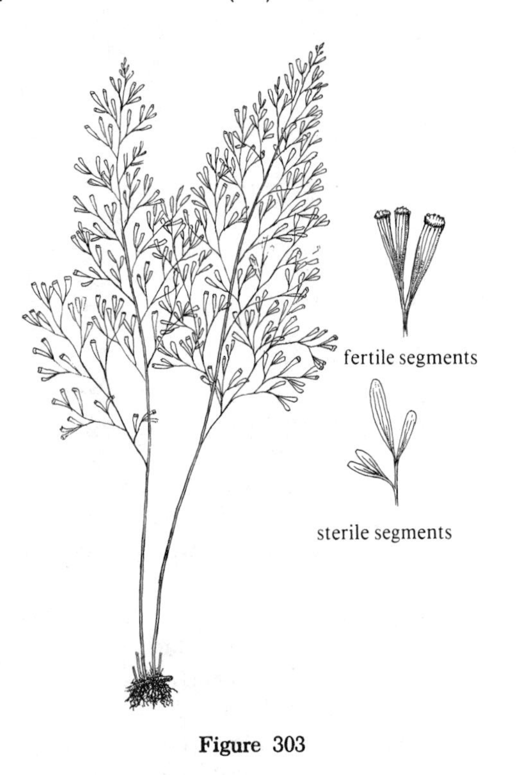

Figure 303

Figure 303 *Sphenomeris clavata*

Frond ten to twenty inches tall, two to five inches wide; stipe slender, green to yellow; segments one-half inch long, one-sixteenth inch or less wide. Limestone rocks. Uncommon. Southernmost Florida; West Indies, southern Mexico.

STENOCHLAENA

Vine Fern

Rhizome long-creeping, climbing on ground and trees, scaly; fronds medium to large size, dimorphic; sterile fronds pinnate; fertile fronds bipinnate, their pinnae very narrow, lacking any green tissue. Five species of Old World tropics. (Fig. 304) .... GREATER VINE-FERN, *Stenochlaena tenuifolia* (Desv.) Moore

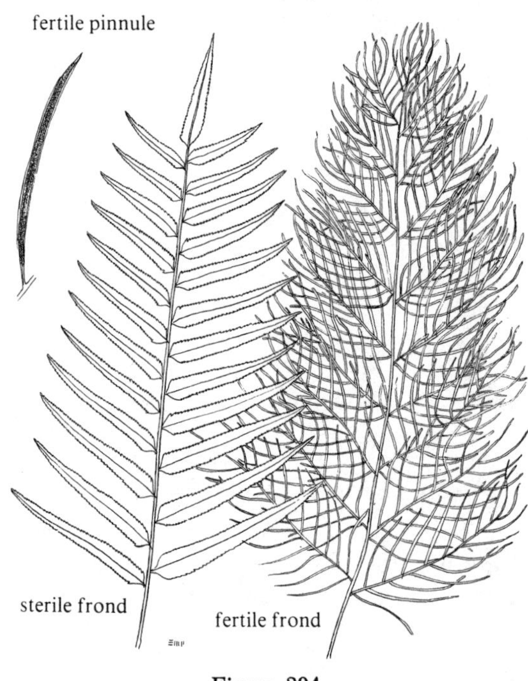

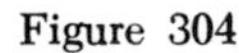

Figure 304

Figure 304 *Stenochlaena tenuifolia*

Fronds one and one-half to three feet long, ten to twenty inches wide; sterile pinnae long and slender, finely toothed. Moist woods. Occasionally introduced in southernmost Florida; native of Old World tropics.

TECTARIA

Halberd Fern

Rhizome ascending, scaly; fronds medium-sized to small, pinnatifid to pinnate-pinnatifid; veins intricately netted with included veinlets; sori medial, round, indusium kidney-shaped or umbrella-shaped. Two hundred fifty species of tropical regions.

1a Plants fourteen to twenty-eight inches tall; indusium umbrella-shaped. (Fig. 305) .... **EARED HALBERD-FERN, *Tectaria heracleifolia* (Willd.) Underw.**

Figure 305

Figure 305 *Tectaria heracleifolia*

Rhizome stout, ascending; fronds fourteen to twenty-eight inches tall, six to twelve inches wide, pinnate with one to four pairs of pinnae, the basal pair of pinnae with greatly exaggerated ears pointing downward, making the frond nearly pentagonal in shape; indusium umbrella-shaped, round. Shaded limestone rocks. Frequent. Southern Florida, western Texas; tropical America.

Tectaria incisa Cav., common in tropical America, has recently been reported in southern Florida. It is generally twenty to fifty inches tall with three to twelve pairs of pinnae, all lacking ears, and the indusia are kidney-shaped.

1b Plants five to ten inches tall; indusium kidney-shaped. (Fig. 306) .... **LEAST HALBERD-FERN, *Tectaria lobata* (Poir.) Morton [*T. minima* Underw.]**

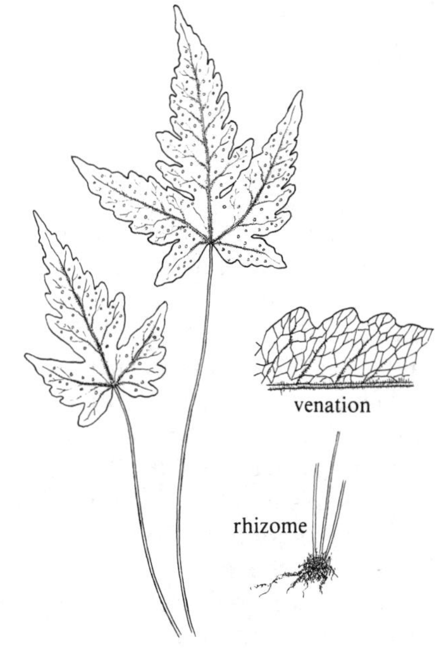

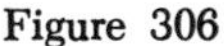

Figure 306

Figure 306 *Tectaria lobata*

Rhizome short-creeping; fronds five to ten inches long, one to two and one-half inches wide, pinnately lobed, occasionally with one pair of distinct basal pinnae, nearly naked; indusium kidney-shaped. Limestone rocks. Uncommon. Southernmost Florida; Cuba.

The HAIRY HALBERD-FERN, *Tectaria coriandrifolia* (Sw.) Underw., is a rare species that resembles the least halberd-fern but differs in having the stipe, rachis, and major veins covered with fine hairs or hair-like scales, more divided blade with two or more pairs of distinct pinnae, the pinnae cut away below to form unequal pinna bases, and the larger fronds having vegetative buds near the tip of the blade. It has been found rarely in Dade Co., Florida, and in the West Indies.

The HYBRID HALBERD FERN, *T.* × *amesiana* A. A. Eaton, is thought to be a hybrid between *T. coriandrifolia* and *T. lobata.* It has scattered scales on the blade, the pinnae with nearly equal bases. This very rare plant has been found in shaded limestone sinks of Dade Co., Florida.

THELYPTERIS

Rhizome creeping or ascending, scaly; fronds of medium to large size; blades pinnate to tripinnate, mostly pinnate-pinnatifid; veins free or netted; blade bearing tiny, needle-shaped hairs unique for this genus; sori medial, round (rarely elongate); indusium kidney-shaped or rarely absent. Nearly one thousand species of tropical, subtropical, and temperate regions.

Most species of *Thelypteris* were at one time placed in *Dryopteris.* Today some botanists divide *Thelypteris* into several segregate genera. I have not used the splinter genera here but have included them in parentheses.

1a Blade pinnate or pinnate-pinnatifid. 2

1b Blade bipinnate to tripinnate. (Fig. 307) *Thelypteris* (*Macrothelypteris*) *torresiana* (Gaud.) Alston [*Dryopteris setigera* of authors, *D. uliginosa* (Kunze) C. Chr.]

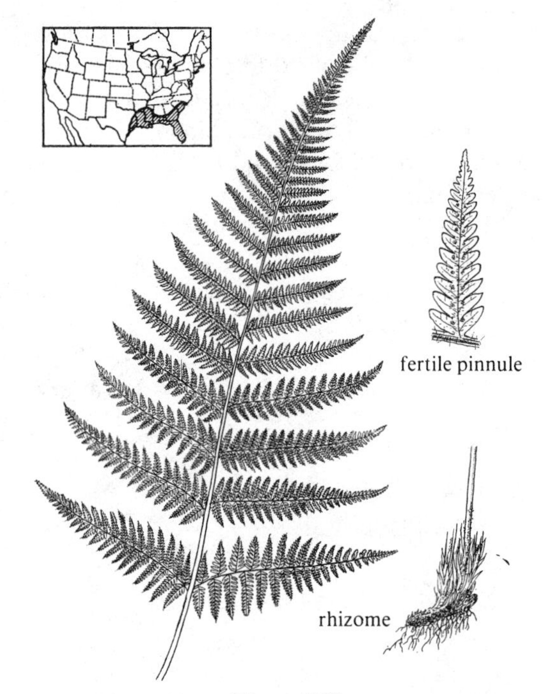

Figure 307

Figure 307 *Thelypteris torresiana*

Rhizome short-creeping, with long, brown scales; fronds twenty-four to forty-two inches tall, ten to twenty inches wide; stipe with a whitish bloom; blade broadly triangular, bipinnate to tripinnate. Moist woods. Frequent. Southeastern United States; native to Old World but now rapidly spreading in America.

2a Blade pinnate-pinnatifid; veins free or with some uniting, but never with more than two rows of areoles between the pinna midvein and the margin. 3

2b **Blade once pinnate; venation a complex network with several rows of areoles between the pinna midvein and the margin. (Fig. 308) ...... *Thelypteris* (*Meniscium*) *serrata* (Cav.) Alston**

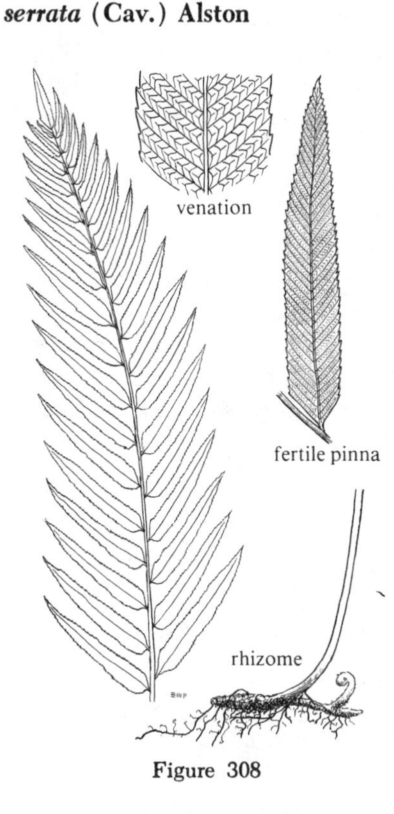

Figure 308

Figure 308 *Thelypteris serrata*

Rhizome short-creeping; fronds twenty-four to forty-eight inches tall; blade pinnate with terminal pinna like the lateral pinnae; margin toothed; main veins run from midveins to margin with minor veins connecting them in a ladder-like pattern; sori elongate on the cross-veins. Wet woods. Rare. Southern Florida; tropical America.

Thelypteris (*Meniscium*) *reticulata* (L.) Proctor, of southernmost Florida and tropical America, has an untoothed margin; very rare.

3a **Blade broadly triangular; rachis winged along most or all of its length; indusium lacking; fronds mostly less than twenty inches tall. 4**

3b **Blade lance-shaped to narrowly triangular; rachis not winged; indusium present or rarely lacking; fronds mostly twenty-five to fifty inches tall. 5**

4a **Lowest pinnae smaller than those above and separate from them, the rachis not winged between them; stipe and rachis scaly. (Fig. 309). .... LONG BEECH-FERN, NORTHERN BEECH-FERN, *Thelypteris* (*Phegopteris*) *phegopteris* (L.) Slosson [*Phegopteris connectilis* (Michx.) Watt]**

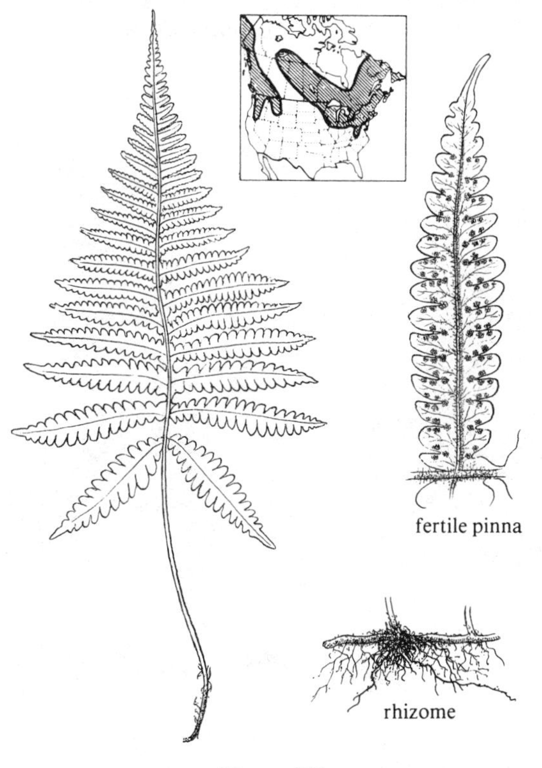

Figure 309

Figure 309 *Thelypteris phegopteris*

Rhizome long-creeping; fronds distant, eight to eighteen inches tall, four to six inches wide; stipe more than half the frond length; the blade pinnate-pinnatifid to bipinnatifid, triangular; upper pinnae connected by a rachis wing but wing lacking between lowest two pairs of pinnae; basal pinnae usually strongly turned downward; stipe and blade covered below with hairs and narrow scales, looking like coarse hairs; sori near the margin; indusium lacking. Moist woods. Common. Northern North America; Europe, Asia.

4b Lowest pinnae larger than those above and connected to them by a wing on the rachis; stipe and rachis with fine hairs. (Fig. 310) .... BROAD BEECH-FERN, SOUTHERN BEECH-FERN, *Thelypteris* (*Phegopteris*) *hexagonoptera* (Michx.) Weath.

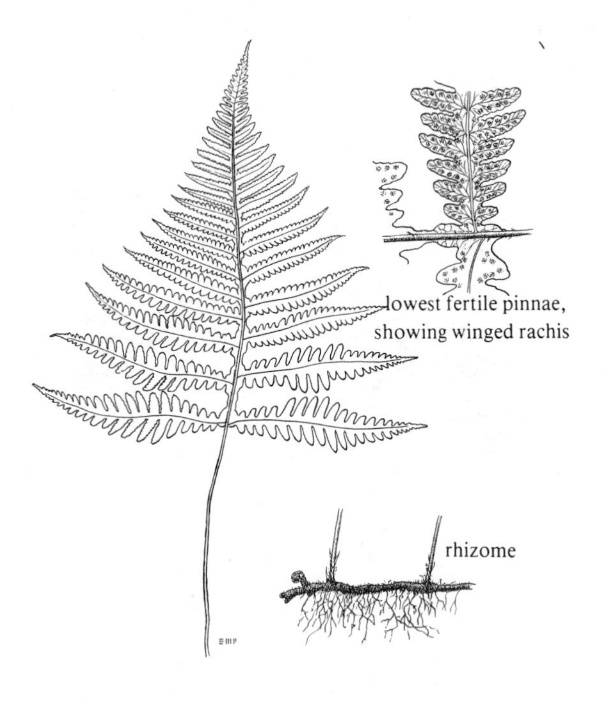

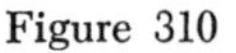

Figure 310

Figure 310 *Thelypteris hexagonoptera*

Rhizome long-creeping; fronds distant, fifteen to twenty-four inches tall, five to twelve inches wide; blade bipinnatifid, broadly triangular; all pinnae connected to those above and below by a wing on the rachis; lower surface with fine hairs; sori near margin; indusium lacking. Common. Moist woods. Northern North America.

5a Basal veins of segments going into the base of the sinuses in the pinna margins (Fig. 311) blades not reduced at the base except in *T. sclerophylla* and slightly in *T. dentata;* veins uniting in some species. 6

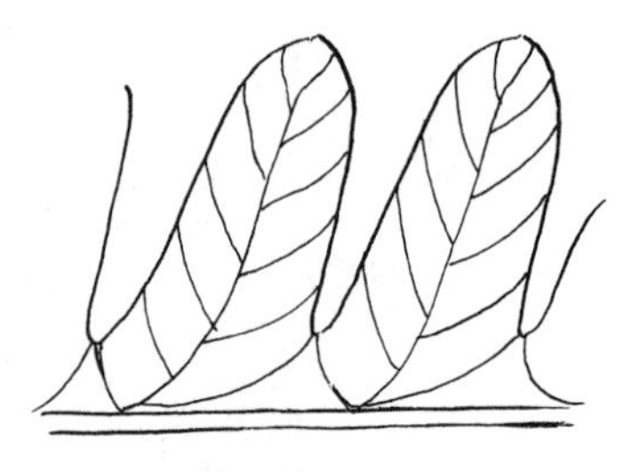

Figure 311

5b Basal veins of segments entering the pinna sinuses above the base of the sinus (Fig. 312); blades usually strongly tapering to the base (except in *T. palustris, T. simulata,* and *T. pilosa*); veins undivided or once-forked, never uniting. 14

Figure 312

6a **Fronds erect, not rooting at the tip; pinnae five to eight inches long, rarely only one and one-half inch long. 7**

6b **Fronds reclining, often rooting at the tip; pinnae one-half to one and one-half inches long (Fig. 313) .... CREEPING FERN, *Thelypteris* (*Goniopteris*) *reptans* (Gmel.) Morton**

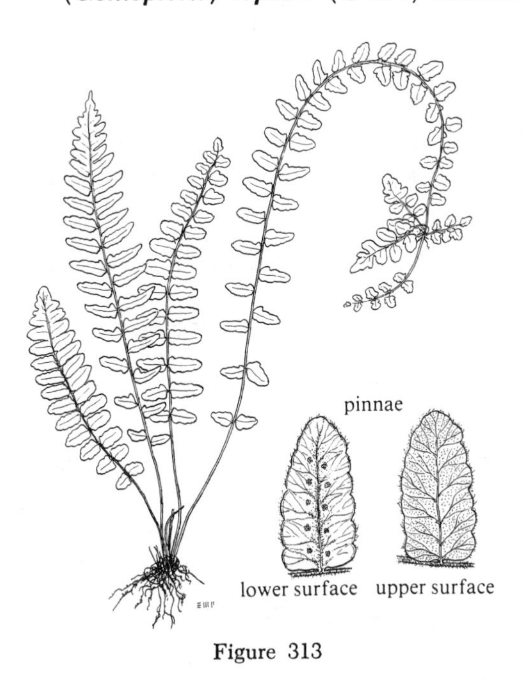

Figure 313

Figure 313 *Thelypteris reptans*

Rhizome short-creeping with clumped fronds; fronds eight to sixteen inches long, one to three inches wide; pinnate to pinnate-pinnatifid, gradually tapering to a pinnatifid tip; pinnae shallowly lobed; stipe very short; blade apex often long and rooting at various points along the rachis to form new plants; frond surface with minute stellate hairs; veins uniting in sinuses between lobes; sori small, round, medial. Shaded limestone rocks. Frequent. Florida; Caribbean region.

7a **Basal veins of adjacent pinna lobes united below the sinus with a vein going toward the sinus. (Fig. 314). 8**

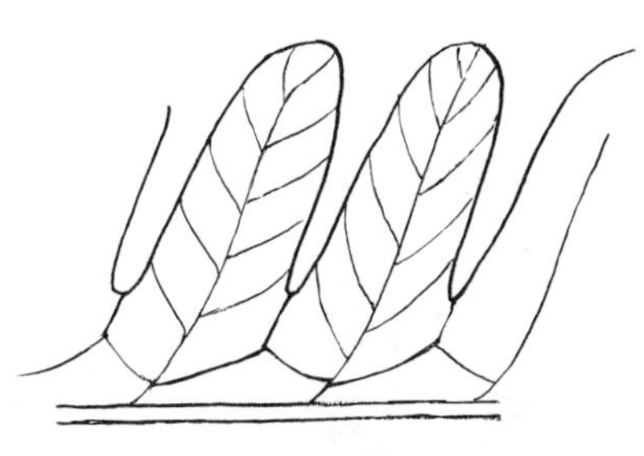

Figure 314

7b **Basal veins of adjacent pinna lobes free from each other or running close together toward the sinus but not uniting. (Fig. 315) 11**

Figure 315

8a **Rhizome short-creeping to nearly erect, scaly; pinnae over one-half inch wide (except in *T. augescens*). 9**

8b **Rhizome long-creeping, black, nearly naked; pinnae only three-eighths to one-half inch wide. (Fig. 316) .... *Thelypteris* (*Cyclosorus*) *interrupta* (Willd.) Iwatsuki [*T. gongylodes* (Schkuhr) Small, *T. totta* (Thunb.) Schelpe]**

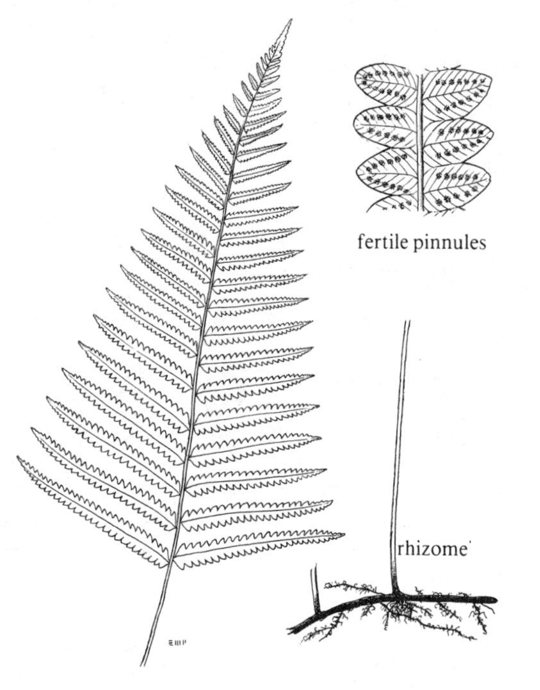

Figure 316

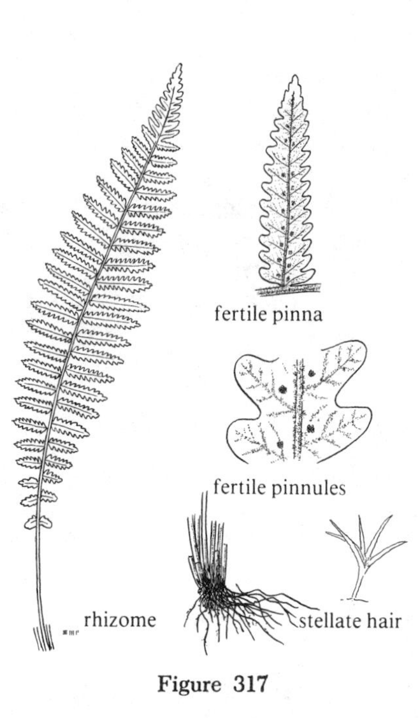

Figure 316 *Thelypteris interrupta*

Rhizome long-creeping, black, nearly naked; fronds distant, thirty to fifty inches tall, eight to twelve inches wide; blade pinnate-pinnatifid; pinnae three-eighths to one-half inch wide, cut shallowly, less than half way to the midvein; naked or finely hairy below; sori much closer to margins than to midvein of pinnule. Swamps and wet ditches. Common. Florida; pantropical.

Hairy and naked forms of this species are known but their significance is uncertain.

Figure 317 *Thelypteris sclerophylla*

Rhizome ascending; fronds clumped, eleven to twenty-two inches tall, three to six inches wide; stipe short, about one-sixth the frond length; blade pinnate-pinnatifid, tapering to the base, the apex gradually reduced, pinnatifid; stipe and rachis with minute stellate hairs. Moist woods. Rare. Southern Florida; West Indies.

Thelypteris (*Goniopteris*) *tetragona* (Sw.) E. P. St. John, of southern Florida and tropical America, differs in the blade not tapering at the base and the blade apex like a terminal pinna, not pinnatifid; rare.

9a Stipe hairs needle-like. 10

9b Stipe hairs minute and stellate. (Fig. 317) .... *Thelypteris* (*Goniopteris*) *sclerophylla* (Poeppig ex Spreng.) Morton

10a Hairs on the lower blade surface of uniform short length, shorter than the sporangia; stipe and rachis purplish. (Fig. 318) .... DOWNY WOOD-FERN, *Thelypteris* (*Cyclosorus*) *dentata* (Forssk.) E. P. St. John

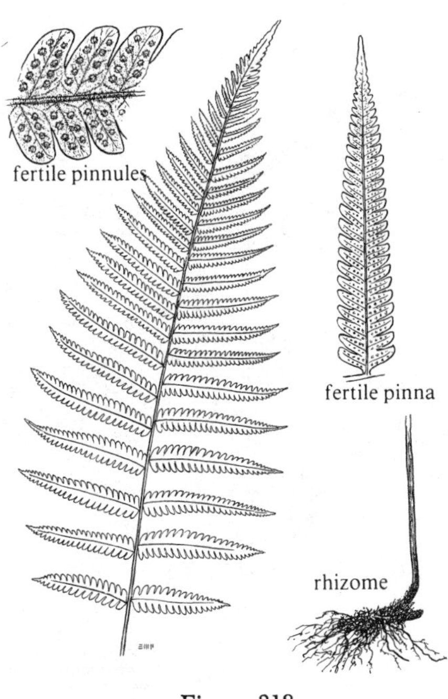

Figure 318

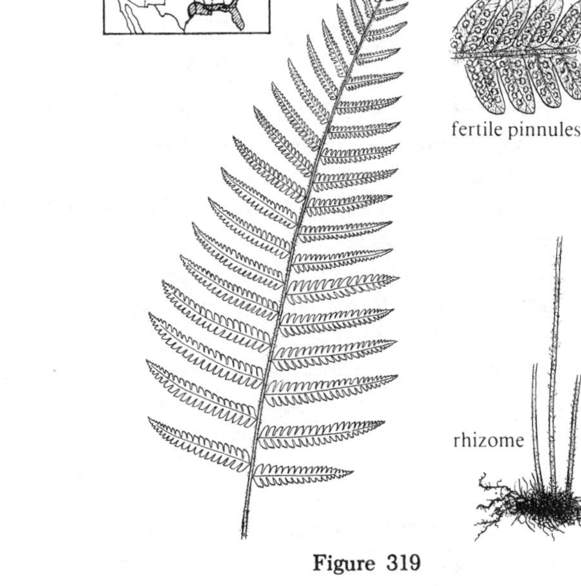

Figure 319

Figure 318 *Thelypteris dentata*

Rhizome short-creeping; fronds twenty-four to fifty inches tall, six to fourteen inches wide; blade pinnate-pinnatifid; pinnae five-eighths to one inch wide, hairy on both upper and lower surfaces; hairs uniformly short, shorter than the sporangia. Moist woods. Frequently escaped from cultivation or naturalized. Gulf states; pantropical.

10b Hairs on lower blade surface of irregular length; many longer than the sporangia, many also short, especially those on the indusium; stipe and rachis straw-colored. (Fig. 319) .... *Thelypteris* (*Cyclosorus*) *quadrangularis* (Fée) Schelpe

Figure 319 *Thelypteris quadrangularis*

Rhizome short-creeping; fronds sixteen to thirty-two inches tall, four to eight inches wide; blade pinnate-pinnatifid; pinnae three-eighths to one inch wide, cut more than half way to midvein; hairy on both upper and lower surfaces; many hairs longer than the sporangia, variable in length, some shorter, especially those on the indusium. Moist woods. Common. Southeastern United States; tropical America.

Our material is var. *versicolor* (R. St. John) A. Reid Smith. Recent evidence suggests that *T. quadrangularis* may be the same as the Old World *T. hispidula* (Decne.) Reed.

11a Veins and midveins on upper surface with stout hairs; scales absent on lower surface of rachis and pinna midveins. 12

11b Veins and midveins on upper surface naked; a few small scales often present on lower surface of rachis and pinna midveins. 13

12a One or two pairs of basal pinnae somewhat reduced; veins variable—basal veins of adjacent pinna lobes either uniting to form a vein going to the sinus or the basal veins run side by side to the sinus; rhizome short-creeping to suberect; veins above always with stout hairs; hairs below both long and short, short on the indusium. (Fig. 319) .... *Thelypteris* (*Cyclosorus*) *quadrangularis* (see above)

12b Lowermost pinnae usually not reduced; basal veins of adjacent pinna lobes run side by side to the sinus without uniting; rhizome short-creeping to long-creeping; veins above with or without long stout hairs; hairs below all long, even those on the indusium (longer than the sporangia). (Fig. 320) .... *Thelypteris* (*Cyclosorus*) *kunthii* (Desv.) Morton [*T. normalis* (C.Chr.) Moxley]

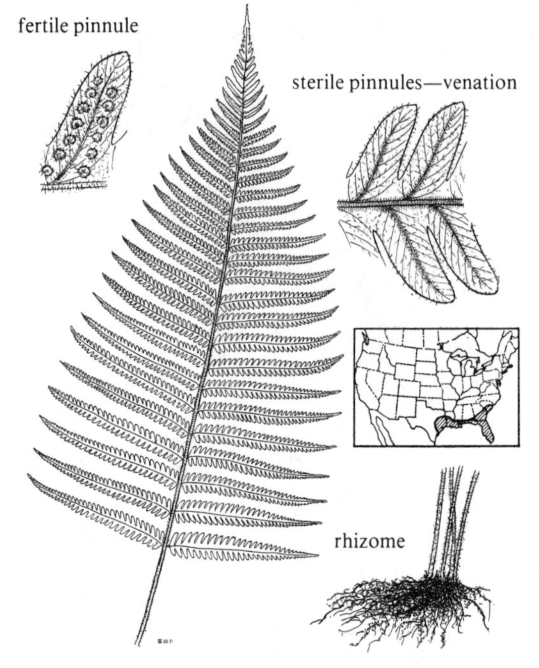

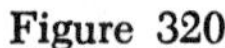

Figure 320

Figure 320 *Thelypteris kunthii*

Rhizome short- to long-creeping; fronds twenty-two to forty-five inches tall, six to twelve inches wide; blade bipinnate-pinnatifid; both upper and lower surfaces very hairy; hairs all nearly equal in length, including those of the indusium, longer than the sporangia. Wooded rocky slopes. Abundant. Southeastern United States; West Indies, Middle America.

Thelypteris (*Cyclosorus*) *patens* (Sw.) Small, of southernmost Florida and tropical America, differs in having an erect rhizome and lacking any hairs on the upper blade surface other than hairs on the pinna midveins; rare.

13a Terminal "pinna" usually at least five times as long as wide; scales on pinna midveins below often numerous; blade leathery; lower pinnae only one-fourth to three-eighths inch wide, cut half way

to pinna midvein. (Fig. 321) ....
...... *Thelypteris (Cyclosorus) augescens* (Link) Munz & Johnston

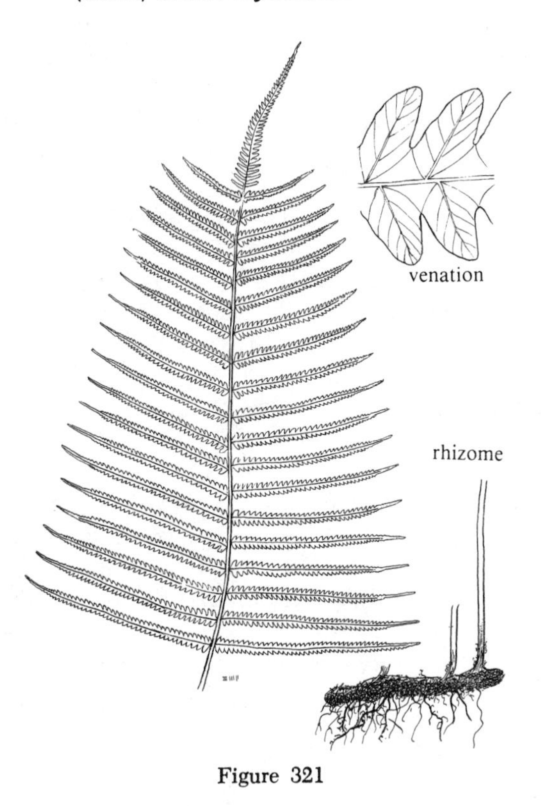

Figure 321

Figure 321 *Thelypteris augescens*

Rhizome short- to long-creeping; fronds thirty to sixty inches tall, ten to twenty inches wide; blade pinnate-pinnatifid, tapering abruptly at the tip to form a very slender terminal "pinna"; pinnae one-fourths to three-eighths inch wide, cut half way to the pinna midvein, lower surface hairy, upper surface naked. Wooded slopes and limestone rocks. Frequent. Southernmost Florida; West Indies.

13b Terminal "pinna" less than five times as long as wide; scales on pinna midveins below lacking or sparse; blade papery to nearly leathery; pinnae about one-half inch wide, cut about three-fourths of the way to the pinna midvein. (Fig. 322) ....
.... *Thelypteris (Cyclosorus) ovata* R. St. John

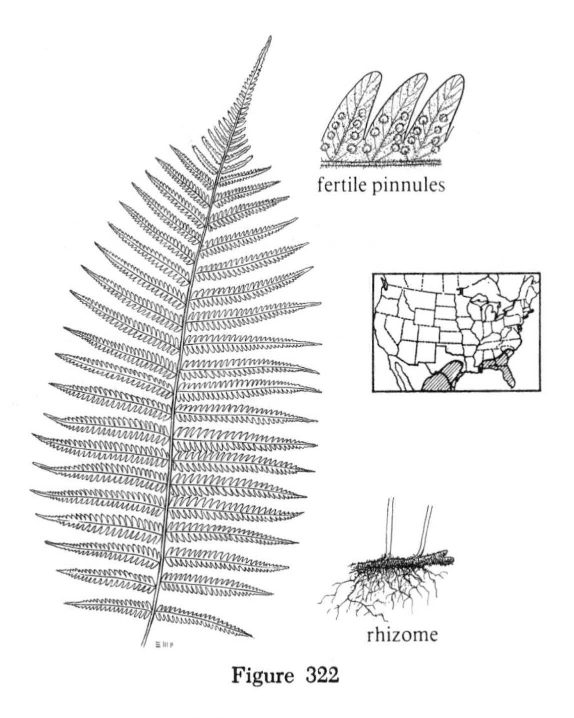

Figure 322

Figure 322 *Thelypteris ovata*

Rhizome short- to long-creeping; fronds twenty-two to fifty inches tall, four to twenty inches wide; blade pinnate-pinnatifid; pinnae about one-half inch wide, cut three-fourths of the way to the pinna midvein; hairy below, naked above. Moist woods. Southeastern United States; Mexico.

East of Texas the scales are usually lacking on the pinna midveins below (var. *ovata*) whereas in Texas there are a few scales present on the lower surface of the pinna midveins (var. *lindheimeri* (C.Chr.) A. Reid Smith).

Thelypteris (Cyclosorus) puberula (Baker) Morton, of Arizona, southern California, and Middle America, may be minutely hairy above (var. *sonorensis* A. Reid Smith) or very rarely naked above (var. *puberula*).

Thelypteris grandis A. Reid Smith var. *grandis* of the West Indies has recently been discovered in Collier County, Florida. It is distinguished by its fronds being over three feet tall and its pinnae an inch or more broad.

14a (5b) Blade not gradually reduced to the base, the pinnae stopping abruptly; stipe one-third to one-half the frond length. 15

14b Blade gradually reduced to the base of the frond; stipe less than one-sixth of the frond length. 16

15a At least some of the veins forked between the pinnule midveins and the margins; plants of a wide variety of wet situations. (Fig. 323). .... MARSH FERN, ... *Thelypteris (Thelypteris) palustris* Schott

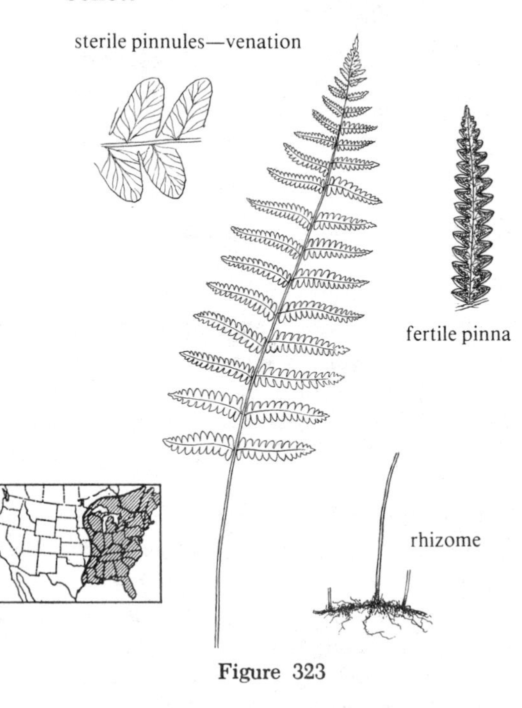

Figure 323

Figure 323 *Thelypteris palustris*

Rhizome long-creeping; fronds distant, eighteen to thirty inches tall; blade pinnate-pinnatifid, narrowly oval; pinnae not tapering toward the base of the blade, stopping abruptly; lowest pinnae more than one-half as long as the longest pinnae; fertile fronds with segments inrolled, appearing more contracted; sori medial but appearing closer to the margin because of the curled margin. Wet meadows and swamps. Abundant. Eastern North America; nearly world-wide.

Thelypteris (Leptogramma) pilosa (Mart. & Gal.) Crawford, only six to nine inches tall, is distinct with its elongate sori that lacks an indusium. The species is widespread in Middle America but is known from a single locality in northern Alabama. The material from this country is sometimes called a separate variety, var. *alabmensis* Crawford.

15b All veins unforked between the pinnule midvein and the margin; plants of acidic coniferous or blueberry swamps. (Fig. 324) .... MASSACHUSETTS FERN, BOG FERN, *Thelypteris (Parathelypteris) simulata* (Davenp.) Nieuwl.

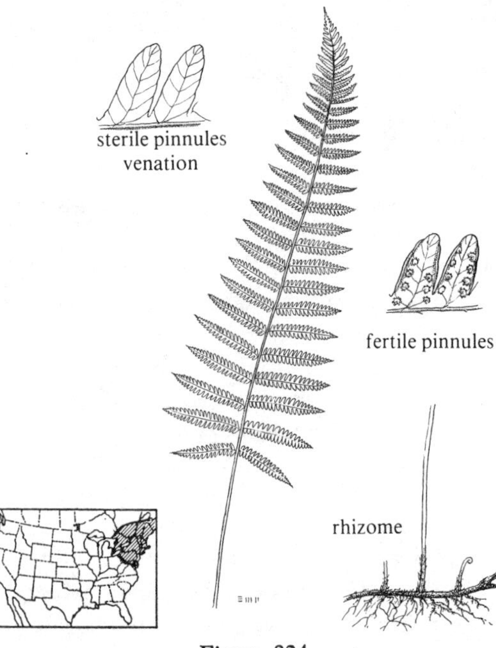

Figure 324

Figure 324 *Thelypteris simulata*

Rhizome long-creeping; fronds distant, eighteen to thirty inches tall; blade pinnate-pinnatifid, lance-shaped; blade not reduced at the base, pinnae stopping abruptly; lowest pinnae more than one-half as long as longest pinnae; sori medial, fertile pinnules not strongly inrolled. Coniferous and blueberry swamps. Rare and local, except in New England. Northeastern North America, with isolated populations in Wisconsin.

16a Plants of eastern North America. (Fig. 325) .... NEW YORK FERN, *Thelypteris* (*Parathelypteris*) *noveboracensis* (L.) Nieuwl.

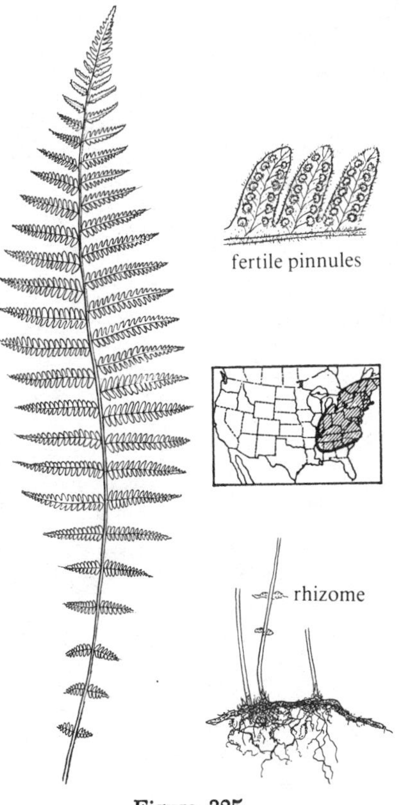

Figure 325

Figure 325 *Thelypteris noveboracensis*

Rhizome long-creeping; fronds usually distant, twelve to twenty-four inches tall; blade pinnate-pinnatifid, tapering gradually to both ends; veins unforked between pinnule midvein and margin; sori closer to margin than the pinnule midvein. Moist woods. Abundant. Eastern North America.

Thelypteris (*Amauropelta*) *resinifera* (Desv.) Proctor, of southern Florida and tropical America, has a stout, short-creeping rhizome and the lower surface is conspicuously resin-dotted. The pinnae are auriculate at the base and the segments lean toward the pinna apex instead of being at nearly right angles to the pinna midvein; rare.

16b Plants of the Northwest. .... NEVADA WOOD-FERN, *Thelypteris* (*Parathelypteris*) *nevadensis* (Baker) Clute ex Morton

Rhizome short- to long-creeping; fronds sixteen to thirty inches tall; blade pinnate-pinnatifid, tapering to both ends; pinnae one-fourth to one-half inch wide; pinnules about one-sixteenth inch wide; rachis not scaly; sori medial. Wooded slopes and wet meadows. Frequent. Northwestern North America.

This species very closely resembles *T. noveboracensis* and is most easily distinguished by geographical range.

Thelypteris (*Oreopteris*) *limbosperma* (All.) H. P. Fuchs [*T. oreopteris* (Ehrh.) Slosson], of northwestern North America, Europe, and Asia, has pinnae one-half to one inch wide, pinnules about one-eighth inch wide, rachis scaly, and sori submarginal.

TRICHOMANES

Filmy Fern

Rhizome thread-like, creeping, hairy; fronds small to minute, very thin, translucent; blade undivided to bipinnate; veins free; sori marginal in tubular pockets. Two hundred seventy-five species of humid tropical regions.

1a **Blade undivided; fronds less than one inch tall; veins spreading fan-like. (Fig. 326) .... PETERS' FILMY-FERN, *Trichomanes petersii* A. Gray**

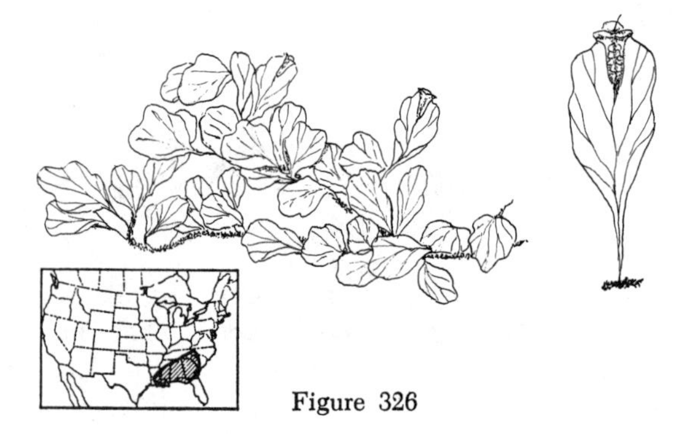

Figure 326

Figure 326 *Trichomanes petersii*

Rhizome long-creeping; fronds spaced, up to one-half inch tall, one-fourth inch wide; sorus without dark lip edge. On moist rocks or tree trunks. Rare. Southeastern United States; Middle America.

Several very rare species of tropical America have undivided fronds, marginal black stellate hairs, and often false veins between the true veins. *Trichomanes lineolatum* (v.d.B.) Hook. (Florida, West Indies and South America) has thick veinlets that are enlarged toward the margin and a sorus lip with a red edge. *Trichomanes punctatum* Poir. (Fig. 327) (Florida, tropical America) has thin veinlets that are not enlarged toward the margin and a sorus lip with a red edge. *Trichomanes membranaceum* L. (Florida; tropical America) has the margin of the leaf clothed with round scales.

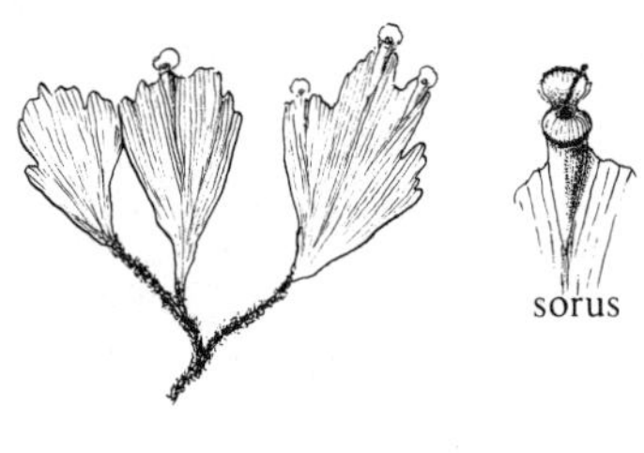

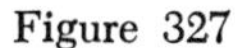

Figure 327

Figure 327 *Trichomanes punctatum*

1b **Blade divided; veins pinnate. (Fig. 328) .... APPALACHIAN FILMY-FERN, *Trichomanes boschianum* Sturm**

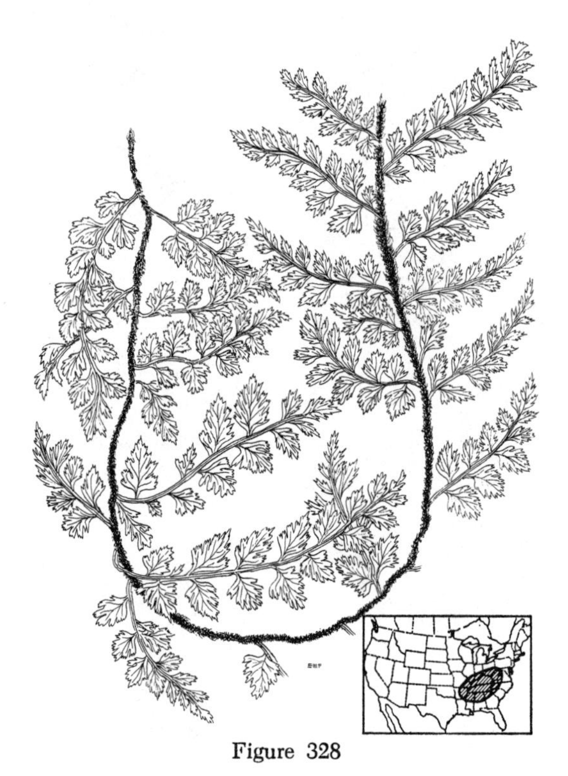

Figure 328

Figure 328 *Trichomanes boschianum*

Rhizome long-creeping; fronds distant, three to six inches long; blade pinnate-pinnatifid; stipe winged. Moist, noncalcareous rock crevices and cave mouths. Rare. Eastern United States.

There are two tropical American species with divided leaves that have been found rarely in Florida. *Trichomanes holopterum* Kunze has an erect rhizome, fronds tufted, two inches long, deeply pinnatifid. This plant is especially interesting in that its gametophytes have creeping, hair-like axes and oval aerial blades which produce gemmae that reproduce it vegetatively. *Trichomanes kraussii* Hook. & Grev. (Fig. 329) (Florida; tropical America) has fronds one to three inches long that are deeply pinnatifid to bipinnatifid with false veins and with black stellate hairs on the margins.

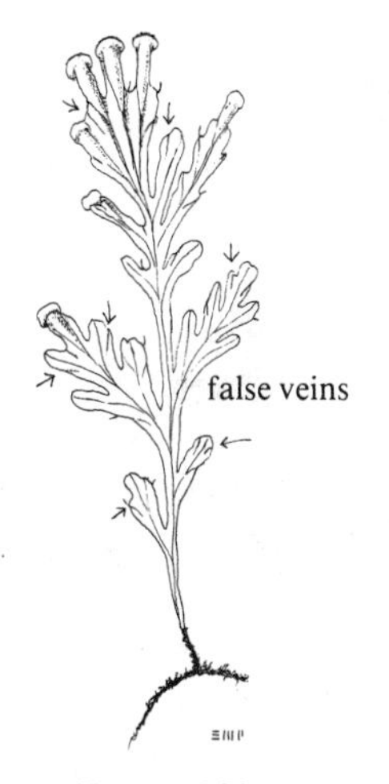

Figure 329

Figure 329 *Trichomanes kraussii*

TRISMERIA

Goldenrod Fern

Rhizome ascending, scaly; fronds large, bipinnate, the pinnae three-parted and spreading three-dimensionally like a goldenrod plant, white waxy beneath; sori running along the veins without an indusium. One species, of American tropics. (Fig. 330) .... GOLDENROD FERN, *Trismeria trifoliata* (L.) Diels

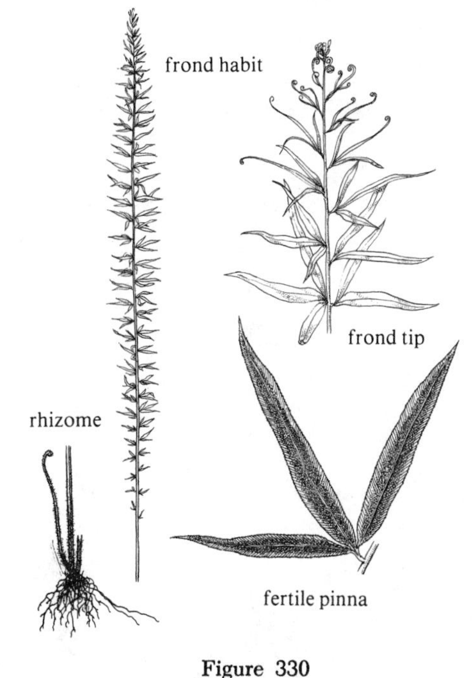

Figure 330

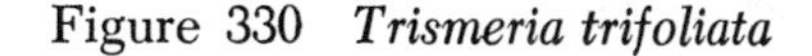
Figure 330 *Trismeria trifoliata*

Fronds three to seven feet tall; pinnae three-parted from the base, each segment one-fourth to three-eighths inch wide, three to seven inches long. Seepage banks and ditches. Rare. Southernmost Florida; tropical America. The appearance of the fronds when pressed belies their three-dimensional character.

VITTARIA

Shoestring Fern

Rhizome short-creeping, scaly; fronds small to medium-sized, shiny, leathery, undivided, linear, very narrow; sori in two grooves running the length of the blade parallel to the midvein. Eighty species, on trees and rocks in tropical and subtropical regions. (Fig. 331) ....
.... SHOESTRING FERN, GRASS FERN, *Vittaria lineata* (L.) J. E. Smith

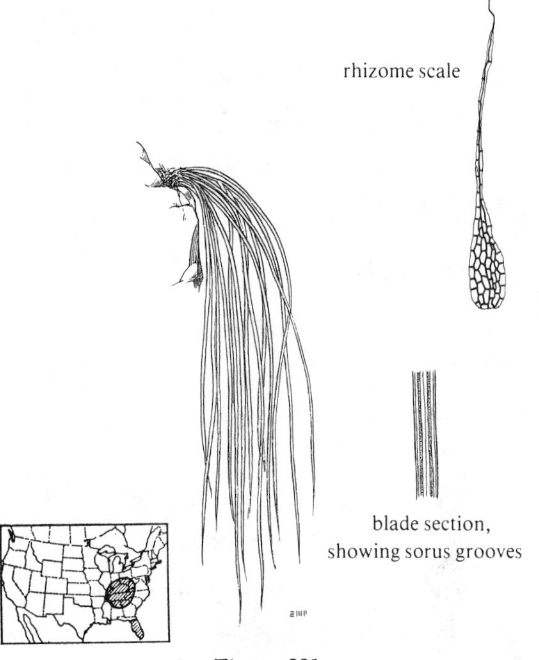

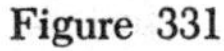

Figure 331

Figure 331 *Vittaria lineata*

Rhizome short-creeping, with iridescent clathrate scales narrowed to hairlike tip only one to two cells wide; fronds twelve to twenty-four inches long, one-sixteenth to one-eighth inch wide. Epiphyte in moist woods on trunks of palmettos and oaks. Common. Florida, once found on rock cliffs in central Georgia; tropical America.

The gametophyte of this or a related species is found in grottoes and at the mouths of caves north to southern Ohio. It produces no sporophytes but reproduces vegetatively by gemmae which produce new gametophytes.

Another species, *V. graminifolia* Kaulf. (*V. filifolia* Fée), is reported from Collier Co., Florida. It is distinguished from *V. lineata* by having rhizome scales that narrow gradually to the tip (more than two cells wide nearly to the end). It is the most common species of *Vittaria* in tropical America.

WOODSIA

Woodsia

Rhizome short-creeping or ascending, scaly; fronds clumped, small, pinnate-pinnatifid to bipinnate-pinnatifid; veins free, sori medial, round; indusium arising beneath the sorus and surrounding it as a series of broad straps or hairs. Twenty-two species of temperate and subtropical regions.

1a **Stipes articulate, breaking off across a distinct, predetermined line, leaving stubble of uniform length; blade not glandular; rhizome scales concolorous, with no dark streaks at all. 2**

1b **Stipes not articulate, old stipe bases irregular in length; blades glandular; rhizome scales tan when young but later developing black streaks. 4**

2a **Stipes yellow. 3**

2b **Stipes brown. (Fig. 332) .... ALPINE WOODSIA, *Woodsia alpina* (Bolton) S. F. Gray**

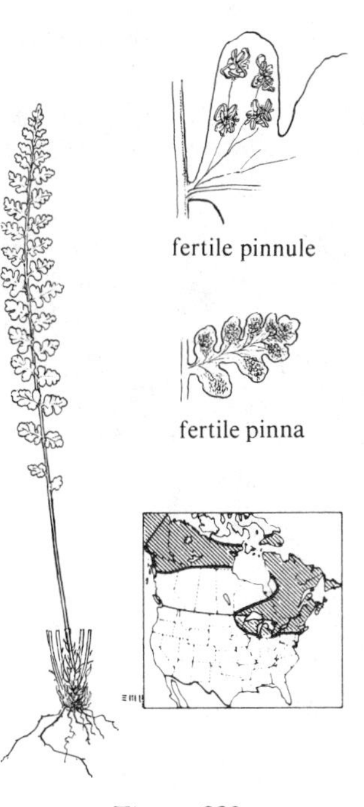

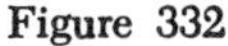

Figure 332

Figure 332 *Woodsia alpina*

Rhizome ascending; scales brown; fronds two to seven inches tall (mostly three to four), one-half to one inch wide; stipes articulate, brown, naked except for a few scales at base, about one-fourth the frond length; blade pinnate-pinnatifid, very narrow, almost linear, nearly naked; indusium of hairs. Dry or moist rocks. Rare. Northern North America; Europe, Asia.

3a Blade and stipe naked. (Fig. 333) .... SMOOTH WOODSIA, *Woodsia glabella* R. Brown

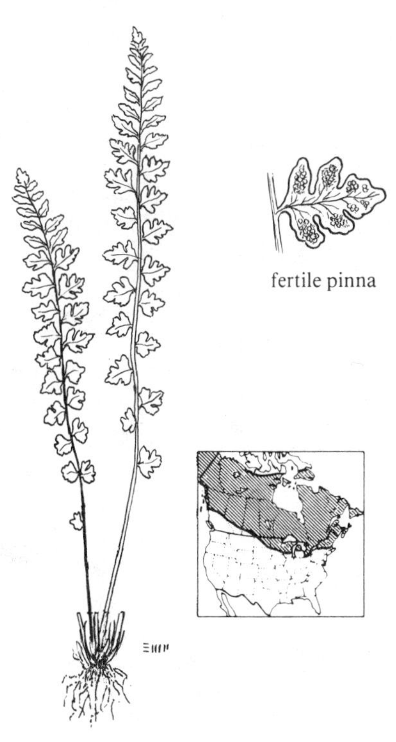

Figure 333

Figure 333 *Woodsia glabella*

Rhizome ascending; scales tan to brown; fronds two to six inches tall (mostly two to three), one-fourth to five-eighths inch wide; stipes articulate, yellow, naked, about one-sixth the frond length; blade pinnate-pinnatifid, linear, naked; indusium of hairs. Moist, shaded, calcareous rocks or cliffs. Rare. Northern North America; Europe, Asia.

3b Blade and stipe with abundant scales and hairs. (Fig. 334) .... RUSTY WOODSIA, *Woodsia ilvensis* (L.) R. Brown

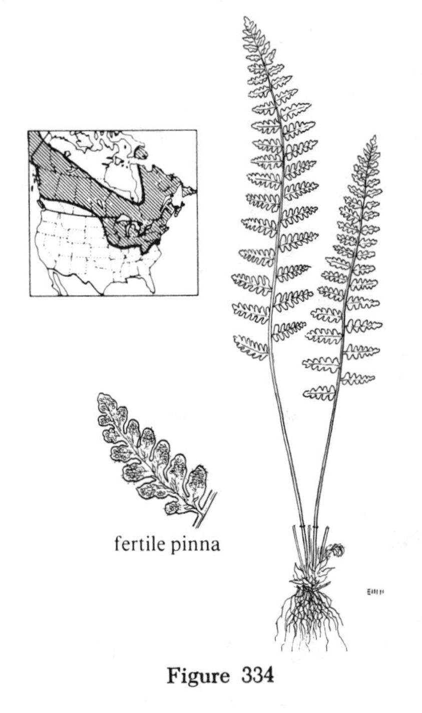

Figure 334

Figure 334 *Woodsia ilvensis*

Rhizome short-creeping; scales light brown; fronds three to eight inches tall (mostly four to six), three-fourths to one and one-fourth inches wide; stipe articulate, yellow, scaly and hairy, one-third to one-half the frond length; blade pinnate-pinnatifid to bipinnate, narrowly oblong; lower surface covered with hairs and scales; indusium of hairs. Dry cliffs and rocky slopes. Frequent to common. Northern North America; Europe, Asia.

Woodsia × abbeae Butters is a hybrid involving *W. ilvensis;* the other parent is still not known for certain; locally abundant in the upper Great Lakes region.

4a Indusium of hairs or very narrow scales; rachis lacking scales. 5

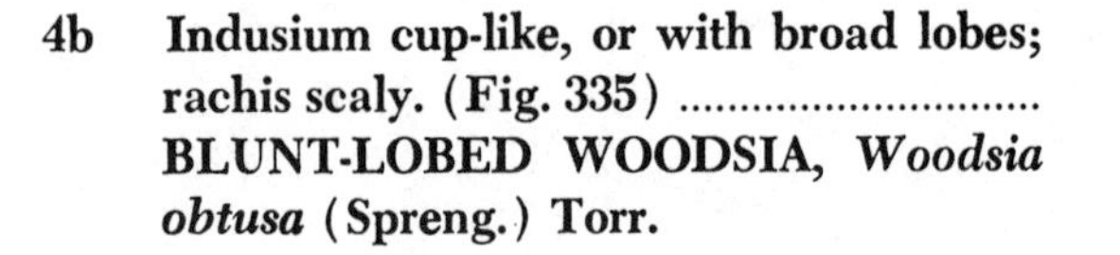
4b Indusium cup-like, or with broad lobes; rachis scaly. (Fig. 335) .... BLUNT-LOBED WOODSIA, *Woodsia obtusa* (Spreng.) Torr.

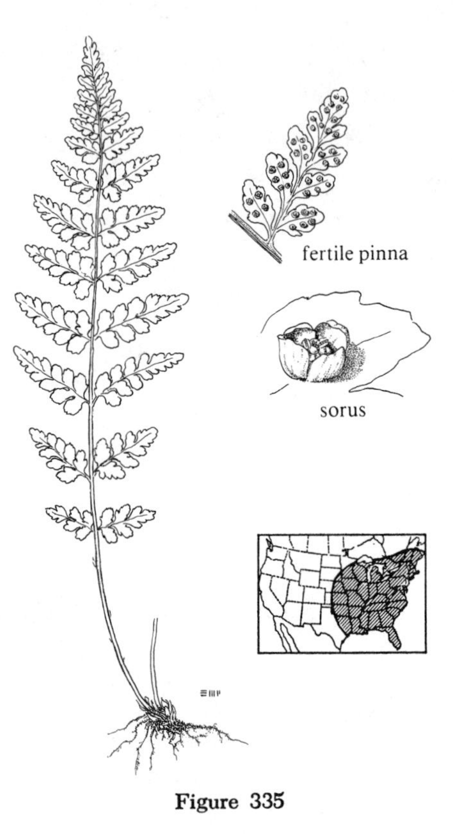

Figure 335

Figure 335 *Woodsia obtusa*

Rhizome short-creeping; scales tan, later with conspicuous dark streaks; fronds five to sixteen inches tall (mostly eight to fourteen), two to four inches wide; stipes not articulate, yellow, with scattered scales to nearly naked, one-fourth to one-half the frond length; blade bipinnate-pinnatifid, oblong, widest near the middle, glandular; indusium of about four broad lobes. Shaded cliffs and rock ledges. Common. Eastern North America.

5a Blade and stipe with minute stalked glands but lacking white hairs. 6

5b Blade and stipe with some glands and also white hairs. (Fig. 336) .... ROCKY MOUNTAIN WOODSIA, *Woodsia scopulina* D. C. Eaton

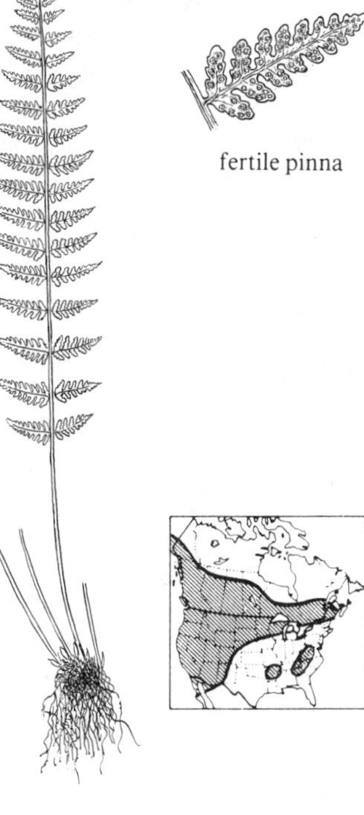

Figure 336

Figure 336 *Woodsia scopulina*

Rhizome ascending; scales tan, older with occasional black streaks; fronds four to twelve inches tall (mostly six to ten); one to one and one-half inch wide; stipes not articulate, brown, especially near the base, to yellow; one-third to one-half the frond length; blade bipinnate to bipinnate-pinnatifid, oblong to linear; stipe and blade with white hairs and short-stalked glands; indusium of narrow, strap-like lobes. Cliffs and rocks. Frequent. Much of northern and mountainous North America.

Woodsia appalachiana Taylor, of the Ozarks and southern Appalachian Mountains, is sometimes distinguished from *W. scopulina* on the basis of broader indusium lobes and narrower rhizome scales, but these characters do not seem to be constant.

6a Blade segments with short white hairs on the teeth. (Fig. 337) .... MEXICAN WOODSIA, *Woodsia mexicana* Fée

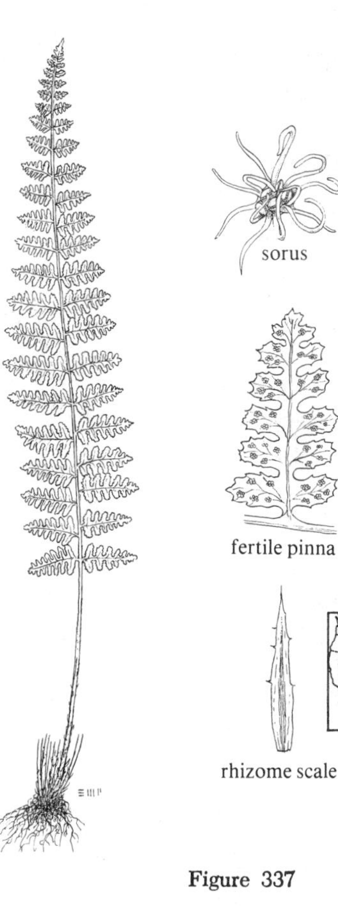

Figure 337

Figure 337 *Woodsia mexicana*

Rhizome short-creeping; scales tan with occasional black streaks; fronds three to fourteen inches tall (mostly seven to eleven inches), one to two inches wide; stipes not articulate, yellow to brown, about one-third the frond length; blade bipinnate to bipinnate-pinnatifid, oblong to oval, glandular; indusium of ciliated lobes. Rock ledges. Frequent. Southwestern United States; Mexico.

Woodsia plummerae Lemmon, also of southwestern United States and Mexico, closely resembles *W. mexicana* but can be distinguished by its conspicuous covering of stalked glands and the indusium of broad lobes.

6b Blade segments lacking short, white hairs on the teeth. (Fig. 338) .... OREGON WOODSIA, *Woodsia oregana* D. C. Eaton

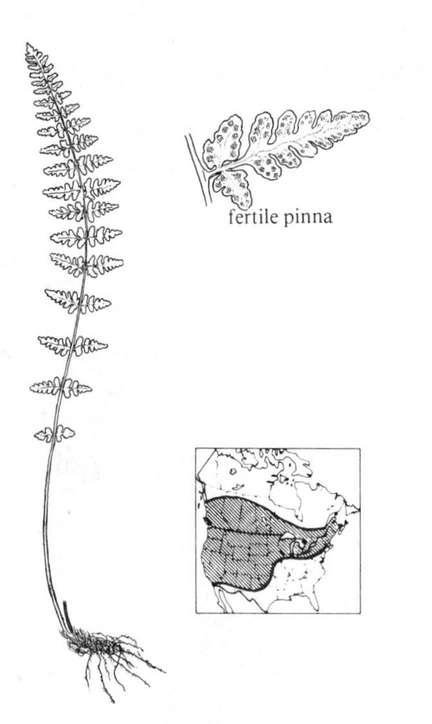

Figure 338

Figure 338 *Woodsia oregana*

Rhizome short-creeping; scales tan with occasional dark streaks; fronds three to ten inches tall (mostly four to eight inches), three-fourths to one and one-fourth inches wide; stipes not articulate, brown near base, yellow above, one-fourth to one-half the frond length; blade pinnate-pinnatifid to bipinnate, oblong to linear; naked or with minute glands; indusium of a few short hairs (about the length of the sporangia). Cliffs and rocks. Frequent. Northern and western North America.

Var. *cathcartiana* (D. C. Eaton) Morton, of Minnesota and Wisconsin, has a very glandular rachis.

WOODWARDIA

Chain Fern

Rhizome creeping or ascending, scaly; fronds medium to large size, pinnatifid to pinnate-pinnatifid; outer veins free or netted; sori originating in areoles along midveins; indusium present, elongate in a broken line, chain-like. Twelve species of temperate regions.

1a Sterile and fertile fronds alike; sterile fronds pinnate-pinnatifid. 2

1b Fronds dimorphic; sterile fronds pinnatifid. (Fig. 339) .... NETTED CHAIN-FERN, *Woodwardia areolata* (L.) Moore [*Lorinseria areolata* (L.) Presl]

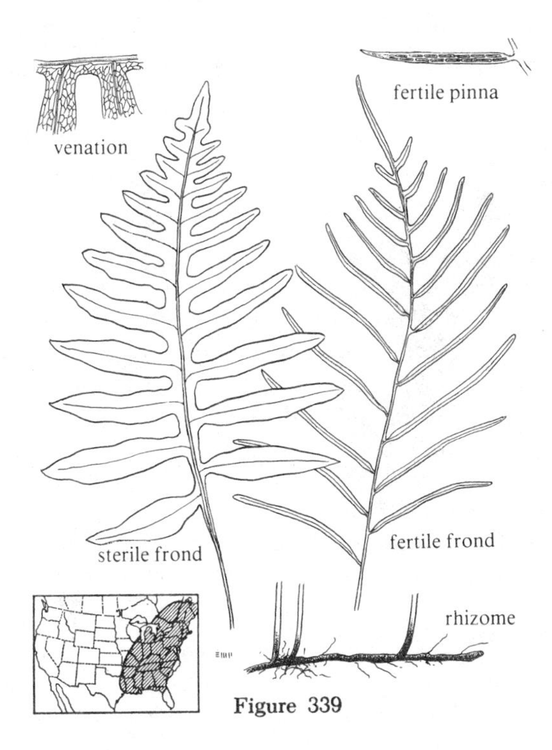

Figure 339

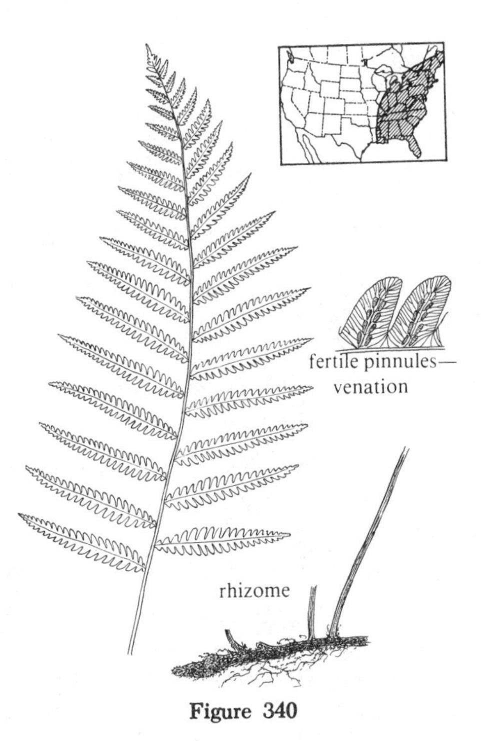

Figure 340

Figure 339 *Woodwardia areolata*

Rhizome long-creeping; fronds scattered, dimorphic; sterile fronds pinnatifid, twelve to twenty-four inches long, four to eight inches wide; all veins netted; fertile fronds as tall as the sterile, the pinnae much more narrow; deciduous. Acidic bogs and swamps. Common. Eastern North America.

The sterile fronds superficially resemble those of the sensitive fern but differ in their venation and more pointed pinnae.

2a Fronds medium-sized, one and one-half to two feet long; pinnae lobes short, blunt, smooth-margined; rhizome creeping, slender; fronds spaced; one row of areoles by the midrib; deciduous; eastern North America. (Fig. 340) .... VIRGINIA CHAIN-FERN, *Woodwardia virginica* (L.) J. E. Smith [Anchistea *virginica* (L.) Presl]

Figure 340 *Woodwardia virginica*

Rhizome long-creeping; fronds spaced along the rhizome; fertile and sterile fronds alike, one and one-half to two feet tall; pinnae pinnatifid, the lobes short and blunt; deciduous. Acidic bogs and swamps. Common. Eastern North America; Bermuda.

2b Fronds large, three to five feet long; pinnae lobes long-pointed, finely toothed; rhizome stout, ascending; fronds clumped; two rows of areoles by the midvein; evergreen; Pacific Northwest. (Fig. 341) .... GIANT CHAIN-FERN, *Woodwardia fimbriata* J. E. Smith [*W. chamissoi* Brack., *W. radicans* of some authors]

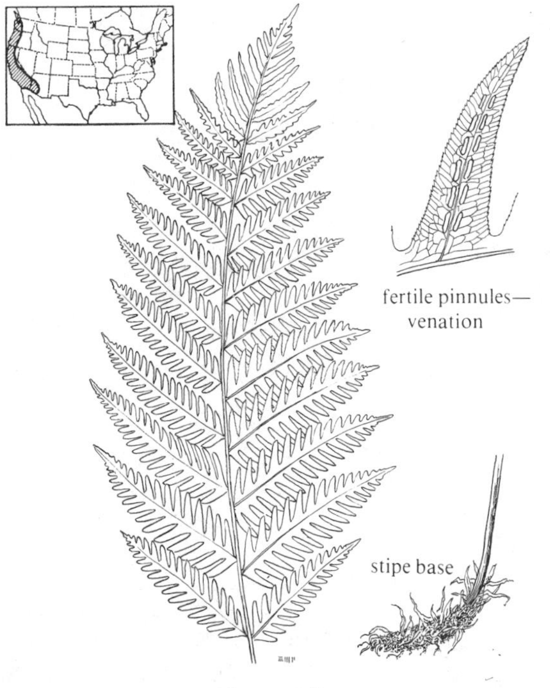

Figure 341

Figure 341 *Woodwardia fimbriata*

Rhizome stout, woody, ascending, densely clothed with large, brown scales; fronds three to five feet tall, sometimes larger, lance-shaped, narrowed toward the base; pinnae pinnatifid, the lobes long and sharp-pointed, with finely toothed margins; evergreen. Moist forests. Common. Western North America.

Woodwardia radicans J. E. Smith, a rare escape in Florida, resembles *W. fimbriata* in size and form, but is distinguished by having a large bud near the frond tip on the under side. This roots and develops into a new plant. Native of southern Europe.

State and Regional Identification Manuals

Northeast:

Cobb, Boughton. A Field Guide to the Ferns. Boston: Houghton Mifflin Co. 1956.

Gleason, Henry A. The New Britton and Brown Illustrated Flora (3 vols.). New York: N. Y. Botanical Garden. 1952.

Wherry, Edgar T. The Fern Guide. New York: Doubleday & Co. 1961; reprinted Philadelphia: Morris Arboretum. 1975.

Southeast:

Small, John Kunkel. Ferns of the Southeastern States. Lancaster, PA: Science Press Printing Co., 1938.

Wherry, Edgar T. The Southern Fern Guide. New York: Doubleday & Co. 1964; reprinted New York: N. Y. Chapter, American Fern Society. 1977.

Northwest:

Taylor, Thomas M. C. Pacific Northwest Ferns and their Allies. Toronto: Univ. of Toronto Press. 1970.

West:

Cronquist, et al. Intermountain Flora: Vascular Plants of the Intermountain West, USA (vol. 1). New York: Hafner Publishing Co. 1972.

Dorn, Robert D. & Jane L. Dorn. The Ferns and Other Pteridophytes of Montana, Wyoming, and the Black Hills of South Dakota. Laramie, WY: by the authors. 1972.

Petrik-Ott, Aleta Jo. A county checklist of the ferns and fern allies of Kansas, Nebraska, South Dakota, and North Dakota. Rhodora 77: 478-511. 1975.

State floras:

Alabama: Dean, Blanche E. Ferns of Alabama. Rev. ed. Birmingham, AL: Southern Univ. Press. 1969.

Alaska: Hulten, Eric. Flora of Alaska and Neighboring Territories. Stanford, CA: Stanford Univ. Press. 1968.

Welsh, Stanley L. Anderson's Flora of Alaska and Adjacent Parts of Canada. Provo, UT: Brigham Young Univ. Press. 1974.

Arizona: Kearney, T. H., & R. H. Peebles. Arizona Flora. Berkeley, CA: Univ. of California Press. 1957.

California: Grillos, Steve J. Ferns and Fern Allies of California. Berkeley, CA: Univ. of California Press. 1966.

Kiefer, Lawrence L., & Barbara Joe. Checklist of California pteridophytes. Madroño 19: 65-73. 1967.

Munz, P. A., & D. D. Keck. A California Flora. Berkeley, CA: Univ. of California Press. 1959.

Carolinas: Evans, A. Murray. Pteridophytes. In: Radford, A., et al. Manual of the Vascular Flora of the Carolinas. Chapel Hill, NC: Univ. of North Carolina Press. 1964.

Colorado: Harrington, H. D., and L. W. Durrell. Colorado Ferns and Fern Allies. Fort Collins, CO: Colorado Agric. Research Foundation. 1950.

Florida: Lakela, Olga, & Robert W. Long. Ferns of Florida. Miami: Banyan Books. 1976.

Small, John Kunkel. The Ferns of Florida. New York: Science Press. 1932.

Georgia: McVaugh, Rogers, & Joseph H. Pyron. Ferns of Georgia. Athens: Univ. of Georgia Press. 1951.

Idaho: Flowers, Seville. A list of the ferns of

Idaho. American Fern Journal 40: 121-131. 1950.
Illinois: Mohlelnbrock, Robert H. The Illustrated Flora of Illinois—Ferns. Carbondale, IL: Southern Illinois Univ. Press. 1967.
Indiana: Clevenger, Sarah. The distribution of the ferns and fern allies in Indiana. Butler Univ. Botanical Studies 10: 1-11. 1951.
Iowa: Cooperrider, T. S. The ferns and other pteridophytes of Iowa. State Univ. of Iowa Studies in Natural History 20 (1): 1-66. 1959.
Kansas: Brooks, R. Ferns in Kansas. Kansas School Nat. 13: 1-15. 1967.
McGregor, Ronald L. Ferns and fern allies in Kansas. American Fern Journal 50: 62-66. 1960.
Louisiana: Brown, C. A., and D. S. Cornell. Ferns and Fern Allies of Louisiana. Baton Rouge: La. State Univ. Press. 1942.
Maine: Ogden, Edith B. The Ferns of Maine. Reprinted from The Maine Bulletin 51, No. 3. Orono, ME: University Press. 1948.
Maryland: Reed, Clyde F. Ferns and Fern Allies of Maryland, Delaware and District of Columbia. Baltimore: Reed Herbarium. 1953.
Mississippi: Jones, S. B., Jr. The pteridophytes of Mississippi. Sida Contributions to Botany 3: 359-364. 1969.
Michigan: Billington, Cecil. Ferns of Michigan. Bloomfield Hills, MI: Cranbrook Institute of Science. 1952.
Minnesota: Tryon, R. M., Jr. The Ferns and Fern Allies of Minnesota. Minneapolis: Univ. of Minnesota Press. 1954.
Missouri: Steyermark, Julian A. Flora of Missouri. Ames, IA: Iowa State University Press. 1963.
New Hampshire: Scamman, Edith. Ferns and Fern Allies of New Hampshire. Durham, NH: Bull. No. 2, N. H. Acad. of Science. 1947.
New Jersey: Chrysler, M A., & J. L. Edwards. Ferns of New Jersey. New Brunswick, NJ: Rutgers Univ. Press. 1947.
New Mexico: Dittmer, Howard J., et al. Ferns and Fern Allies of New Mexico. Univ. of New Mexico Publication in Biology No. 6. Albuquerque: Univ. of New Mexico Press. 1954.
New York: Small, John Kunkel. Ferns of the Vicinity of New York. Lancaster, PA: Science Press Printing Co. 1935.
Ohio: Vannorsdall, H. H. Ferns of Ohio. Wilmington, OH: by the author. 1956.
Oklahoma: Waterfall, U. T. A Catalogue of the Flora of Oklahoma. Stillwater, OK: Research Foundation. 1952.
Pennsylvania: Canan, Elsie D. A Key to the Ferns of Pennsylvania. Lancaster, PA: Science Press. 1946.
Puerto Rico: Kepler, Kay. Common Ferns of Luquillo Forest. Hato Rey, PR: Inter American Univ. Press. 1975.
Rhode Island: Crandall, Dorothy L. County distribution of ferns and fern allies in Rhode Island. American Fern Journal 55: 97-112. 1965.
South Dakota: Van Bruggen, T. The pteridophytes of South Dakota. Proc. South Dakota Acad. Sci. 46: 126-144. 1967.
Tennessee: Shaver, Jesse M. Ferns of Tennessee. Nashville: George Peabody College for Teachers. 1954.
Texas: Correll, Donovan S. Ferns and Fern Allies of Texas. Renner: Texas Research Foundation. 1956.
Virginia: Massy, A. B. The Ferns and Fern Allies of Virginia (2nd ed.). VPI Agricultural Extension Service Bull. 256. Blacksburg, VA. 1958.
Wisconsin: Tryon, R. M., Jr., et al. The Ferns and Fern Allies of Wisconsin (2nd ed.). Madison: Univ. of Wisconsin Press. 1953.
West Virginia: Brooks, M. G., and A. S. Margolin. The Pteridophytes of West Virginia. West Virginia Univ. Bull. Series 39, No. 2. 1938.
Strausbaugh, P. D., & Earl L. Core. Flora of West Virginia (Pt. 1). West Virginia Univ. Bull., Ser. 52, No. 12-2, Morgantown, W. VA. 1952.
Wyoming: Porter, C. L. A Flora of Wyoming, Part 1. Univ. of Wyoming Agricultural Experiment Station Bull. 402. 1962.
Mexico: Knobloch, Irving W., & Donovan S. Correll. Ferns and Fern Allies of Chihuahua. Renner, TX: Texas Research Foundation. 1962.

Common Names of Ferns

Weatherby, Una F. The English names of North American ferns. American Fern Journal 42: 134-151. 1952.

Index

Broun, Maurice. Index to North American Ferns. Orleans, MA: by the author. 1938.

Checklist of North American Ferns and Fern Allies

____ Acrostichum aureum
____ Acrostichum danaeifolium

____ Adiantum capillus-veneris
____ Adiantum hispidulum
____ Adiantum jordanii
____ Adiantum melanoleucum
____ Adiantum pedatum
____ Adiantum tenerum
____ Adiantum × tracyi
____ Adiantum tricholepis

____ Anemia adiantifolia
____ Anemia cicutaria
____ Anemia mexicana
____ Anemia wrightii

____ Asplenium abscissum
____ Asplenium adiantum-nigrum
____ Asplenium auritum
____ Asplenium × biscayneanum
____ Asplenium bradleyi
____ Asplenium × clermontiae
____ Asplenium cristatum
____ Asplenium × curtissii
____ Asplenium dentatum
____ Asplenium exiguum
____ Asplenium heterochroum
____ Asplenium heteroresiliens
____ Asplenium monanthes
____ Asplenium montanum
____ Asplenium myriophyllum
____ Asplenium palmeri
____ Asplenium platyneuron
____ Asplenium × plenum
____ Asplenium pumilum
____ Asplenium resiliens
____ Asplenium ruta-muraria
____ Asplenium septentrionale
____ Asplenium serratum
____ Asplenium trichomanes
____ Asplenium vespertinum
____ Asplenium × virginicum
____ Asplenium viride
____ Asplenium × wherryi

____ Asplenosorus × ebenoides
____ Asplenosorus × gravesii
____ Asplenosorus × herb-wagneri
____ Asplenosorus × inexpectatus
____ Asplenosorus × kentuckiensis
____ Asplenosorus pinnatifidus
____ Asplenosorus × trudellii

____ Athyrium distentifolium
____ Athyrium filix-femina
____ Athyrium pycnocarpon
____ Athyrium thelypterioides

____ Azolla caroliniana
____ Azolla filiculoides
____ Azolla mexicana

____Blechnum occidentale
____Blechnum serrulatum
____Blechnum spicant

____Bommeria hispida

____Botrychium alabamense
____Botrychium biternatum
____Botrychium boreale
____Botrychium dissectum
____Botrychium dusenii
____Botrychium lanceolatum
____Botrychium lunaria
____Botrychium lunarioides
____Botrychium matricariifolium
____Botrychium minganense
____Botrychium multifidum
____Botrychium oneidense
____Botrychium pumicola
____Botrychium simplex
____Botrychium ternatum
____Botrychium virginianum

____Camptosorus rhizophyllus

____Ceratopteris pteridoides
____Ceratopteris richardii
____Ceratopteris thalictroides

____Ceterach dalhousiae

____Cheilanthes aemula
____Cheilanthes alabamensis
____Cheilanthes aliena
____Cheilanthes arizonica
____Cheilanthes aschenborniana
____Cheilanthes bonariensis
____Cheilanthes californica
____Cheilanthes cancellata
____Cheilanthes candida
____Cheilanthes carolotta-halliae
____Cheilanthes clevelandii
____Cheilanthes cochisensis
____Cheilanthes cooperae
____Cheilanthes covillei
____Cheilanthes davenportii
____Cheilanthes dealbata
____Cheilanthes deserti
____Cheilanthes eatonii
____Cheilanthes feei
____Cheilanthes fendleri
____Cheilanthes fibrillosa
____Cheilanthes gracillima
____Cheilanthes grayi
____Cheilanthes greggii
____Cheilanthes horridula
____Cheilanthes integerrima
____Cheilanthes intertexta
____Cheilanthes jonesii
____Cheilanthes kaulfussii
____Cheilanthes lanosa
____Cheilanthes lemmonii
____Cheilanthes lendigera
____Cheilanthes leucopoda
____Cheilanthes limitanea
____Cheilanthes lindheimeri
____Cheilanthes microphylla
____Cheilanthes nealleyi
____Cheilanthes neglecta
____Cheilanthes newberryi
____Cheilanthes notholaenoides
____Cheilanthes × parishii
____Cheilanthes parryi
____Cheilanthes parvifolia
____Cheilanthes pringlei
____Cheilanthes siliquosa
____Cheilanthes sinuata
____Cheilanthes standleyi
____Cheilanthes tomentosa
____Cheilanthes villosa
____Cheilanthes viscida
____Cheilanthes wootonii
____Cheilanthes wrightii

____Cryptogramma acrostichoides
____Cryptogramma stelleri

____Ctenitis sloanei
____Ctenitis submarginalis

____Cyrtomium falcatum
____Cyrtomium fortunei

____Cystopteris bulbifera
____Cystopteris fragilis
____Cystopteris laurentiana
____Cystopteris montana
____Cystopteris protrusa
____Cystopteris tennesseensis

____Dennstaedtia bipinnata
____Dennstaedtia globulifera
____Dennstaedtia punctilobula

____Dicranopteris flexuosa

____Diplazium esculentum
____Diplazium japonicum
____Diplazium lonchophyllum

____Dryopteris × algonquinensis
____Dryopteris arguta
____Dryopteris × australis
____Dryopteris × benedictii
____Dryopteris × boottii
____Dryopteris × burgessii
____Dryopteris campyloptera
____Dryopteris celsa
____Dryopteris clintoniana
____Dryopteris cristata
____Dryopteris × dowellii
____Dryopteris expansa
____Dryopteris filix-mas
____Dryopteris fragans
____Dryopteris goldiana
____Dryopteris intermedia
____Dryopteris × leedsii
____Dryopteris ludoviciana
____Dryopteris marginalis
____Dryopteris × neo-wherryi
____Dryopteris patula
____Dryopteris × pittsfordensis
____Dryopteris × separabilis
____Dryopteris × slossonae
____Dryopteris spinulosa
____Dryopteris × triploidea
____Dryopteris × uliginosa

____Equisetum arvense
____Equisetum ×ferrissii
____Equisetum fluviatile
____Equisetum hyemale
____Equisetum laevigatum
____Equisetum × litorale
____Equisetum × nelsonii
____Equisetum palustre
____Equisetum pratense
____Equisetum ramosissimum
____Equisetum scirpoides
____Equisetum sylvaticum
____Equisetum telmateia
____Equisetum × trachyodon
____Equisetum variegatum

____Grammatis nimbata

____Gymnocarpium dryopteris
____Gymnocarpium × heterosporum
____Gymnocarpium robertianum

____Hymenophyllum tunbrigense
____Hymenophyllum wrightii
____Hymenophyllum (Sphaerocionium) sp.

____Hypolepis repens

____Isoetes bolanderi
____Isoetes butleri
____Isoetes eatonii
____Isoetes echinospora
____Isoetes engelmannii
____Isoetes flaccida
____Isoetes foveolata
____Isoetes howellii
____Isoetes lithophila
____Isoetes louisianensis
____Isoetes macrospora
____Isoetes melanopoda
____Isoetes melanospora
____Isoetes nuttallii
____Isoetes occidentalis
____Isoetes orcuttii
____Isoetes riparia

____Isoetes tuckermanii
____Isoetes virginica

____Lomariopsis kunzeana

____Lycopodium appressum
____Lycopodium alopecuroides
____Lycopodium alpinum
____Lycopodium annotinum
____Lycopodium carolinianum
____Lycopodium cernuum
____Lycopodium clavatum
____Lycopodium complanatum
____Lycopodium dendroideum
____Lycopodium dichotomum
____Lycopodium flabelliforme
____Lycopodium × habereri
____Lycopodium inundatum
____Lycopodium × issleri
____Lycopodium lucidulum
____Lycopodium obscurum
____Lycopodium porophilum
____Lycopodium prostratum
____Lycopodium sabinifolium
____Lycopodium selago
____Lycopodium sitchense
____Lycopodium tristachyum

____Lygodium japonicum
____Lygodium microphyllum
____Lygodium palmatum

____Marsilea macropoda
____Marsilea mexicana
____Marsilea quadrifolia
____Marsilea tenuifolia
____Marsilea uncinata
____Marsilea vestita

____Matteuccia struthiopteris

____Maxonia apiifolia

____Nephrolepis biserrata
____Nephrolepis cordifolia
____Nephrolepis exaltata
____Nephrolepis multiflora
____Nephrolepis pectinata

____Neurodium lanceolatum

____Onoclea sensibilis

____Ophioglossum crotalophoroides
____Ophioglossum dendroneuron
____Ophioglossum engelmannii
____Ophioglossum lusitanicum
____Ophioglossum nudicaule
____Ophioglossum palmatum
____Ophioglossum petiolatum
____Ophioglossum vulgatum

____Osmunda cinnamomea
____Osmunda claytoniana
____Osmunda regalis
____Osmunda × ruggii

____Pellaea andromedifolia
____Pellaea atropurpurea
____Pellaea brachyptera
____Pellaea breweri
____Pellaea bridgesii
____Pellaea cardiomorpha
____Pellaea glabella
____Pellaea intermedia
____Pellaea mucronata
____Pellaea ovata
____Pellaea ternifolia
____Pellaea truncata
____Pellaea wrightiana

____Phanerophlebia auriculata
____Phanerophlebia umbonata

____Phyllitis scolopendrium

____Pilularia americana

____Pityrogramma calomelanos
____Pityrogramma triangularis

____Polypodium angustifolium
____Polypodium astrolepis
____Polypodium aureum
____Polypodium australe
____Polypodium californicum
____Polypodium costatum
____Polypodium dispersum
____Polypodium erythrolepis
____Polypodium glycyrrhiza
____Polypodium hesperium
____Polypodium heterophyllum
____Polypodium latum
____Polypodium phyllitidis
____Polypodium plumula
____Polypodium polypodioides
____Polypodium ptilodon
____Polypodium scolopendrium
____Polypodium scouleri
____Polypodium thyssanolepis
____Polypodium triseriale
____Polypodium virginianum

____Polystichum acrostichoides
____Polystichum aleuticum
____Polystichum andersonii
____Polystichum braunii
____Polystichum californicum
____Polystichum dudleyi
____Polystichum imbricans
____Polystichum kruckebergii
____Polystichum lemmonii
____Polystichum lonchitis
____Polystichum microchlamys
____Polystichum munitum
____Polystichum muricatum
____Polystichum scopulinum
____Polystichum setigerum

____Psilotum nudum

____Pteridium aquilinum

____Pteris bahamensis
____Pteris cretica
____Pteris ensiformis
____Pteris grandifolia
____Pteris longifolia
____Pteris multifida
____Pteris tripartita
____Pteris vittata

____Salvinia minima

____Schizaea germanii
____Schizaea pusilla

____Selaginella apoda
____Selaginella arenicola
____Selaginella arizonica
____Selaginella asprella
____Selaginella bigelovii
____Selaginella braunii
____Selaginella cinerascens
____Selaginella densa
____Selaginella douglasii
____Selaginella eatonii
____Selaginella eclipes
____Selaginella eremophila
____Selaginella hansenii
____Selaginella kraussiana
____Selaginella lepidophylla
____Selaginella leucobryoides
____Selaginella ludoviciana
____Selaginella mutica
____Selaginella × neomexicana
____Selaginella oregana
____Selaginella peruviana
____Selaginella pilifera
____Selaginella rupestris
____Selaginella rupincola
____Selaginella selaginoides
____Selaginella sibirica
____Selaginella tortipila
____Selaginella uncinata
____Selaginella underwoodii
____Selaginella utahensis
____Selaginella viridissima
____Selaginella wallacei
____Selaginella watsonii
____Selaginella weatherbeana

____Selaginella willdenovii
____Selaginella wrightii

____Sphenomeris clavata

____Stenochlaena tenuifolia

____Tectaria × amesiana
____Tectaria coriandrifolia
____Tectaria heracleifolia
____Tectaria incisa
____Tectaria lobata

____Thelypteris augescens
____Thelypteris dentata
____Thelypteris grandis
____Thelypteris hexagonoptera
____Thelypteris interrupta
____Thelypteris kunthii
____Thelypteris limbosperma
____Thelypteris nevadensis
____Thelypteris noveboracensis
____Thelypteris ovata
____Thelypteris palustris
____Thelypteris patens
____Thelypteris phegopteris
____Thelypteris pilosa
____Thelypteris puberula
____Thelypteris quadrangularis
____Thelypteris reptans
____Thelypteris resinifera
____Thelypteris reticulata
____Thelypteris sclerophylla
____Thelypteris serrata
____Thelypteris simulata
____Thelypteris tetragona
____Thelypteris torresiana

____Trichomanes boschianum
____Trichomanes holopterum
____Trichomanes kraussii
____Trichomanes lineolatum
____Trichomanes membranaceum
____Trichomanes petersii
____Trichomanes punctatum

____Trismeria trifoliata

____Vittaria graminifolia
____Vittaria lineata

____Woodsia × abbeae
____Woodsia alpina
____Woodsia glabella
____Woodsia ilvensis
____Woodsia mexicana
____Woodsia obtusa
____Woodsia oregana
____Woodsia plummerae
____Woodsia scopulina

____Woodwardia areolata
____Woodwardia fimbriata
____Woodwardia radicans
____Woodwardia virginica

Index and Glossary

D

E

F

T

V

W

X

NOTES

NOTES

NOTES

NOTES

NOTES

NOTES

NOTES